MATHEMATICS PLUS

Teacher's Guide • Testing Program

- **Performance Assessment**
  - Interview/Task Tests
  - Portfolio
  - Problem-Solving Assessment
  - Checklists
- **Standardized Format Tests**
- **Free Response Format Tests**
- **Management Forms**

PROPERTY OF
RUTHERFORD CO. BOARD OF EDUCATION
SP. ED. DEPT.

Harcourt Brace Jovanovich, Inc.
Orlando Austin San Diego Chicago Dallas New York

Copyright © 1992 by Harcourt Brace Jovanovich, Inc.

All rights reserved. No part of this publication may be reproduced or transmitted in any form or by any means, electronic or mechanical, including photocopy, recording, or any information storage and retrieval system, without permission in writing from the publisher.

Permission is hereby granted to reproduce the Copying Masters in this publication in complete pages, with the copyright notice, for instructional use and not for resale by any teacher using classroom quantities of the related student textbook.

Printed in the United States of America

ISBN 0-15-301015-0

5 6 7 8 9 10   082   95   94   93   92

Tab I. **Performance Assessment**

Mathematics Portfolio	1–3
Build a Portfolio	4–5
Evaluating Interview/Task Test Items	7
Evaluation of Interview Task Test	9–25
Performance Assessment • Class Record Form	27–31
Performance Checklists	33
Teacher Observation Checklist	
Cooperative Learning	34
Self-Evaluation Checklists	
Attitude Survey	35
Group Work	36

Tab II. **Formal Assessment**

- Standardized Tests • Copying Masters
- **A.** Standardized Test Answers
- Free Response Tests • Copying Masters
- **B.** Free Response Test Answers

Tab III. **Management Forms**

- Test Answer Sheet
- Grading Made Easy
- Individual Record Form
- Formal Assessment Class Record Form
- Cumulative Record Form

I
Performance Assessment

It has been the custom in the past to use narrow approaches to testing, such as tests composed of multiple-choice questions to be answered by students in a fixed time period. Many have been uneasy about the process and have recognized that this testing format, while efficient, has little to do with what the students have learned and the ways in which they have been taught.

Multiple means of assessment for all aspects of mathematical knowledge and its connections are provided in the Performance Assessment methods described in this section. Through the evaluation of students' portfolios, problem-solving processes, and interview/task items, a broader vision of what the student knows will emerge. A Class Record Form for Performance Assessment is provided to record the results of the interview/task items. These methods, together with the formal assessment, provide an accurate and thorough indication of the mathematics that students know.

The Mathematics Portfolio

What Is a Mathematics Portfolio?

The portfolio is a collection of each student's work gathered over an extended period of time. A portfolio illustrates the growth, talents, achievements, and reflections of the mathematics learner and provides a means for the teacher to assess the student's performance.

An effective portfolio will

- include items collected over the entire school year.
- show the "big picture"—providing a broad understanding of a student's mathematics language and feelings through words, diagrams, checklists, and so on.
- give students opportunities to collaborate with their peers.
- give students a chance to experience success, to develop pride in their work, and to develop positive attitudes toward mathematics.

Building a Portfolio

There are many opportunities to collect student's work throughout the year as you use *Mathematics Plus*. A list of suggested portfolio items for each chapter is given on pages 4 and 5. These suggestions are also listed on page F of each Chapter Overview in the Teacher's Edition. Students may also appreciate a chance to select some of their work to be included in the portfolio. Attitudes and performance can be recorded on checklists provided at the end of this section of the *Teacher's Guide * Testing Program*.

To begin:

- Provide a file folder for each student with the student's name clearly marked on the tab or folder.
- Explain to students that throughout the year they will save some of their work in the folder. Sometimes it will be their individual work; sometimes it will be group reports and projects, or completed checklists.
- Assign a fun activity to the entire class that can be placed into the portfolio by all students. EXAMPLE: Ask students to draw a map of the school playground in which they use at least three different shapes.
- Comment positively on the maps and reinforce the process. Have students place their maps into their portfolios.

Performance Assessment

Evaluating a Portfolio

The ultimate purpose of assessment is to enable students to evaluate themselves. Portfolios have the potential to create authentic portraits of what students learn and offer an alternative means for documenting growth, change, and risk-taking in mathematics learning. Evaluating their own growth in mathematics will be a new experience for most students. The following points made with regular portfolio evaluation will encourage growth in self-evaluation.

- Discuss the contents of the portfolio with each student as you examine it at regular intervals during the school year.
- Examine each portfolio on the basis of the growth the student has made rather than in comparison with other portfolios.
- Ask the student questions as you examine the portfolio.
- Point out the strengths and weaknesses in the work.
- Encourage and reward students by emphasizing the growth you see, the original thinking, and the completion of tasks.
- Reinforce and adjust instruction of the broad goals you want to accomplish as you evaluate the portfolios.

What Should the Teacher Look For?

Growth in mathematics is shown by:

- the non-standard responses that students make.
- the ways students solve a problem or attempt to solve a problem. Note the strategies they used and the reasons they succeeded or became confused.
- the ways students communicate their understanding of math problems. Note whether they use words, pictures, or abstract algorithms.
- unique solutions or ways of thinking.

Placing comments such as the following on student's work can have instructional benefits.

"Interesting approach to solving this problem."
"Is this the only solution to the problem?"
"Where would you take it from here?"
"Can you think of a related problem?"
"What about . . . ?"

What Should the Student Look For?

Have students look through their samples and describe the kinds of learning experiences they had. Point out to students the following statements—which are likely to represent some of the ways they will feel.

I enjoyed the problems.
I learned something new.
This repeated things I already knew.
Now I'm getting the hang of it.
I was challenged.
I discovered something new myself.

Sharing the Portfolios

- Examine the portfolio with parents or guardians to share concrete examples of the work the student is doing. Emphasize the growth you see as well as the expectations you have.
- Examine portfolios with your students to emphasize their experiences of success and to develop pride in their work and positive attitudes toward mathematics.
- Examine portfolios with your supervisor to share the growth your students have made and to show the ways you have developed the curriculum objectives.

The Benefits of Mathematics Portfolios

Portfolios can be the basis for informed change in mathematics classrooms because they:

- send positive messages to students about successful processes rather than end results.
- give you better insights as to how students understand and work problems.
- focus on monitoring the development of reasoning skills.
- help students become responsible for their own learning.
- promote teacher-student dialogue.
- focus on the student rather than on the assignment.
- focus on the development of conceptual understandings rather than applications of skills and procedures.

Build a Portfolio

CHAPTER 1

Activity 1 Use after Lesson 1.2 on PE pages 5–6.
Children trace a row of 9 counters. They color the first 6 counters red, blue, blue, red, blue, blue. Then they extend the pattern by coloring the next three counters.

Activity 2 Use after Lesson 1.13 on PE pages 29–30.
Have children fold a sheet of paper in half and write 8 in one section and 10 in the other. They trace that number of counters in each section and ring the number that is less and underline the number that is greater.

CHAPTER 2

Activity 1 Use after Lesson 2.2 on PE pages 43–44.
Have children draw 3 red counters and 2 blue counters. Have them write an addition sentence for the counters.

Activity 2 Use after Lesson 2.5 on PE pages 49–50.
Have children make up a story and then draw a picture for the addition sentence $2 + 3 = \underline{\ ?\ }$.

Activity 3 Use after Lesson 2.7 on PE pages 55–56.
Have children use counters to help them find seven ways to make a sum of 6. Have them record each way on a sheet of paper.

CHAPTER 3

Activity 1 Use after Lesson 3.3 on PE pages 73–74.
Have children trace 6 counters onto a sheet of paper and cross out 2 of the counters. Then have them write a subtraction sentence for the counters.

CHAPTER 3 (continued)

Activity 2 Use after Lesson 3.5 on PE pages 77–78.
Have children draw a picture for the addition sentence $2 + 4 = \underline{\ ?\ }$. Have them draw a picture for the subtraction sentence $6 - 4 = \underline{\ ?\ }$. Then have them complete the sentences.

CHAPTER 4

Activity 1 Use after Lesson 4.4 on PE pages 107–108.
Have children draw a picture of 3 toys with price tags showing 3¢, 4¢, and 5¢. Have them choose pairs of toys to buy and write addition sentences to show how much they would spend.

Activity 2 Use after Lesson 4.8 on PE pages 117–118.
Have children write $3 + 4 + 1 = \underline{\ ?\ }$ on a sheet of paper. Have them draw a picture or trace counters to go with the addition sentence.

CHAPTER 5

Activity 1 Use after Lesson 5.3 on PE pages 135–136.
Give children a sheet of paper with four number lines. Have them show how they would count back to solve these:
$9 - 1 = \underline{\ ?\ }$, $8 - 3 = \underline{\ ?\ }$, $7 - 2 = \underline{\ ?\ }$, $10 - 3 = \underline{\ ?\ }$.

Activity 2 Use after Lesson 5.4 on PE pages 137–138.
Have children draw a picture for this problem.
There are 9 red balls and 4 blue balls. How many more red balls are there than blue balls?
Then have them solve.

CHAPTER 6

Activity 1 Use after Lesson 6.3 on PE pages 163–164.
Have children draw a picture of an object at home or school that has each of these shapes: cone, sphere, cube, cylinder.

Activity 2 Use after Lesson 6.4 on PE pages 165–166.
Have children trace a circle, a square, a triangle, and a rectangle onto a sheet of paper. For each shape, have them put a / inside the shape, an O outside the shape, and an X on the shape.

CHAPTER 7

Activity 1 Use after Lesson 7.5 on PE pages 203–204.
Have children fold a sheet of paper in half and write 24 in one section and 42 in the other section. Have them put place-value models in the sections to illustrate the numbers and trace the models.

Activity 2 Use after Lesson 7.10 on PE pages 215–216.
Have children write the following numbers in a column down the center of a piece of paper.
9 25 78 43 37
Have them write in front of each number the number that comes before. Have them write after each number the number that comes after.

CHAPTER 8

Activity 1 Use after Lesson 8.6 on PE pages 245–246.
Have children find out the favorite sport of a group of ten classmates. Have them make a tally table to show the responses.

CHAPTER 8 (continued)

Activity 2 Use after Lesson 8.10 on PE pages 255–256.
Have children fold a sheet of paper in half twice to make four sections. In each section, have them write one of the facts in the fact family for 9, 2, 11. Have them draw a picture to illustrate each fact.

CHAPTER 9

Activity 1 Use after Lesson 9.3 on PE pages 275–276.
Have children fold a sheet of paper in half. Write *Less than 6 inches* and *More than 6 inches*. Then have them draw pictures of objects that belong in each section.

Activity 2 Use after Lesson 9.14 on PE pages 299–300.
Have children fold a sheet of paper in half. In the first section, have them write $\frac{1}{2}$, trace 2 counters, and color $\frac{1}{2}$ of the counters. In the second section, have them write $\frac{1}{4}$, trace 4 counters, and color $\frac{1}{4}$.

CHAPTER 10

Activity 1 Use after Lesson 10.11 on PE pages 335–336.
Have children trace and label a punch-out quarter at the top of a sheet of paper. Then have them trace and label other combinations of coins that make 25¢.

Activity 2 Use after Lesson 10.13 on PE pages 339–340.
Have children draw a picture of a small toy with a price tag showing 37¢. Then have them trace and label punch-out coins to show how they could pay for the toy.

Performance Assessment

CHAPTER 11

Activity 1 Use after Lesson 11.5 on PE pages 361–362.

Have children label a sheet of paper to match Workmat 3. Have them use punch-out tens and ones to model 23 + 35. Have them trace pieces and record the sum.

Activity 2 Use after Lesson 11.8 and PE pages 369–370.

Have children label a sheet of paper to match Workmat 3. Have them use punch-out tens and ones to model 57 − 24. Then have them trace the punch-out pieces and record the difference.

CHAPTER 12

Activity 1 Use after Lesson 12.9 on PE pages 403–404.

Have children fold a sheet of paper in half and write 15 − 9 = ? and 17 − 8 = ? in the sections. Have them write under each subtraction fact an addition fact they could use to help find the difference. Then have them draw a picture to illustrate each fact.

Activity 2 Use after Lesson 12.11 on PE pages 407–408.

Have children arrange 12 counters in equal groups in as many ways as they can. Have them trace and record each way.

Evaluating Interview/Task Items

The interview/task test items are designed to provide an optional instrument to evaluate each student's level of accomplishment for each tested objective in the *Mathematics Plus* program. These items provide opportunities for students to verbalize or write about their thinking or to use manipulatives or other pictorial representations to represent their thinking. They test students at the concrete and pictorial levels, where appropriate, so that you can assess each student's progress toward functioning at the abstract level. The items will enable you to analyze the student's thought processes as they work on different types of problems and will enable you to plan instruction that will meet your students' needs.

You may wish to use these test items as you work through the content in the chapter to determine whether students are ready to move on or whether they need additional teaching or reinforcement activities. You may also wish to use these test items with students who did not successfully pass the standardized format or free response test for the chapter to determine what types of reteaching activities are appropriate. These test items may also be used with students who have difficulty reading written material or who have learning disabilities.

The test items are designed to focus on evaluating how students think about mathematics and how they work at solving problems rather than on whether they can get the correct answer. The evaluation criteria given for each test item will help you pinpoint the errors in the students' thinking processes as they work through the problem.

A checklist of possible responses is provided to record each student's thinking processes. The Class Record Form can be used to show satisfactory completion of interview/task test items.

Evaluation of Interview/Task Test

DATE _____

CHILD'S NAME _____ CLASS _____

TEST ITEM	EVALUATE WHETHER CHILD
1-A Show the child two rows of connecting cubes in these color patterns: —red, blue, red, blue, red, blue —green, yellow, yellow, green, yellow, yellow Have the child name each color pattern and then continue the pattern by adding 3 cubes to it.	_____ names each pattern correctly. _____ continues the patterns by adding red, blue, red to the first pattern and green, yellow, yellow to the second pattern.
1-B Show the child number cards for 0, 2, 4, 6, and 8. For each number, have the child show an appropriate number of counters and tell how many. Then have the child add 1 more counter to each group and tell how many counters are in the new group.	shows the correct number of counters _____ for 0. _____ for 2. _____ for 4. _____ for 6. _____ for 8. adds 1 counter and identifies the number of counters in the new groups as _____ 1. _____ 3. _____ 5. _____ 7. _____ 9.
1-C Give the child a set of shuffled number cards for 0–10. Have the child arrange the cards in order from least to greatest. Then have the child name a number that is less than 4 and a number that is greater than 7.	_____ arranges the cards in order from least to greatest. _____ identifies 0, 1, 2, or 3 as a number less than 4. _____ identifies 8, 9, or 10 as a number greater than 7.
1-D Have the child look at Exercise 4 on PE page 18. Have the child explain how he or she would draw a picture to continue the pattern.	_____ identifies a flag as the next object in the pattern. _____ explains why a flag is the next object in the pattern.

Performance Assessment (See Teacher's Edition, p. 1F)

Interview/Task Test • Chapter 1 *continued*

TEST ITEM	EVALUATE WHETHER CHILD
1-D Show the child a set of 4 red cubes, 5 blue cubes, 2 yellow cubes, and 1 green cube. Have the child look at the grid on PE page 31 and explain how to use the grid and squares cut from red, blue, yellow, and green construction paper to complete the graph for the cubes.	_____ makes the pictograph correctly. _____ explains how the graph was made.

Performance Assessment (See Teacher's Edition, p. 1F)

Evaluation of Interview/Task Test

DATE _____

CHILD'S NAME _____ CLASS _____

TEST ITEM	EVALUATE WHETHER CHILD
2-A Have the child make a set of 4 counters and a set of 1 counter. Have the child tell an addition sentence for the counters.	_____ uses the correct number of counters for each set. _____ states the addition sentence as $4 + 1 = 5$.
2-B Have the child use counters or draw pictures to show each addition fact. Then have the child give the sum. $3 + 2 = \underline{?}$ $2 + 3 = \underline{?}$ $4 + 0 = \underline{?}$	_____ shows the facts correctly with counters or pictures. _____ identifies the sums as 5, 5, and 4.
2-C Give the child 5 counters. Have the child use the counters to find six ways to make a sum of 5. Have the child tell or write the ways he or she found.	_____ uses the counters to show pairs of addends with a sum of 5. _____ identifies the addition sentences as $5 + 0 = 5, 4 + 1 = 5, 3 + 2 = 5,$ $0 + 5 = 5, 1 + 4 = 5, 2 + 3 = 5.$

2-D Read the following story problem.
   There are 3 kittens in a basket.
   There are 2 kittens on a rug.
   How many kittens are there in all?
Have the child *draw a picture* to solve the problem.
	_____ draws an appropriate picture for the story.
	
	_____ identifies the answer as 5 kittens.

2-D Read the following story problem.
   Four crayons are on the desk.
   Two crayons are on the floor.
   How many crayons are there?
Have the child use crayons to *act out* the problem. Then have the child write a number sentence to solve the problem.
	_____ *acts out* the problem by putting 4 crayons on the desk and 2 crayons on the floor.
	
	_____ identifies the number sentence as $4 + 2 = 6$.

Performance Assessment (See Teacher's Edition, p. 39F)

Evaluation of Interview/Task Test

DATE _____

CHILD'S NAME _____ CLASS _____

TEST ITEM	EVALUATE WHETHER CHILD
3-A Show the child a set of 6 counters. Have the child take away 2 counters and tell how many are left.	_____ takes away 2 counters. _____ identifies the number left as 4.
3-B Have the child use counters or draw pictures to show each subtraction fact. Then have the child tell how many are left. $4 - 2 = \underline{?}$ $5 - 1 = \underline{?}$ $6 - 3 = \underline{?}$	_____ shows the facts correctly with counters or pictures. _____ identifies the numbers left as 2, 4, and 3.
3-C Have the child use cubes to show $3 + 2 = \underline{?}$ and $5 - 2 = \underline{?}$. Have the child give the answers and then explain what is the same and what is different about the sentences.	_____ uses cubes to model the sentences. _____ identifies the answers as 5 and 3. _____ states at least one way in which the sentences are alike (same numbers are in both sentences) and one way in which they are different (one is addition; the other is subtraction).
3-D Have the child look at Exercises 5 and 6 on PE page 78. Have the child give a story and then *write a number sentence* for each picture.	_____ tells an appropriate story for each picture. _____ writes an appropriate number sentence for each story.

3-D Read the following story.
  Joy catches 3 fish.
  Then she catches 1 more fish.
Have the child give a question he or she could ask for the story and then give a number sentence he or she could use to answer the question.
	_____ identifies an appropriate question (How many fish did she catch in all?).
	_____ identifies the number sentence as $3 + 1 = 4$.

Performance Assessment (See Teacher's Edition, p. 67F)

Evaluation of Interview/Task Test

DATE _____

CHILD'S NAME _____ CLASS _____

TEST ITEM	EVALUATE WHETHER CHILD
4-A Show the child addition fact cards for 5 + 2 = __?__, 8 + 1 = __?__, and 4 + 4 = __?__. Have the child explain how he or she would find each sum.	____ explains a strategy that could be used to solve each fact; for example, 5 + 2 = __?__ → counting on. 8 + 1 = __?__ → counting on. 4 + 4 = __?__ → doubles. ____ identifies the sums as 7, 9, and 8.
4-B Show the child the addition sentence 3 + 2 + 4 = __?__. Have the child explain how to find the sum.	____ explains that you add two addends and then add the third addend. ____ identifies the sum as 9.
4-C Have the child look at the pictures of the toys at the top of PE page 107. Have the child explain how to write an addition sentence to show how much the child would spend if he or she bought a shovel and a bucket.	____ explains that the addition sentence is 6¢ + 2¢ = 8¢. ____ identifies the amount spent as 8¢.

4-C Read the following story problem.
    Max sees 3 rabbits.
    He sees 4 ducks.
    Then he sees 2 more rabbits.
    How many rabbits does he see?
Have the child tell the sentence that is not needed to answer the question. Then have the child *write a number sentence* to solve the problem.
	____ identifies the sentence "He sees 4 ducks" as not needed.
	 ____ writes the number sentence correctly (3 + 2 = 5).
	 ____ identifies the solution as 5 rabbits.

Performance Assessment (See Teacher's Edition, p. 99F)   Chapter 4 • 13

Evaluation of Interview/Task Test

DATE _____

CHILD'S NAME _____ CLASS _____

TEST ITEM	EVALUATE WHETHER CHILD
5-A Show the child subtraction fact cards for $8 - 2 = \underline{?}$, $10 - 3 = \underline{?}$, and $9 - 9 = \underline{?}$. Have the child explain how he or she would find each difference and then give each difference.	_____ explains a strategy that could be used to solve each fact; for example, $8 - 2 = \underline{?} \rightarrow$ counting back. $10 - 3 = \underline{?} \rightarrow$ counting back. $9 - 9 = \underline{?} \rightarrow$ zero fact. _____ identifies the differences as 6, 7, and 0.
5-B Show the child 4 red cubes and 1 blue cube. Have the child tell the two addition and the two subtraction sentences in the fact family for these cubes.	_____ identifies the two addition sentences as $4 + 1 = 5$ and $1 + 4 = 5$. _____ identifies the two subtraction sentences as $5 - 1 = 4$ and $5 - 4 = 1$.

5-C Read the following story problem.
   There are 7 birds on a branch.
   Then 3 birds fly away.
   How many birds are left?
 Have the child show how he or she would use counters to solve the problem.
	_____ shows 7 counters and then takes away 3 counters.
	 _____ identifies the answer as 4 birds.

5-C Read the following clues.
   The sum is 7.
   The difference is 1.
 Have the child explain how he or she would *guess and check* to find the number pair and then give the number pair.
	_____ explains how to guess a number pair and then check it with the clues.
	 _____ identifies the number pair as 4 and 3.

14 • Chapter 5   Performance Assessment (See Teacher's Edition, p. 129F)

Evaluation of Interview/Task Test

DATE _____

CHILD'S NAME _____ CLASS _____

TEST ITEM	EVALUATE WHETHER CHILD
6-A Display a set of solid shapes that include a sphere, cube, cone, prism, pyramid, and cylinder. Have the child point to the shapes that will stack, will roll, and will slide. Then have the child use the shapes to identify a flat face that is a square, a triangle, a rectangle, and a circle.	identifies shapes that will _____ stack (cube, prism, cylinder). _____ roll (sphere, cone, cylinder). _____ slide (cube, cone, prism, pyramid, cylinder). identifies _____ a square. _____ a triangle. _____ a rectangle. _____ a circle.
6-B Have the child draw an open figure and a closed figure. Then have the child draw a red X *inside* the closed figure, a blue X *outside* the closed figure, and a green X *on* the closed figure.	_____ draws an open figure and a closed figure. _____ draws the correctly colored X's inside, outside, and on the closed figure.
6-C Show the child a triangle and a square. Have the child tell how many sides and corners each figure has.	_____ states that the triangle has 3 sides and the square has 4 sides. _____ states that the triangle has 3 corners and the square has 4 corners.
6-D Show the child a square or a triangle on a geoboard or on dot paper. Then have the child make a figure on another geoboard or on dot paper that is the same size and shape as your figure. Have the child explain how he or she knows that the figures are the same size and shape.	_____ makes a figure that is the same size and shape as your figure. _____ explains how he or she knows the figures are the same size and shape.
6-E Use connecting cubes to show the child these patterns. —red, green, green, red, green, green, red, green, green —red, green, blue, red, green, blue, red, green, blue Have the child describe the patterns by naming the colors. Then have the child add 3 more blocks to continue each pattern.	_____ names the colors in the patterns. _____ continues the patterns by adding red, green, green to the first pattern and red, green, blue to the second pattern.

Performance Assessment (See Teacher's Edition, p. 157F)

Chapter 6 • 15

Interview/Task Test • Chapter 6 *continued*

TEST ITEM	EVALUATE WHETHER CHILD
6-F Show the child a set of shapes that include 5 squares, 7 circles, 4 triangles, and 8 rectangles. Have the child look at the grid on PE page 172 and explain how as he or she uses the grid to complete a bar graph for the shapes.	_____ explains how to fill in the grid to show the number of each shape. _____ completes the bar graph correctly.
6-F Use shapes to show the child these patterns, each with a missing shape: —square, circle, square, circle, _?_, circle —circle, triangle, circle, circle, triangle, circle, _?_, triangle, circle Have the child read each pattern and add the missing shape.	_____ reads the patterns correctly. _____ identifies the missing shape in the first pattern as a square and in the second pattern as a circle.

Performance Assessment (See Teacher's Edition, p. 157F)

Evaluation of Interview/Task Test

DATE _____

CHILD'S NAME _____ CLASS _____

TEST ITEM	EVALUATE WHETHER CHILD
7-A Have the child arrange 37 connecting cubes into groups of tens and ones. Have the child tell the number of tens and ones and then tell how many in all.	_____ arranges the cubes into 3 groups of tens and 1 group of 7 ones. _____ identifies the number in all as 37.
7-B Have the child use place-value models to show 32 and 26. Then have him or her explain how to compare the models to tell which number is greater.	_____ models the numbers correctly. _____ explains that 32 is the greater number because 32 has more tens than 26.
7-C Make a cube train with ten cubes. The first cube should be red; all the other cubes should be blue or yellow. Place the cube train in front of the child with the red cube to the left. Tell him or her that the red cube is *first*. Have him or her tell the position of each blue (or yellow) cube.	_____ identifies the correct ordinal numbers for the cubes.
7-D Have the child count by twos from 2 to 20, by fives from 5 to 50, and by tens from 10 to 100.	_____ counts by twos correctly. _____ counts by fives correctly. _____ counts by tens correctly.
7-E Use place-value models to show the child the number 38. Ask the child to explain how he or she would decide whether 20, 40, or 50 is the best estimate for the number shown.	_____ explains that he or she would think about groups of ten to determine the best estimate. _____ identifies the best estimate as 40.
7-E Show the child a row of stacks with 2, 4, 6, and 8 connecting cubes in them. Have the child explain how he or she would decide how many cubes to put in the next stack to continue the pattern.	_____ identifies that there are 2, 4, 6, and 8 cubes in the stacks. _____ states that the next stack would have 10 cubes.

Performance Assessment (See Teacher's Edition, p. 193F)

Evaluation of Interview/Task Test

DATE _____

CHILD'S NAME _____ CLASS _____

TEST ITEM	EVALUATE WHETHER CHILD
8-A Show the child addition fact cards for $8 + 2 = \underline{?}$, $1 + 9 = \underline{?}$, $6 + 6 = \underline{?}$, and $5 + 6 = \underline{?}$. Have the child explain how he or she would find each sum.	_____ explains a strategy that could be used to find each sum; for example, $8 + 2 = \underline{?} \rightarrow$ counting on. $1 + 9 = \underline{?} \rightarrow$ counting on. $6 + 6 = \underline{?} \rightarrow$ doubles. $5 + 6 = \underline{?} \rightarrow$ doubles plus one. _____ identifies the sums as 10, 10, 12, and 11.
8-B Show the child subtraction fact cards for $11 - 2 = \underline{?}$, $12 - 9 = \underline{?}$, and $10 - 5 = \underline{?}$. Have the child explain how to find each difference.	_____ explains a strategy that could be used to solve each fact; for example, $11 - 2 = \underline{?} \rightarrow$ counting back. $12 - 9 = \underline{?} \rightarrow$ counting up. $10 - 5 = \underline{?} \rightarrow$ related addition fact: $5 + 5 = 10$. _____ identifies the differences as 9, 3, and 5.
8-C Have the child look at the pictures at the top of PE page 258. Have him or her explain how to answer these questions: How much would you spend if you bought Little Pig and Little Miss Muffet? How much would you have left if you had 11¢ and bought Jack in the Beanstalk?	_____ adds to find the amount spent ($7¢ + 4¢ = 11¢$). _____ subtracts to find the amount left ($11¢ - 9¢ = 2¢$). _____ identifies the answers as 14¢ and 2¢.
8-D Show the child a train of 8 red connecting cubes and 3 blue connecting cubes and fact cards for $8 + 3 = \underline{?}$, $3 + 8 = \underline{?}$, $11 - 3 = \underline{?}$, and $11 - 8 = \underline{?}$. Have the child explain what is the same and what is different about the facts and how to find the sums or differences.	_____ explains at least one way in which the facts are alike (same numbers). _____ explains at least one way in which the facts are different (two are addition; two are subtraction). _____ identifies the sums as 11 and the differences as 8 and 3.

Performance Assessment (See Teacher's Edition, p. 233F)

Interview/Task Test • Chapter 8 *continued*

TEST ITEM	EVALUATE WHETHER CHILD
8-E Have the child look at the grid on PE page 246 and explain how to make tally marks and use the grid to make a graph to show whether a kitten, puppy, bird, or hamster was the pet liked best by a group of ten classmates.	_____ explains that he or she would ask the ten classmates which pet they liked best. _____ explains that he or she would use tally marks to count. _____ explains that he or she would color a section in the grid for each tally mark.

8-E Read the following story problem.
    Matt has 11 red marbles.
    He has 6 green marbles.
    How many more red marbles than green marbles does he have?
Have the child explain how to decide whether to add or subtract to solve the problem. Then have the child write a number sentence to solve the problem.
	_____ explains that he or she would subtract to find how many more.
	
	_____ writes the number sentence $11 - 6 = \underline{\ ?\ }$.
	
	_____ identifies the answer as 5 more red marbles.

Performance Assessment (See Teacher's Edition, p. 233F)

Evaluation of Interview/Task Test

CHILD'S NAME _____ DATE _____ CLASS _____

TEST ITEM	EVALUATE WHETHER CHILD
9-A Cut strips of paper that are 5 inches long and 8 centimeters long. Have the child use inch and centimeter rulers to measure the lengths of the strips.	_____ aligns rulers with edges of strips. _____ identifies the lengths as 5 inches and 8 centimeters.
9-B Show the child a chalkboard eraser and a reading book. Have him or her tell which object is heavier. Then have him or her use a balance and cubes to tell how many cubes it takes to balance the eraser.	_____ states that the book is heavier than the eraser. _____ weighs the eraser on the balance and tells the number of cubes it weighs.
9-C Show the child a sheet of paper folded in half twice to make four sections. Have the child color $\frac{1}{4}$ of the sheet of paper. Then show him or her a set of 3 counters. Have him or her give you $\frac{1}{3}$ of the counters in the set.	_____ colors 1 of the 4 sections of the sheet of paper to show $\frac{1}{4}$. _____ gives you 1 of the 3 counters in the set to show $\frac{1}{3}$.
9-D Have the child look at the table on PE page 288. Have him or her tell or write a question about the table. Then have the child explain how to use the data in the table to answer the question.	_____ tells or writes an appropriate question for the table. _____ identifies the correct answer to the question.
9-D Have the child look at the pizzas at the bottom of PE page 301. Have the child explain which picture shows a pizza that would give four children equal shares.	_____ identifies the correct pizza.

Performance Assessment (See Teacher's Edition, p. 269F)

Evaluation of Interview/Task Test

DATE _____

CHILD'S NAME _____ CLASS _____

TEST ITEM	EVALUATE WHETHER CHILD
10-A Have the child look at Exercises 3 and 4 on PE page 315. For each exercise, have the child tell which activity takes less time and explain why it takes less time.	_____ chooses the first photo in Exercise 3 and the second photo in Exercise 4. _____ explains why these 2 activities take less time.
10-B Show these times on an analog clock: 5:00, 8:30, 12:00. Have the child tell each time.	identifies the correct time for _____ 5:00. _____ 8:30. _____ 12:00.
10-C Have the child look at a calendar for the current month and answer these questions about the calendar. What is today's date? What day was it yesterday? How many Sundays are in this month?	_____ identifies today's date. _____ identifies yesterday's date. _____ identifies the number of Sundays in this month.
10-D Show the child 1 quarter, 1 dime, 1 nickel, and 3 pennies. Have him or her count on from 25¢ to find the total amount. Then have him or her write the amount.	_____ counts on to find the total amount: 25¢, 35¢, 40¢, 41¢, 42¢, 43¢. _____ writes the total amount as 43¢.
10-E Show the child 2 dimes. Have him or her use other coins to show a different way to make the same amount.	_____ uses coins to show 20¢.

Performance Assessment (See Teacher's Edition, p. 311F)

Interview/Task Test • Chapter 10 *continued*

TEST ITEM	EVALUATE WHETHER CHILD
10-F Read the following story problem. The soccer game started at 3:00. It lasted for 1 hour. At what time did the game end? Have the child show how to use a clock to solve the problem. Then have him or her give the answer.	____ sets the clock at 3:00. ____ moves the hands to show 1 hour later (4:00). ____ identifies the ending time as 4:00.
10-F Read the following story problem. Kim has 3 dimes and 1 nickel. He loses 2 of the dimes. How much money does Kim have now? Have the child show how to use punch-out coins to *act out* the problem and solve it.	____ models the problem correctly. ____ identifies the answer as 15¢.

Performance Assessment (See Teacher's Edition, p. 311F)

Evaluation of Interview/Task Test

DATE _____

CHILD'S NAME _____ CLASS _____

TEST ITEM	EVALUATE WHETHER CHILD
11-A Have the child use 12 counters and Workmat 2 to show and name at least 3 pairs of numbers that have a sum of 12.	_____ use the counters to show pairs of numbers with a sum of 12 _____ writes or names at least 3 pairs of numbers with a sum of 12 (12, 0; 11, 1; 10, 2; 9, 3; 8, 4; 7, 5; or 6, 6).
11-B Show the child 27 connecting cubes. Have the child put the cubes into trains of ten cubes each and then tell how many cubes in all.	_____ makes 2 trains of 10 cubes each. _____ identifies the number of cubes as 27.
11-C Have the child use punch-out tens and ones and Workmat 3 to show the sum 36 + 23 = __?__ .	_____ shows 36 as 3 tens and 6 ones; shows 23 as 2 tens and 3 ones. _____ combines the ones and tens to show the sum. _____ identifies the sum as 59.
11-D Have the child use punch-out tens and ones and Workmat 3 to show the difference 47 − 21 = __?__ .	_____ shows 47 as 4 tens and 7 ones. _____ takes away 2 tens and 1 one to show the difference. _____ identifies the difference as 26.

11-E Read the following story problem.
    Carl had 25¢.
    He lost 5¢.
    How much money does he have left?
Have the child explain whether 20¢, 40¢, or 60¢ is the most reasonable answer to the question and how he or she knows.
	_____ chooses 20¢ as the most reasonable answer.
	
	 _____ explains why 20¢ is the most reasonable answer.

11-E Read the following story problem.
    Lori bought a marker for 23¢ and a pencil for 11¢.
    About how much did she spend?
Have the child explain how to decide whether Lori spent about 20¢ or about 30¢. Then have him or her find the exact answer.
	_____ explains that he or she would add the tens to find about how much Lori spent (2 tens + 1 ten = 20 + 10 = 30).
	
	 _____ identifies that Lori spent about 30¢.
	
	 _____ states that Lori spent exactly 34¢.

Performance Assessment (See Teacher's Edition, p. 351F)

Evaluation of Interview/Task Test

DATE _____

CHILD'S NAME _____ CLASS _____

TEST ITEM	EVALUATE WHETHER CHILD
12-A Show the child addition fact cards for $7 + 7 = \underline{\ ?\ }$, $7 + 8 = \underline{\ ?\ }$, $7 + 6 = \underline{\ ?\ }$, $9 + 5 = \underline{\ ?\ }$, and $3 + 4 + 6 = \underline{\ ?\ }$. Have the child explain how to find each sum and then give each sum.	_____ explains a strategy that could be used to solve each fact; for example, $7 + 7 = \underline{\ ?\ } \rightarrow$ doubles. $7 + 8 = \underline{\ ?\ } \rightarrow$ doubles plus one. $7 + 6 = \underline{\ ?\ } \rightarrow$ doubles minus one. $9 + 5 = \underline{\ ?\ } \rightarrow$ make a ten. $3 + 4 + 6 = \underline{\ ?\ } \rightarrow$ add 2 addends; then add the third. _____ identifies the sums as 14, 15, 13, 14, and 13.
12-B Show the child subtraction fact cards for $14 - 5 = \underline{\ ?\ }$ and $17 - 9 = \underline{\ ?\ }$. Have the child explain how to find the differences and then give the differences.	_____ explains a strategy that could be used to solve each fact; for example, using related addition facts: $5 + 9 = 14$ and $9 + 8 = 17$. _____ identifies the differences as 9 and 8.
12-C Have the child write the two addition and the two subtraction sentences in the fact family for 9, 6, 15.	_____ identifies the two addition sentences as $9 + 6 = 15$ and $6 + 9 = 15$. _____ identifies the two subtraction sentences as $15 - 9 = 6$ and $15 - 6 = 9$.
12-D Have the child look at the table on PE page 398. Have the child answer these questions about the table. How many children in Grade 2 like bananas best? How many children in Grades 2 and 3 like apples best?	_____ finds the information in the table to answer the questions. _____ states that 3 children in Grade 2 like bananas best. _____ states that 14 children in Grades 2 and 3 like apples best.

Performance Assessment (See Teacher's Edition, p. 383F)

Interview/Task Test • Chapter 12 *continued*

TEST ITEM	EVALUATE WHETHER CHILD
12-D Read the following story problem. There are 12 pennies. There are 4 children. Each child gets the same number of pennies. How many pennies does each child get? Have the child use counters to model the problem and then give the answer.	_____ shows 4 equal groups with 3 counters in each group. _____ identifies the answer as 3 pennies each.
12-D Read the following story problem. There are 2 boxes. There are 5 apples in each box. How many apples are in both boxes? Have the child explain how to *draw a picture* or use counters to solve the problem.	_____ draws an appropriate picture or uses counters to show 2 sets of 5 each. _____ identifies the answer as 10 apples.

Performance Assessment (See Teacher's Edition, p. 383F)

Performance Assessment

MATHEMATICS PLUS
Grade 1

Class Record Form

TEACHER _____

Page 1

1-A	To sort by two attributes and to extend a simple pattern											
1-B	To identify groups of 0 to 10 objects											
1-C	To compare and order numbers through 10											
1-D	To solve problems by drawing a picture and by reading a pictograph											
2-A	To identify an addition sentence represented by a model											
2-B	To add basic facts with sums to 6 in horizontal and vertical format											
2-C	To identify combinations of addends with sums to 6											
2-D	To solve problems by using a picture and by acting out problems											
3-A	To identify a subtraction sentence represented by a model											
3-B	To subtract basic facts to 6 in horizontal and vertical format											
3-C	To identify families of facts											
3-D	To solve problems by choosing the operation or choosing an appropriate question											

continued

Performance Assessment

MATHEMATICS PLUS
Grade 1

Class Record Form

Page 2

TEACHER _____

- 4-A To add basic facts to 10
- 4-B To add three addends with sums to 10
- 4-C To solve problems by using pictures and identifying irrelevant information
- 5-A To subtract basic facts through 10
- 5-B To use inverse operations; to identify families of facts; to recall sums and differences through 10
- 5-C To solve problems by guessing and checking and using models
- 6-A To identify plane and solid figures
- 6-B To identify whether an object is inside, outside, or on a plane figure; to identify open and closed figures
- 6-C To count the number of sides and corners of plane figures
- 6-D To identify congruent figures; to identify lines of symmetry
- 6-E To identify and extend patterns
- 6-F To solve problems by using graphs and identifying a pattern

continued

HBJ material copyrighted under notice appearing earlier in this work.

Performance Assessment

MATHEMATICS PLUS
Grade 1

Page 3

Class Record Form

TEACHER _____

7-A	To identify numbers to 99														
7-B	To compare and order numbers to 100														
7-C	To identify ordinal numbers first through tenth														
7-D	To skip-count by twos, fives, and tens														
7-E	To solve problems by using estimation and by using patterns														
8-A	To recall basic facts with sums to 12														
8-B	To recall basic facts with differences from 12														
8-C	To add and subtract pennies; to count pennies to find the total amount and to count change														
8-D	To use inverse operations; to identify families of facts														
8-E	To solve problems by using bar graphs and choosing the operation														

continued

Performance Assessment

MATHEMATICS PLUS
Grade 1

Page 4

Class Record Form

TEACHER _____

- 9-A To estimate and measure length, using nonstandard units, inches, and centimeters
- 9-B To estimate weight, capacity, and temperature
- 9-C To identify equal parts, halves, thirds, and fourths of a region or group of objects
- 9-D To solve problems by using tables and visualizing results
- 10-A To sequence events; to estimate which event takes more or less time
- 10-B To tell time to the hour and half-hour
- 10-C To read a calendar
- 10-D To find the value of a group of coins: pennies, nickels, dimes, and quarters
- 10-E To determine equivalent groups of coins
- 10-F To solve problems by using a picture and by using a model

continued

HBJ material copyrighted under notice appearing earlier in this work.

Performance Assessment

MATHEMATICS PLUS
Grade 1

Page 5

Class Record Form

TEACHER _____

11-A To add and subtract basic facts with sums to 12												
11-B To identify a two-digit number represented by groups of tens and ones												
11-C To add two-digit numbers without regrouping												
11-D To subtract two-digit numbers without regrouping												
11-E To solve problems by using estimation and choosing a sensible answer												
12-A To add basic facts with sums to 18												
12-B To subtract basic facts with sums to 18												
12-C To identify fact families with sums to 18												
12-D To solve problems by using a model and by choosing the strategy												

Performance Checklists

Two types of checklists are included in this section: observation by the teacher and self-evaluation by the student. Observation checklists help evaluate student participation in cooperative learning groups. Student self-evaluation checklists give students a chance to reflect upon their own work in a group and their attitude about what they are learning.

These checklists give you information about students' confidence, flexibility, willingness to persevere, interest, curiosity, and inventiveness, inclination to monitor and reflect on their own thinking and doing, and appreciation of the role of mathematics in our culture.

Student's Name _____

Cooperative Learning Checklist

Ring the response that best describes each student's behavior.

Never — Behavior is not observable.

Sometimes — Behavior is sometimes, but not always, observable when appropriate.

Always — Behavior is observable throughout the activity or whenever appropriate.

The student

• is actively involved in the activity.	Never	Sometimes	Always
• shares materials with others.	Never	Sometimes	Always
• helps others in the group.	Never	Sometimes	Always
• seeks the teacher's help only when all group members need help.	Never	Sometimes	Always
• fulfills their assigned role in the group.	Never	Sometimes	Always
• dominates the activity of the group.	Never	Sometimes	Always
• shares ideas with others.	Never	Sometimes	Always
• tolerates different views within the group about how to solve problems.	Never	Sometimes	Always

> Use this checklist to discuss each student's successful cooperative learning experiences and ways in which he or she can become a more effective group member.

How I Feel About Math

Color the happy face if you agree.
Color the sad face if you disagree.

1. I like math.
2. I like doing math alone.
3. I like doing math with my classmates.
4. I like to do math at home.
5. Math is my favorite subject in school.
6. I like numbers.
7. I like figuring things out.
8. I like solving puzzles.
9. I like using blocks to figure out problems.
10. The best part of math is doing it in your head.
11. I like to talk about math.
12. My class goes too fast in math.
13. I would like math more if there were no story problems.
14. I think math is easy.
15. I like to work on problems that take a long time to solve.

Student Self-Evaluation

How Well Did I Work in My Group Today?

Color the happy face if you agree.
Color the sad face if you disagree.

1. I solve problems better when I work with a group than when I work alone.

2. I shared my ideas with my group.

3. I listened to the ideas of others in my group.

4. I told others in my group when I liked their ideas.

5. I can explain the problem my group worked on and the answer to others.

II
Formal Assessment

Two types of formal assessment are provided in this section. The multiple-choice format is provided to assess mastery of the broad objectives of the program. These tests assess concepts, skills, and problem solving. The two forms of the test can be used as Pretest/Posttest or as two forms of the Posttest. The use of this test format helps prepare students for the standardized achievement tests.

The free-response tests, also given in two forms, are useful diagnostic tools. The work the student performs provides information about what this student understands about the concepts and/or procedures so that appropriate reteaching can be chosen from the many options in the program.

Name _____
Inventory Test

PAGE 1

1.

2.

3.
   Ⓐ ◯   Ⓑ △

4.

5.

6.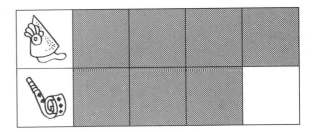

Which one (1) is the right hand; (2) is the same; (3) is the same shape; (4) has the same number; (5) has more; (6) has fewer?

Name _____

Inventory Test

PAGE 2

7.

8.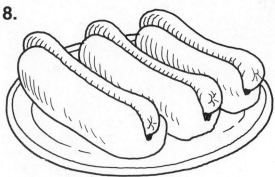

Ⓐ Ⓑ

9.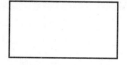

Ⓐ   Ⓑ

10.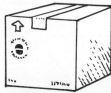

Ⓐ Ⓑ

11.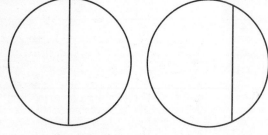

Ⓐ Ⓑ

12.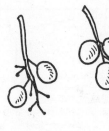

Ⓐ Ⓑ

What number (7) tells how many? Which one (8) has fewer; (9) is a square; (10) is a cone; (11) shows halves; (12) comes next?

 **GO ON**

Name _____

Inventory Test

PAGE 3

13.

Ⓐ 6   Ⓑ 7

14.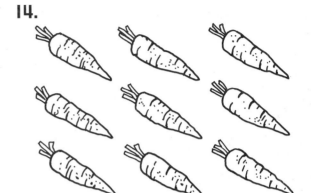

Ⓐ 8   Ⓑ 9

15.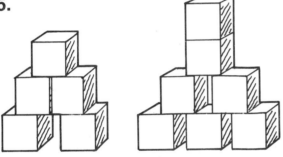

Ⓐ   Ⓑ

16.

Ⓐ ○   Ⓑ △

17.

Ⓐ 12   Ⓑ 13

18.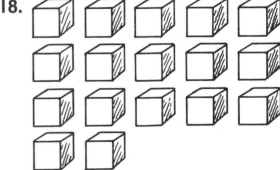

Ⓐ 17   Ⓑ 18

What number (13–14) tells how many? Which one (15) has more; (16) comes next? What number (17–18) tells how many?

GO ON

Name _____

Inventory Test

PAGE 4

19.

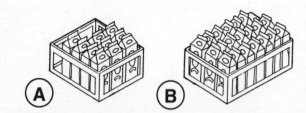

20.

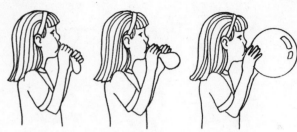

21.

Ⓐ 2 o'clock
Ⓑ 12 o'clock

22.

Ⓐ 1 o'clock
Ⓑ 10 o'clock

23.

Ⓐ   Ⓑ

24.

Ⓐ 1¢   Ⓑ 10¢

Which (19) one has fewer; (20) happened last? What (21–22) time is shown?
Which (23) one is a penny? How (24) many cents are there?

GO ON

Name _____

Inventory Test

PAGE 5

25.

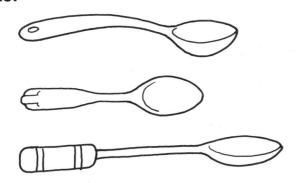

Ⓐ  Ⓑ

26.

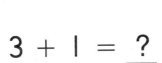

Ⓐ   Ⓑ

27.

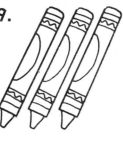

$3 + 1 = \underline{?}$

Ⓐ 3   Ⓑ 4

28.

$5 - 3 = \underline{?}$

Ⓐ 2   Ⓑ 3

29.

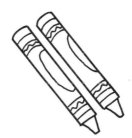

Ⓐ $3 + 2 = 5$
Ⓑ $2 + 1 = 3$

30.

Ⓐ $2 - 1 = 1$
Ⓑ $3 - 1 = 2$

Which (25) one is the longest; (26) holds more? How many (27) in all; (28) are left? Which number sentence (29–30) tells the story?

Name _____

Pretest

CHAPTER 1
PAGE 1

Choose the correct answer.

1. Which one belongs?

Ⓐ

Ⓑ

Ⓒ

Ⓓ

2. Which one belongs?

Ⓐ

Ⓑ

Ⓒ

Ⓓ

3. What comes next?

 ?

Ⓐ 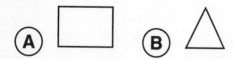 Ⓑ

Ⓒ  Ⓓ

4. What comes next?

 ?

Ⓐ Ⓑ

Ⓒ Ⓓ

GO ON

Name _____

Pretest

CHAPTER 1
PAGE 2

5. How many □ are there?

- (A) 4
- (B) 5
- (C) 6
- (D) not here

6. How many □ are there?

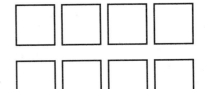

- (A) 4
- (B) 7
- (C) 8
- (D) 9

7. How many □ are there?

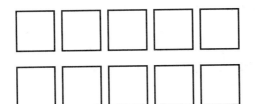

- (A) 5
- (B) 8
- (C) 9
- (D) 10

8. How many □ are there?

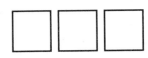

- (A) 2
- (B) 3
- (C) 4
- (D) not here

9. How many △ are there?

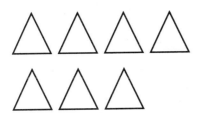

- (A) 5
- (B) 6
- (C) 7
- (D) 8

10. How many ○ are there?

- (A) 0
- (B) 2
- (C) 3
- (D) 8

Name _____

Pretest

CHAPTER 1
PAGE 3

11. How many ☐ are there?

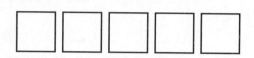

- (A) 2
- (B) 3
- (C) 4
- (D) not here

12. How many ☆ are there?

- (A) 6
- (B) 7
- (C) 9
- (D) 10

13. Which group has more than this?

- (A)
- (B)
- (C) ☆☆☆

14. Which group has fewer than this?

- (A)
- (B)
- (C) ☆☆☆

15. What comes next?

5, 6, 7, ___?

- (A) 4
- (B) 8
- (C) 9
- (D) 10

16. What comes next?

0, 1, 2, ___?

- (A) 3
- (B) 4
- (C) 5
- (D) not here

GO ON

Name _____

Pretest

CHAPTER 1
PAGE 4

Use the pictograph below to answer questions 17–20.

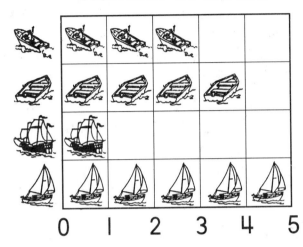

---

17. Which group has more, or ?

18. Which group has fewer, 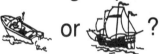 ?

---

19. Which group has more than 4?

20. Which group has fewer than 3?

Name _____

Posttest

CHAPTER 1
PAGE 1

Choose the correct answer.

1. Which one belongs?

 A

 B

 C

 D

2. Which one belongs?

 A

 B

 C

 D

3. What comes next?

 A   B

 C  D

4. What comes next?

 A   B

 C   D

Posttest

CHAPTER 1
PAGE 2

5. How many □ are there?

   (A) 3   (B) 4
   (C) 5   (D) 6

6. How many □ are there?

   (A) 1   (B) 2
   (C) 4   (D) not here

7. How many □ are there?

   (A) 6   (B) 8
   (C) 9   (D) not here

8. How many □ are there?

   (A) 2   (B) 8
   (C) 9   (D) 10

9. How many ○ are there?

   (A) 6   (B) 7
   (C) 8   (D) 9

10. How many □ are there?

    (A) 0   (B) 1
    (C) 2   (D) 10

Name _____

Posttest

CHAPTER 1
PAGE 3

11. How many ☆ are there?

- (A) 0
- (B) 6
- (C) 7
- (D) 8

12. How many ○ are there?

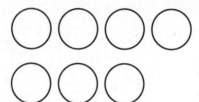

- (A) 4
- (B) 5
- (C) 6
- (D) not here

13. Which group has fewer than this?

- (A)
- (B)
- (C) ☆☆☆☆☆

14. Which group has more than this?

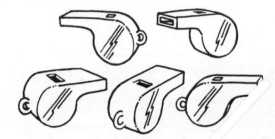

- (A)
- (B) ○○○○○
- (C)

15. What comes next?

2, 3, 4, ___?

- (A) 5
- (B) 6
- (C) 8
- (D) 9

16. What comes next?

6, 7, 8, ___?

- (A) 4
- (B) 5
- (C) 9
- (D) 10

Name _____

Posttest

CHAPTER 1
PAGE 4

Use the pictograph below to answer questions 17–20.

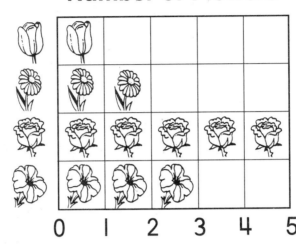

17. Which group has fewer, or ?

     (A)   (B)

18. Which group has more,  ?

    (A)    (B)

19. Which group has more than 4?

    (A)   (B)
    (C)   (D)

20. Which group has fewer than 2?

    (A)   (B)
    (C)   (D)

Name _____

Pretest

> CHAPTER 2
> PAGE 1

Choose the correct answer.

1. Which addition sentence tells how many are there?

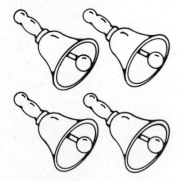

- Ⓐ 4 + 2 = 6
- Ⓑ 4 + 1 = 5
- Ⓒ 1 + 3 = 4
- Ⓓ 2 + 1 = 3

2. Which addition sentence tells how many are there?

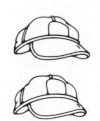

- Ⓐ 2 + 1 = 3
- Ⓑ 3 + 3 = 6
- Ⓒ 2 + 3 = 5
- Ⓓ 3 + 1 = 4

3. Which addition sentence tells how many are there?

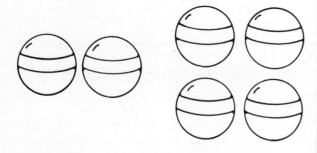

- Ⓐ 2 + 4 = 6
- Ⓑ 2 + 2 = 4
- Ⓒ 2 + 3 = 5
- Ⓓ 1 + 1 = 2

4. Which addition sentence tells how many are there?

- Ⓐ 3 + 3 = 6
- Ⓑ 1 + 3 = 4
- Ⓒ 1 + 1 = 2
- Ⓓ 1 + 2 = 3

GO ON

Name _____

Pretest

CHAPTER **2**
PAGE 2

5. Which addition sentence tells how many are there?

- (A) 2 + 2 = 4
- (B) 1 + 2 = 3
- (C) 1 + 1 = 2
- (D) 2 + 3 = 5

6. Which addition sentence tells how many are there?

- (A) 1 + 3 = 4
- (B) 2 + 1 = 3
- (C) 2 + 3 = 5
- (D) 1 + 1 = 2

7. 3 + 1 = ?

- (A) 3
- (B) 4
- (C) 5
- (D) not here

8. 3 + 3 = ?

- (A) 1
- (B) 3
- (C) 5
- (D) not here

9. 0 + 2 = ?

- (A) 0
- (B) 2
- (C) 3
- (D) 4

10. 3 + 2 = ?

- (A) 2
- (B) 3
- (C) 5
- (D) 6

GO ON

Pretest

CHAPTER 2
PAGE 3

11. 1
    +5
    ───

   Ⓐ 1   Ⓑ 4
   Ⓒ 5   Ⓓ not here

12. 2
    +2
    ───

   Ⓐ 0   Ⓑ 2
   Ⓒ 3   Ⓓ 4

13. 5
    +0
    ───

   Ⓐ 0   Ⓑ 1
   Ⓒ 5   Ⓓ 6

14. 1
    +1
    ───

   Ⓐ 0   Ⓑ 1
   Ⓒ 2   Ⓓ not here

15. Which is a name for 3?

   Ⓐ 3 + 0
   Ⓑ 1 + 3
   Ⓒ 3 + 3
   Ⓓ 0 + 6

16. Which is a name for 5?

   Ⓐ 1 + 5
   Ⓑ 2 + 3
   Ⓒ 3 + 3
   Ⓓ 3 + 1

17. Which is a name for 6?

   Ⓐ 1 + 3
   Ⓑ 3 + 2
   Ⓒ 3 + 3
   Ⓓ 5 + 0

18. Which is a name for 2?

   Ⓐ 1 + 1
   Ⓑ 2 + 1
   Ⓒ 1 + 3
   Ⓓ 2 + 2

Name _____

Pretest

CHAPTER 2
PAGE 4

19. Which is a name for 1?

(A) 1 + 1
(B) 1 + 2
(C) 2 + 2
(D) 1 + 0

20. Which is a name for 4?

(A) 3 + 1
(B) 4 + 1
(C) 3 + 3
(D) 0 + 3

21.

4 cats sleep.
2 cats play.

How many cats are there?

(A) 2 cats (B) 4 cats
(C) 5 cats (D) 6 cats

22.

1 girl plays.
3 more girls play.

How many girls play?

(A) 1 girl (B) 3 girls
(C) 4 girls (D) 5 girls

23.

1 dog sits.
1 dog runs.

How many dogs are there?

(A) 1 dog   (B) 2 dogs
(C) 3 dogs (D) 4 dogs

24.

3 boys read.
3 boys write.

How many boys are there?

(A) 3 boys (B) 4 boys
(C) 5 boys (D) 6 boys

STOP

Name _____

Posttest

CHAPTER 2
PAGE 1

Choose the correct answer.

1. Which addition sentence tells how many are there?

- (A) $4 + 2 = 6$
- (B) $3 + 2 = 5$
- (C) $2 + 2 = 4$
- (D) $2 + 1 = 3$

2. Which addition sentence tells how many are there?

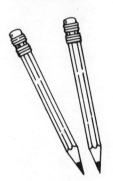 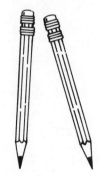

- (A) $1 + 1 = 2$
- (B) $2 + 3 = 5$
- (C) $2 + 2 = 4$
- (D) $4 + 2 = 6$

3. Which addition sentence tells how many are there?

 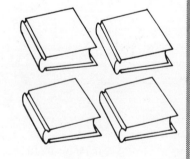

- (A) $1 + 4 = 5$
- (B) $2 + 4 = 6$
- (C) $2 + 2 = 4$
- (D) $1 + 3 = 4$

4. Which addition sentence tells how many are there?

- (A) $2 + 1 = 3$
- (B) $3 + 1 = 4$
- (C) $3 + 3 = 6$
- (D) $3 + 2 = 5$

GO ON

Name _____

Posttest

CHAPTER 2
PAGE 2

5. Which addition sentence tells how many are there?

Ⓐ 3 + 3 = 6
Ⓑ 1 + 2 = 3
Ⓒ 2 + 2 = 4
Ⓓ 3 + 2 = 5

6. Which addition sentence tells how many are there?

Ⓐ 1 + 3 = 4
Ⓑ 1 + 2 = 3
Ⓒ 2 + 3 = 5
Ⓓ 1 + 1 = 2

7. 2 + 3 = ?

Ⓐ 3   Ⓑ 4
Ⓒ 5   Ⓓ 6

8. 0 + 4 = ?

Ⓐ 0   Ⓑ 3
Ⓒ 4   Ⓓ 5

9. 1 + 1 = ?

Ⓐ 1   Ⓑ 2
Ⓒ 3   Ⓓ not here

10. 3 + 3 = ?

Ⓐ 3   Ⓑ 4
Ⓒ 5   Ⓓ not here

Name _____

Posttest

CHAPTER 2
PAGE 3

11. 2
    +2

   Ⓐ 2   Ⓑ 3
   Ⓒ 4   Ⓓ 5

12. 5
    +1

   Ⓐ 1   Ⓑ 4
   Ⓒ 5   Ⓓ 6

13. 3
    +0

   Ⓐ 0   Ⓑ 3
   Ⓒ 4   Ⓓ not here

14. 2
    +4

   Ⓐ 2   Ⓑ 3
   Ⓒ 5   Ⓓ not here

15. Which is a name for 4?

   Ⓐ 3 + 1
   Ⓑ 2 + 3
   Ⓒ 3 + 3
   Ⓓ 0 + 3

16. Which is a name for 1?

   Ⓐ 0 + 1
   Ⓑ 1 + 1
   Ⓒ 1 + 2
   Ⓓ 2 + 2

17. Which is a name for 5?

   Ⓐ 1 + 3
   Ⓑ 3 + 2
   Ⓒ 3 + 3
   Ⓓ 4 + 0

18. Which is a name for 6?

   Ⓐ 1 + 5
   Ⓑ 5 + 0
   Ⓒ 3 + 2
   Ⓓ 2 + 2

GO ON

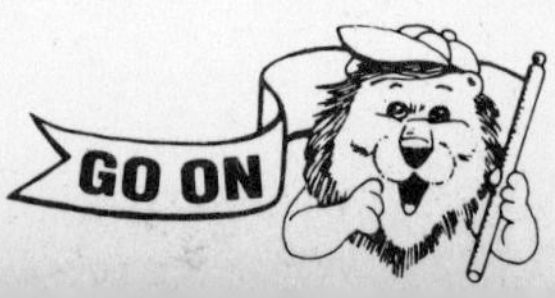

Name _____

Posttest

CHAPTER 2
PAGE 4

19. Which is a name for 3?
   - (A) 1 + 1
   - (B) 0 + 2
   - (C) 2 + 2
   - (D) 3 + 0

20. Which is a name for 2?
   - (A) 3 + 1
   - (B) 0 + 2
   - (C) 1 + 0
   - (D) 2 + 2

21.

2 frogs jump.
3 frogs sit.

How many frogs are there?

- (A) 2 frogs
- (B) 4 frogs
- (C) 5 frogs
- (D) 6 frogs

22.

1 girl plays.
1 girl reads.

How many girls are there?

- (A) 0 girls
- (B) 1 girl
- (C) 2 girls
- (D) 3 girls

23.

4 boys smile.
2 more boys smile.

How many boys smile in all?

- (A) 2 boys
- (B) 4 boys
- (C) 5 boys
- (D) 6 boys

24.

1 cat sleeps.
5 cats drink.

How many cats are there?

- (A) 1 cat
- (B) 4 cats
- (C) 5 cats
- (D) 6 cats

Name _____

Pretest

CHAPTER 3
PAGE 1

Choose the correct answer.

1. Which subtraction sentence tells the story?

- Ⓐ $6 - 1 = 5$
- Ⓑ $6 - 6 = 0$
- Ⓒ $6 - 3 = 3$
- Ⓓ $6 - 2 = 4$

2. Which subtraction sentence tells the story?

- Ⓐ $4 - 1 = 3$
- Ⓑ $4 - 0 = 4$
- Ⓒ $4 - 2 = 2$
- Ⓓ $4 - 1 = 3$

3. Which subtraction sentence tells the story?

- Ⓐ $4 - 0 = 4$
- Ⓑ $4 - 1 = 3$
- Ⓒ $5 - 1 = 4$
- Ⓓ $5 - 5 = 0$

4. Which subtraction sentence tells the story?

- Ⓐ $3 - 3 = 0$
- Ⓑ $3 - 0 = 3$
- Ⓒ $6 - 5 = 1$
- Ⓓ $6 - 3 = 3$

GO ON

Name _____

Pretest

CHAPTER 3
PAGE 2

5.

$4 - 2 = \underline{?}$

- Ⓐ 1
- Ⓑ 2
- Ⓒ 3
- Ⓓ 6

6.

$5 - 4 = \underline{?}$

- Ⓐ 0
- Ⓑ 2
- Ⓒ 5
- Ⓓ not here

7.

$3 - 3 = \underline{?}$

- Ⓐ 0
- Ⓑ 1
- Ⓒ 3
- Ⓓ 5

8.

$2 - 1 = \underline{?}$

- Ⓐ 0
- Ⓑ 1
- Ⓒ 2
- Ⓓ 3

9. $4 - 3 = \underline{?}$

- Ⓐ 1
- Ⓑ 2
- Ⓒ 3
- Ⓓ not here

10. $3 - 0 = \underline{?}$

- Ⓐ 0
- Ⓑ 1
- Ⓒ 2
- Ⓓ 3

11. $3 - 1 = \underline{?}$

- Ⓐ 0
- Ⓑ 1
- Ⓒ 2
- Ⓓ 3

12. $6 - 5 = \underline{?}$

- Ⓐ 1
- Ⓑ 2
- Ⓒ 3
- Ⓓ 5

GO ON

Name _____

Pretest

CHAPTER 3
PAGE 3

13. 6
    −4

Ⓐ 0   Ⓑ 1
Ⓒ 2   Ⓓ 10

14. 1
    −1

Ⓐ 0   Ⓑ 1
Ⓒ 2   Ⓓ not here

15. 5
    −3

Ⓐ 2   Ⓑ 3
Ⓒ 5   Ⓓ 8

16. 4
    −2

Ⓐ 0   Ⓑ 3
Ⓒ 4   Ⓓ not here

17. Which subtraction sentence goes with the addition sentence?

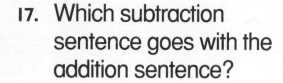

3 + 2 = 5

Ⓐ 3 − 3 = 0
Ⓑ 3 − 2 = 1
Ⓒ 5 − 1 = 4
Ⓓ 5 − 2 = 3

18. Which subtraction sentence goes with the addition sentence?

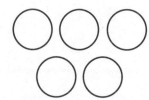

5 + 1 = 6

Ⓐ 5 − 1 = 4
Ⓑ 6 − 1 = 5
Ⓒ 6 − 3 = 4
Ⓓ 3 − 3 = 0

GO ON

Name _____

Pretest

CHAPTER 3
PAGE 4

19. Which number sentence does not belong?

 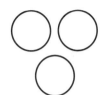

- (A) $1 + 3 = 4$
- (B) $4 - 3 = 1$
- (C) $4 - 1 = 3$
- (D) $3 - 1 = 2$

20. Which number sentence does not belong?

 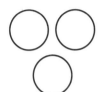

- (A) $3 - 2 = 1$
- (B) $5 - 2 = 3$
- (C) $5 - 3 = 2$
- (D) $3 + 2 = 5$

21.

What do you do to find how many 🐖 in all?

(A) Add (B) Subtract

22.

What do you do to find how many  are left?

(A) Add (B) Subtract

Name _____

Pretest

CHAPTER 3
PAGE 5

23. Sara sees 4 🦆. 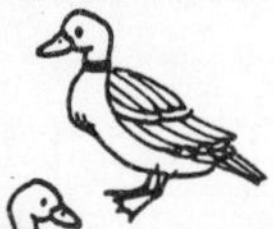
Then 1 🦆 swims away.
Which is the correct question?

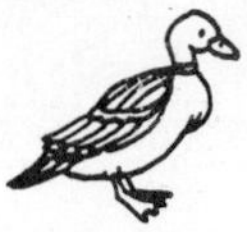

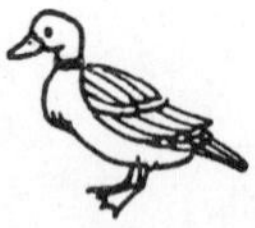

Ⓐ How many 🦆 are left?

Ⓑ How many 🦆 are there in all?

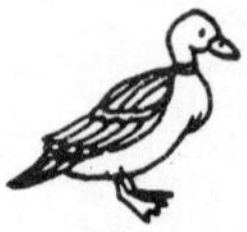

24. Tom sees 1 ✈️. 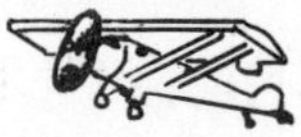
Then he sees 1 more ✈️. 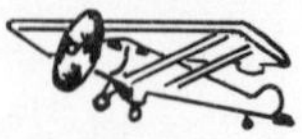
Which is the correct question?

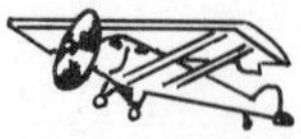

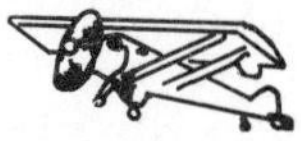

Ⓐ How many ✈️ are left?

Ⓑ How many ✈️ are there in all?

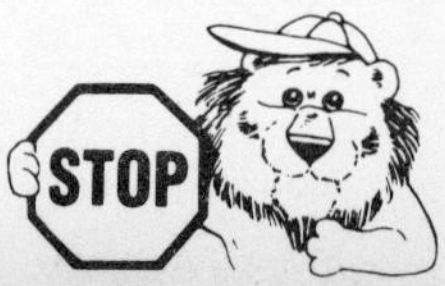

Name _____

Posttest

CHAPTER 3
PAGE 1

Choose the correct answer.

1. Which subtraction sentence tells the story?

- Ⓐ $3 - 2 = 1$
- Ⓑ $3 - 1 = 2$
- Ⓒ $4 - 1 = 3$
- Ⓓ $4 - 2 = 2$

2. Which subtraction sentence tells the story?

- Ⓐ $5 - 1 = 4$
- Ⓑ $5 - 2 = 3$
- Ⓒ $3 - 2 = 1$
- Ⓓ $3 - 3 = 0$

3. Which subtraction sentence tells the story?

- Ⓐ $6 - 3 = 3$
- Ⓑ $3 - 1 = 2$
- Ⓒ $3 - 3 = 0$
- Ⓓ $4 - 1 = 3$

4. Which subtraction sentence tells the story?

- Ⓐ $4 - 1 = 3$
- Ⓑ $5 - 4 = 1$
- Ⓒ $5 - 1 = 4$
- Ⓓ $6 - 1 = 5$

GO ON

Name _____

Posttest

CHAPTER 3
PAGE 2

5.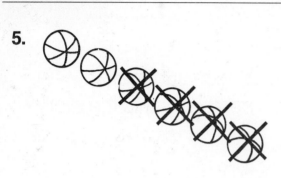

$6 - 4 = \underline{?}$

- Ⓐ 1
- Ⓑ 2
- Ⓒ 4
- Ⓓ not here

6.

$5 - 3 = \underline{?}$

- Ⓐ 2
- Ⓑ 3
- Ⓒ 5
- Ⓓ 8

7.

$4 - 3 = \underline{?}$

- Ⓐ 1
- Ⓑ 3
- Ⓒ 4
- Ⓓ not here

8.

$2 - 2 = \underline{?}$

- Ⓐ 0
- Ⓑ 1
- Ⓒ 2
- Ⓓ 4

9. $5 - 4 = \underline{?}$

- Ⓐ 1
- Ⓑ 2
- Ⓒ 3
- Ⓓ 4

10. $6 - 2 = \underline{?}$

- Ⓐ 0
- Ⓑ 2
- Ⓒ 4
- Ⓓ 6

11. $2 - 1 = \underline{?}$

- Ⓐ 0
- Ⓑ 1
- Ⓒ 2
- Ⓓ 3

12. $4 - 0 = \underline{?}$

- Ⓐ 0
- Ⓑ 2
- Ⓒ 3
- Ⓓ not here

GO ON

Name _____

Posttest

CHAPTER 3
PAGE 3

13. 4
    −4

(A) 0   (B) 1
(C) 2   (D) 4

14. 6
    −3

(A) 0   (B) 2
(C) 3   (D) 6

15. 5
    −1

(A) 2   (B) 3
(C) 5   (D) not here

16. 6
    −5

(A) 0   (B) 1
(C) 2   (D) 3

17. Which subtraction sentence goes with the addition sentence?

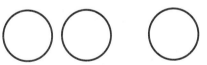

2 + 1 = 3

(A) 2 − 1 = 1
(B) 3 − 3 = 0
(C) 2 − 2 = 0
(D) 3 − 1 = 2

18. Which subtraction sentence goes with the addition sentence?

2 + 2 = 4

(A) 2 − 0 = 2
(B) 4 − 2 = 2
(C) 3 − 3 = 0
(D) 3 − 2 = 1

Name _____

Posttest

CHAPTER 3
PAGE 4

19. Which number sentence does not belong?

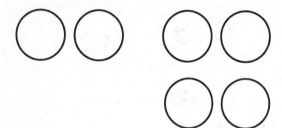

Ⓐ $4 + 2 = 6$
Ⓑ $6 - 4 = 2$
Ⓒ $4 - 2 = 2$
Ⓓ $6 - 2 = 4$

20. Which number sentence does not belong?

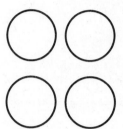

Ⓐ $5 - 1 = 4$
Ⓑ $4 + 1 = 5$
Ⓒ $5 - 4 = 1$
Ⓓ $4 - 1 = 3$

21.

What do you do to find how many  in all?

Ⓐ Add   Ⓑ Subtract

22.

What do you do to find how many are left?

Ⓐ Add   Ⓑ Subtract

Name _____

Posttest

CHAPTER 3
PAGE 5

23. There are 5 in this . Then 2 fly away.

    Which is the correct question?

    Ⓐ How many are left?

    Ⓑ How many are there in all?

24. Sara sees 5 turtles in a pond. Then she sees 1 more turtle.

    Which is the correct question?

    Ⓐ How many are left?

    Ⓑ How many are there in all?

Name _____

Cumulative Test

CHAPTERS 1–3
PAGE 1

Choose the letter of the correct answer.

1. How many ○ are there?

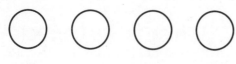

- (A) 3
- (B) 6
- (C) 7
- (D) 8

2. How many ▢ are there?

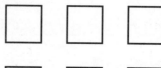
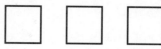

- (A) 5
- (B) 6
- (C) 7
- (D) 9

3. How many △ are there?

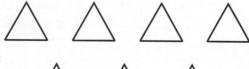

- (A) 5
- (B) 6
- (C) 7
- (D) 8

4. How many ☆ are there?

- (A) 0
- (B) 1
- (C) 9
- (D) not here

5. How many ◇ are there?

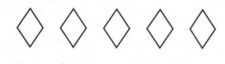

- (A) 6
- (B) 8
- (C) 9
- (D) not here

6. How many ▭ are there?

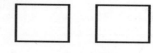

- (A) 3
- (B) 4
- (C) 5
- (D) 6

Name _____

Cumulative Test

CHAPTERS 1-3
PAGE 2

7. Which group has fewer than this?

A

B

C

8. Which group has fewer than this?

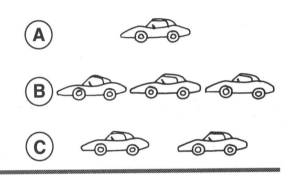
A
B
C

9. Which group has more than this?

A

B

C

10. Which group has more than this?

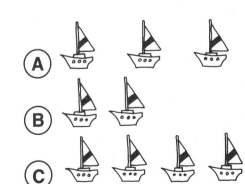
A
B
C

GO ON

Cumulative Test

CHAPTERS 1–3
PAGE 3

11. Which comes next?

    3, 4, 5, ?

    - (A) 1
    - (B) 2
    - (C) 6
    - (D) 9

12. Which comes next?

    7, 8, 9, ?

    - (A) 0
    - (B) 1
    - (C) 6
    - (D) 10

13. Which comes next?

    0, 1, 2, ?

    - (A) 4
    - (B) 5
    - (C) 6
    - (D) not here

14. Which comes next?

    4, 5, 6, ?

    - (A) 3
    - (B) 5
    - (C) 7
    - (D) 9

15. Which addition sentence tells how many in all?

    - (A) 2 + 1 = 3
    - (B) 3 + 2 = 5
    - (C) 4 + 2 = 6
    - (D) 3 + 1 = 4

16. Which addition sentence tells how many in all?

    - (A) 2 + 2 = 4
    - (B) 2 + 4 = 6
    - (C) 4 + 4 = 8
    - (D) 2 + 3 = 5

Name _____

Cumulative Test

CHAPTERS **1–3**
PAGE 4

17. Which addition sentence tells how many in all?

- Ⓐ 2 + 2 = 4
- Ⓑ 1 + 1 = 2
- Ⓒ 3 + 0 = 3
- Ⓓ 4 + 4 = 8

18. Which addition sentence tells how many in all?

- Ⓐ 1 + 2 = 3
- Ⓑ 1 + 1 = 2
- Ⓒ 2 + 3 = 5
- Ⓓ 2 + 2 = 4

19. Which addition sentence tells how many in all?

- Ⓐ 2 + 3 = 5
- Ⓑ 1 + 5 = 6
- Ⓒ 1 + 0 = 1
- Ⓓ 2 + 5 = 7

20. Which addition sentence tells how many in all?

- Ⓐ 2 + 2 = 4
- Ⓑ 4 + 1 = 5
- Ⓒ 1 + 1 = 2
- Ⓓ 3 + 3 = 6

Name _____

Cumulative Test

CHAPTERS 1–3
PAGE 5

21. $5 + 1 =$ ___?___
 - (A) 1
 - (B) 4
 - (C) 5
 - (D) 6

22. $2 + 0 =$ ___?___
 - (A) 0
 - (B) 1
 - (C) 2
 - (D) 3

23. $2 + 2 =$ ___?___
 - (A) 0
 - (B) 1
 - (C) 2
 - (D) 4

24. $1 + 4 =$ ___?___
 - (A) 1
 - (B) 4
 - (C) 5
 - (D) 6

25. $\begin{array}{r} 0 \\ +1 \\ \hline \end{array}$
 - (A) 0
 - (B) 1
 - (C) 2
 - (D) not here

26. $\begin{array}{r} 3 \\ +3 \\ \hline \end{array}$
 - (A) 0
 - (B) 1
 - (C) 3
 - (D) not here

27. $\begin{array}{r} 2 \\ +3 \\ \hline \end{array}$
 - (A) 5
 - (B) 6
 - (C) 7
 - (D) 8

28. Which subtraction sentence tells the story?

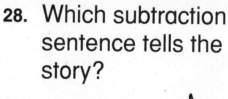

 - (A) $4 - 2 = 2$
 - (B) $6 - 2 = 4$
 - (C) $6 - 1 = 5$
 - (D) $4 - 3 = 1$

GO ON

Name _____

Cumulative Test

CHAPTERS 1–3
PAGE 6

29. Which subtraction sentence tells the story?

(A) 5 − 2 = 3
(B) 5 − 1 = 4
(C) 3 − 2 = 1
(D) 3 − 3 = 0

30. Which subtraction sentence tells the story?

(A) 3 − 1 = 2
(B) 4 − 3 = 1
(C) 4 − 2 = 2
(D) 3 − 3 = 0

31. Which subtraction sentence tells the story?

(A) 2 − 1 = 1
(B) 1 − 1 = 0
(C) 4 − 1 = 3
(D) 3 − 1 = 2

32. 4 − 1 = ___?

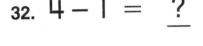

(A) 1   (B) 2
(C) 3   (D) 4

33. 6 − 3 = ___?

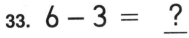

(A) 0   (B) 3
(C) 6   (D) not here

34. 5 − 0 = ___?

(A) 0   (B) 1
(C) 5   (D) 6

GO ON

Name _____

Cumulative Test

CHAPTERS 1–3
PAGE 7

35. $4 - 4 = \underline{\ ?\ }$

~~🍎~~ ~~🍎~~ ~~🍎~~ ~~🍎~~

- Ⓐ 0
- Ⓑ 1
- Ⓒ 2
- Ⓓ 4

36. $\begin{array}{r} 5 \\ -3 \\ \hline \end{array}$

- Ⓐ 2
- Ⓑ 3
- Ⓒ 5
- Ⓓ 8

37. $6 - 4 = \underline{\ ?\ }$

- Ⓐ 0
- Ⓑ 1
- Ⓒ 2
- Ⓓ 3

38. $2 - 1 = \underline{\ ?\ }$

- Ⓐ 1
- Ⓑ 2
- Ⓒ 3
- Ⓓ 4

39. $4 - 0 = \underline{\ ?\ }$

- Ⓐ 0
- Ⓑ 1
- Ⓒ 4
- Ⓓ 5

40. $6 - 5 = \underline{\ ?\ }$

- Ⓐ 0
- Ⓑ 1
- Ⓒ 5
- Ⓓ 6

41. $3 - 3 = \underline{\ ?\ }$

- Ⓐ 0
- Ⓑ 1
- Ⓒ 3
- Ⓓ 6

42. $4 - 1 = \underline{\ ?\ }$

- Ⓐ 0
- Ⓑ 1
- Ⓒ 2
- Ⓓ not here

43. Jack sees 3 🐟. Then he sees 2 more 🐟. Which is the correct question?

- Ⓐ How many 🐟 are there in all?
- Ⓑ How many 🐟 are left?

44. Jane sees 5 🐦. Then 4 🐦 fly away. Which is the correct question?

- Ⓐ How many 🐦 are there in all?
- Ⓑ How many 🐦 are left?

GO ON

Name _____

Cumulative Test

CHAPTERS 1–3
PAGE 8

Use the pictograph below to answer questions 45–50.

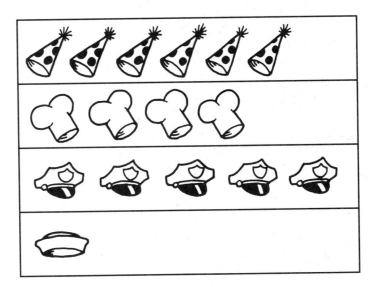

45. Which group has fewer, or ?

   A   B

46. Which group has more, or ?

   A   B

47. Which group has more than 5?

   A   B
   C   D

48. Which group has fewer than 4?

   A   B
   C   D

49. Which group has more than 5?

   A
   B

50. Which group has fewer, or ?

   A
   B

Name _____

Pretest

CHAPTER 4
PAGE 1

Choose the correct answer.

1.

$5 + 3 = \underline{?}$

Ⓐ 5   Ⓑ 7
Ⓒ 8   Ⓓ 9

2.

$6 + 4 = \underline{?}$

Ⓐ 2   Ⓑ 6
Ⓒ 9   Ⓓ 10

3.
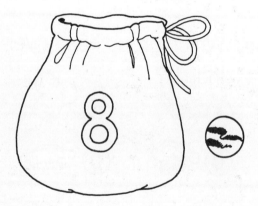

$8 + 1 = \underline{?}$

Ⓐ 7   Ⓑ 8
Ⓒ 9   Ⓓ 10

4.

$5 + 2 = \underline{?}$

Ⓐ 6   Ⓑ 7
Ⓒ 8   Ⓓ 9

Name _____

Pretest

CHAPTER 4
PAGE 2

5.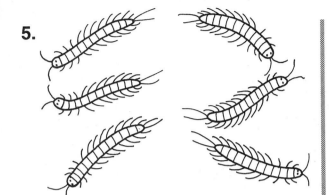

3 + 3 = ?

- Ⓐ 0
- Ⓑ 5
- Ⓒ 6
- Ⓓ 9

6.

5 + 5 = ?

- Ⓐ 5
- Ⓑ 6
- Ⓒ 9
- Ⓓ 10

7. 0 + 2 = ?
- Ⓐ 0
- Ⓑ 2
- Ⓒ 3
- Ⓓ not here

8. 5 + 1 = ?
- Ⓐ 5
- Ⓑ 6
- Ⓒ 7
- Ⓓ 9

9. 4 + 4 = ?
- Ⓐ 0
- Ⓑ 4
- Ⓒ 8
- Ⓓ 9

10. 2 + 7 = ?
- Ⓐ 5
- Ⓑ 7
- Ⓒ 8
- Ⓓ 9

11. 4 + 3 = ?
- Ⓐ 7
- Ⓑ 8
- Ⓒ 9
- Ⓓ not here

12. 3 + 2 = ?
- Ⓐ 1
- Ⓑ 3
- Ⓒ 4
- Ⓓ not here

13. 8 + 2 = ?
- Ⓐ 6
- Ⓑ 8
- Ⓒ 9
- Ⓓ 10

14. 2 + 2 = ?
- Ⓐ 0
- Ⓑ 2
- Ⓒ 4
- Ⓓ 5

GO ON

Name _____

Pretest

CHAPTER 4
PAGE 3

15. Which is a name for 9?

   (A) 1 + 6   (B) 4 + 3
   (C) 5 + 2   (D) 9 + 0

16. Which is a name for 10?

   (A) 4 + 4   (B) 6 + 3
   (C) 7 + 0   (D) 8 + 2

17. Which addition sentence tells how many in all?

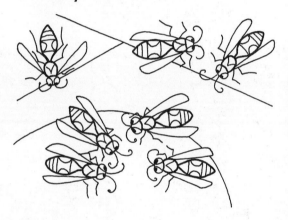

   (A) 5 + 1 + 1 = 7
   (B) 4 + 2 + 1 = 7
   (C) 4 + 1 = 5
   (D) 2 + 1 = 3

18. Which addition sentence tells how many in all?

   (A) 5 + 2 = 7
   (B) 2 + 2 = 4
   (C) 5 + 2 + 2 = 9
   (D) 4 + 2 + 2 = 8

19. 3 + 0 + 3 = ?

   (A) 0   (B) 3
   (C) 6   (D) 9

20. 4 + 1 + 3 = ?

   (A) 4   (B) 5
   (C) 8   (D) not here

Name _____

Pretest

CHAPTER 4
PAGE 4

21. How much did Kara spend?

Ⓐ 3¢   Ⓑ 4¢
Ⓒ 7¢   Ⓓ 10¢

22. How much did Karl spend?

Ⓐ 2¢   Ⓑ 5¢
Ⓒ 6¢   Ⓓ 7¢

23. Which sentence is not needed?

Lin sees 5 birds.
Then she sees 3 more birds.
Then she sees 2 cats.
How many birds does she see in all?

Ⓐ Lin sees 5 birds.
Ⓑ Then she sees 3 more birds.
Ⓒ Then she sees 2 cats.

24. Which sentence is not needed?

Tom feeds 4 dogs.
Tom feeds 3 cats.
Then he feeds 2 more cats.
How many cats does he feed in all?

Ⓐ Tom feeds 4 dogs.
Ⓑ Tom feeds 3 cats.
Ⓒ Then he feeds 2 more cats.

STOP

Name _____

Posttest

CHAPTER 4
PAGE 1

Choose the correct answer.

1.

$7 + 1 = \underline{?}$

- Ⓐ 6
- Ⓑ 7
- Ⓒ 8
- Ⓓ 9

2.

$8 + 2 = \underline{?}$

- Ⓐ 6
- Ⓑ 8
- Ⓒ 9
- Ⓓ 10

3.

$6 + 3 = \underline{?}$

- Ⓐ 3
- Ⓑ 6
- Ⓒ 7
- Ⓓ 9

4.

$5 + 2 = \underline{?}$

- Ⓐ 3
- Ⓑ 7
- Ⓒ 8
- Ⓓ 9

GO ON

Name _____

Posttest

CHAPTER 4
PAGE 2

5.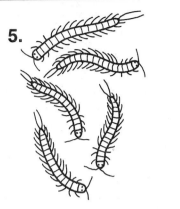

5 + 5 =

(A) 0   (B) 5
(C) 9   (D) 10

6.

4 + 4 = ?

(A) 4   (B) 7
(C) 8   (D) 9

7. 0 + 3 = ?

(A) 0   (B) 1
(C) 3   (D) not here

8. 1 + 4 = ?

(A) 1   (B) 4
(C) 5   (D) 6

9. 6 + 2 = ?

(A) 2   (B) 4
(C) 7   (D) 8

10. 5 + 3 = ?

(A) 7   (B) 8
(C) 9   (D) not here

11. 6 + 4 = ?

(A) 2   (B) 8
(C) 9   (D) not here

12. 7 + 0 = ?

(A) 0   (B) 1
(C) 7   (D) 8

13. 4 + 5 = ?

(A) 0   (B) 1
(C) 9   (D) 10

14. 3 + 3 = ?

(A) 0   (B) 3
(C) 6   (D) not here

Posttest

CHAPTER 4
PAGE 3

15. Which is a name for 8?

   Ⓐ 8 + 1   Ⓑ 3 + 3
   Ⓒ 6 + 3   Ⓓ 4 + 4

16. Which is a name for 7?

   Ⓐ 1 + 8   Ⓑ 3 + 2
   Ⓒ 4 + 1   Ⓓ 5 + 2

17. Which addition sentence tells how many in all?

   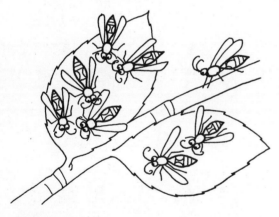

   Ⓐ 5 + 1 = 6
   Ⓑ 5 + 1 + 2 = 8
   Ⓒ 4 + 2 + 2 = 8
   Ⓓ 1 + 2 = 3

18. Which addition sentence tells how many in all?

   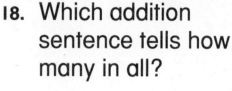

   Ⓐ 4 + 2 + 2 = 8
   Ⓑ 2 + 2 + 2 = 6
   Ⓒ 2 + 0 = 2
   Ⓓ 2 + 2 = 4

19. 4 + 0 + 4 = ___

   Ⓐ 0   Ⓑ 4
   Ⓒ 8   Ⓓ 9

20. 3 + 2 + 5 = ___

   Ⓐ 5   Ⓑ 7
   Ⓒ 9   Ⓓ 10

GO ON

Name _____

Posttest

CHAPTER 4
PAGE 4

21. How much did Kathy spend?

- (A) 2¢
- (B) 6¢
- (C) 9¢
- (D) 10¢

22. How much did Carlos spend?

- (A) 1¢
- (B) 8¢
- (C) 9¢
- (D) 10¢

23. Which sentence is <u>not</u> needed?

Nan sees 4 cows.
Then she sees 3 more cows.
Then she sees 2 dogs.
How many cows does she see in all?

- (A) Nan sees 4 cows.
- (B) Then she sees 3 more cows.
- (C) Then she sees 2 dogs.

24. Which sentence is <u>not</u> needed?

Sam feeds 6 fish.
Sam feeds 4 birds.
Then he feeds 2 more birds.
How many birds does he feed in all?

- (A) Sam feeds 6 fish.
- (B) Sam feeds 4 birds.
- (C) Then he feeds 2 more birds.

STOP

Name _____

Pretest

CHAPTER 5
PAGE 1

Choose the correct answer.

1.

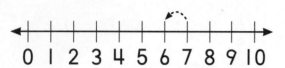

7 − 1 = _?_

- Ⓐ 5
- Ⓑ 6
- Ⓒ 7
- Ⓓ 8

2.

8 − 3 = _?_

- Ⓐ 4
- Ⓑ 5
- Ⓒ 6
- Ⓓ 7

3.

9 − 3 = _?_

- Ⓐ 6
- Ⓑ 7
- Ⓒ 8
- Ⓓ 9

4.

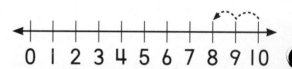

10 − 2 = _?_

- Ⓐ 5
- Ⓑ 6
- Ⓒ 7
- Ⓓ 8

5.

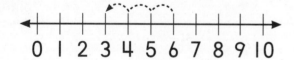

6 − 3 = _?_

- Ⓐ 3
- Ⓑ 4
- Ⓒ 5
- Ⓓ 9

6.

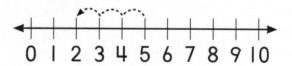

5 − 3 = _?_

- Ⓐ 2
- Ⓑ 3
- Ⓒ 4
- Ⓓ 8

GO ON

Name _____

Pretest

CHAPTER 5
PAGE 2

7. 7
   −2

   (A) 5   (B) 6
   (C) 8   (D) 9

8. 10
   − 7

   (A) 2   (B) 3
   (C) 4   (D) not here

9. 8
   −4

   (A) 0   (B) 3
   (C) 5   (D) not here

10. 6
    −6

   (A) 0   (B) 1
   (C) 6   (D) 9

11. 9 − 2 = ?

   (A) 7   (B) 8
   (C) 12   (D) not here

12. 8 − 6 = ?

   (A) 1   (B) 3
   (C) 4   (D) not here

13. 7 − 0 = ?

   (A) 0   (B) 1
   (C) 7   (D) 8

14. 9 − 8 = ?

   (A) 1   (B) 2
   (C) 4   (D) 9

GO ON

Pretest

CHAPTER 5
PAGE 3

15. Use counters to solve.

Dale has 6 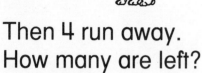.

Then 4 run away.
How many are left?

Ⓐ 2
Ⓑ 4
Ⓒ 6
Ⓓ 10

16. Use counters to solve.

There are 8 🚗.

Then 1 leaves.
How many are left?

Ⓐ 6
Ⓑ 7
Ⓒ 8
Ⓓ 9

17.

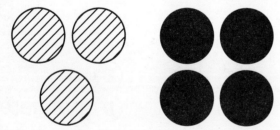

3 + 4 = 7
7 − 4 = ?

Ⓐ 1 Ⓑ 2
Ⓒ 3 Ⓓ 4

18.

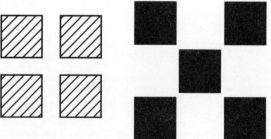

4 + 5 = 9
9 − 5 = ?

Ⓐ 1 Ⓑ 2
Ⓒ 3 Ⓓ 4

Name _____

Pretest

CHAPTER 5
PAGE 4

19. Which subtraction fact goes with this addition fact?

$$2 + 7 = 9$$

- (A) $7 - 5 = 2$
- (B) $7 - 2 = 5$
- (C) $9 - 5 = 4$
- (D) $9 - 7 = 2$

20. Which subtraction fact goes with this addition fact?

$$6 + 4 = 10$$

- (A) $6 - 4 = 2$
- (B) $4 - 2 = 2$
- (C) $10 - 6 = 4$
- (D) $10 - 2 = 8$

21. Which is the number pair?

The sum is 6.
The difference is 4.

- (A) 4 and 2
- (B) 3 and 3
- (C) 5 and 1
- (D) 6 and 2

22. Which is the number pair?

The sum is 7.
The difference is 7.

- (A) 4 and 3
- (B) 7 and 0
- (C) 7 and 7
- (D) 6 and 1

GO ON

Name _____

Pretest

CHAPTER 5
PAGE 5

23.

There are 5 .

There is 1 .

How many more than are there?

Ⓐ 3 more

Ⓑ 4 more

Ⓒ 5 more

Ⓓ 6 more

24.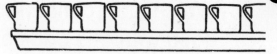

There are 8 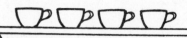.

There are 4 .

How many more 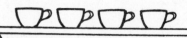 than are there?

Ⓐ 4 more

Ⓑ 6 more

Ⓒ 8 more

Ⓓ 10 more

Name _____

Posttest

CHAPTER 5
PAGE 1

Choose the correct answer.

1.

8 − 1 = ?

- Ⓐ 6
- Ⓑ 7
- Ⓒ 8
- Ⓓ 9

2.

7 − 4 = ?

- Ⓐ 3
- Ⓑ 4
- Ⓒ 5
- Ⓓ 7

3.

10 − 4 = ?

- Ⓐ 4
- Ⓑ 5
- Ⓒ 6
- Ⓓ 7

4.

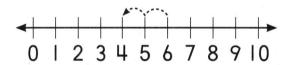

6 − 2 = ?

- Ⓐ 4
- Ⓑ 5
- Ⓒ 6
- Ⓓ 8

5.

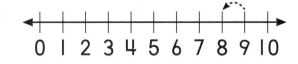

9 − 1 = ?

- Ⓐ 1
- Ⓑ 7
- Ⓒ 8
- Ⓓ 10

6.

8 − 3 = ?

- Ⓐ 5
- Ⓑ 6
- Ⓒ 7
- Ⓓ 8

GO ON

Posttest

CHAPTER 5 PAGE 2

7. 9
   −2

   Ⓐ 7 Ⓑ 8
   Ⓒ 9 Ⓓ not here

8. 10
   − 8

   Ⓐ 0 Ⓑ 1
   Ⓒ 8 Ⓓ not here

9. 7
   −5

   Ⓐ 1 Ⓑ 2
   Ⓒ 3 Ⓓ 7

10. 5
    −5

   Ⓐ 0 Ⓑ 1
   Ⓒ 5 Ⓓ 10

11. 10 − 3 = ?

   Ⓐ 6 Ⓑ 7
   Ⓒ 8 Ⓓ not here

12. 8 − 5 = ?

   Ⓐ 2 Ⓑ 3
   Ⓒ 4 Ⓓ not here

13. 6 − 0 = ?

   Ⓐ 0 Ⓑ 1
   Ⓒ 6 Ⓓ 7

14. 9 − 6 = ?

   Ⓐ 3 Ⓑ 4
   Ⓒ 5 Ⓓ 6

GO ON

Name _____

Posttest

CHAPTER 5
PAGE 3

15. Use counters to solve.

There are 5 .

Then 1 leaves.
How many are left?

Ⓐ 1
Ⓑ 4
Ⓒ 5
Ⓓ 6

16. Use counters to solve.

There are 9 🐦.

Then 4 fly away.
How many are left?

Ⓐ 4
Ⓑ 5
Ⓒ 6
Ⓓ 7

17.

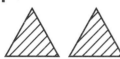 

$4 + 3 = 7$
$7 - 3 = \underline{?}$

Ⓐ 1   Ⓑ 4
Ⓒ 5   Ⓓ 10

18.

$3 + 5 = 8$
$8 - 5 = \underline{?}$

Ⓐ 1   Ⓑ 2
Ⓒ 3   Ⓓ 4

GO ON

Name _____

Posttest

CHAPTER 5
PAGE 4

19. Which subtraction fact goes with this addition fact?

$$6 + 1 = 7$$

Ⓐ $6 - 1 = 5$
Ⓑ $1 - 1 = 0$
Ⓒ $8 - 1 = 7$
Ⓓ $7 - 1 = 6$

20. Which subtraction fact goes with this addition fact?

$$5 + 5 = 10$$

Ⓐ $5 - 0 = 5$
Ⓑ $5 - 5 = 0$
Ⓒ $10 - 5 = 5$
Ⓓ $10 - 0 = 10$

21. Which is the number pair?

The sum is 5.
The difference is 3.

Ⓐ 3 and 3
Ⓑ 5 and 0
Ⓒ 3 and 2
Ⓓ 4 and 1

22. Which is the number pair?

The sum is 9.
The difference is 9.

Ⓐ 8 and 1
Ⓑ 6 and 3
Ⓒ 3 and 3
Ⓓ 9 and 0

GO ON

Name _____

Posttest

CHAPTER 5
PAGE 5

23.

There are 6 🚚.

There are 3 🚗.

How many more 🚚 than 🚗 are there?

- Ⓐ 3 more 🚚
- Ⓑ 4 more 🚚
- Ⓒ 5 more 🚚
- Ⓓ 8 more 🚚

24.

There are 4 🐝.

There are 2 🌹.

How many more 🐝 than 🌹 are there?

- Ⓐ 1 more 🐝
- Ⓑ 2 more 🐝
- Ⓒ 3 more 🐝
- Ⓓ 6 more 🐝

Name _____

Pretest

CHAPTER 6
PAGE 1

Choose the correct answer.

1. Which one is the same shape?

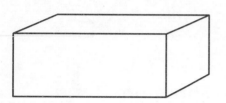

 A  B

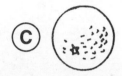 C   D

2. Which one is the same shape?

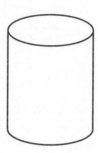

 A   B

 C   D

3. Which plane shape matches the face of the solid shape?

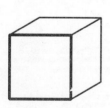

 A    B

 C   D

4. Which plane shape matches the face of the solid shape?

 A    B

 C   D

Name _____

Pretest

CHAPTER 6
PAGE 2

5. Which picture shows the star inside the square?

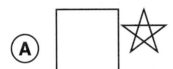
Ⓐ

Ⓑ

Ⓒ

6. Which picture shows the circle on the triangle?

Ⓐ

Ⓑ

Ⓒ

7. Which figure is open?

Ⓐ   Ⓑ

Ⓒ    Ⓓ

8. Which figure is closed?

Ⓐ    Ⓑ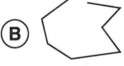

Ⓒ   Ⓓ

9. How many corners are there?

Ⓐ 4   Ⓑ 5
Ⓒ 6   Ⓓ 7

10. How many corners are there?

Ⓐ 1   Ⓑ 2
Ⓒ 3   Ⓓ 4

Pretest

CHAPTER 6
PAGE 3

11. How many sides are there?

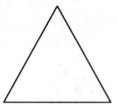

- Ⓐ 1
- Ⓑ 2
- Ⓒ 3
- Ⓓ 4

12. How many sides are there?

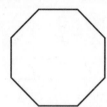

- Ⓐ 4
- Ⓑ 6
- Ⓒ 8
- Ⓓ 16

13. Which figure is the same size and shape as this one?

- Ⓐ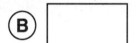
- Ⓑ
- Ⓒ
- Ⓓ

14. Which figure is the same size and shape as this one?

- Ⓐ
- Ⓑ
- Ⓒ
- Ⓓ

15. Which picture shows two parts that match?

- Ⓐ
- Ⓑ

16. Which picture shows two parts that match?

- Ⓐ
- Ⓑ

GO ON

Name _____

Pretest

CHAPTER 6
PAGE 4

17. What is the pattern?

A

B

C

D

18. What is the pattern?

A

B

C

D

19. Which shape comes next in the pattern?

A △

B □

C ○

20. Which shape comes next in the pattern?

A △

B ○

C □

Name _____

Pretest

CHAPTER 6
PAGE 5

21. Which is the missing shape?

☐ ○ ☐ ○ ? ○

Ⓐ ☐   Ⓑ ○

22. Which is the missing shape?

△ △ ○ △ ? ○

Ⓐ ○   Ⓑ △

Use the graph to answer questions 23–24.

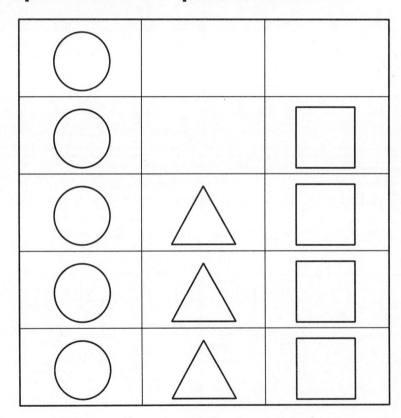

23. How many ○ are there?

Ⓐ 2   Ⓑ 3
Ⓒ 4   Ⓓ 5

24. How many ☐ are there?

Ⓐ 3   Ⓑ 4
Ⓒ 5   Ⓓ 6

Name _____

Posttest

CHAPTER 6
PAGE 1

Choose the correct answer.

1. Which one is the same shape?

2. Which one is the same shape?

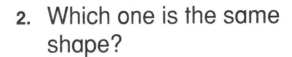

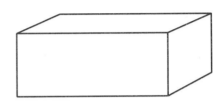

3. Which plane shape matches the face of the solid shape?

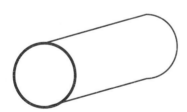

4. Which plane shape matches the face of the solid shape?

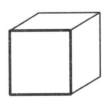

Name _____

Posttest

CHAPTER 6
PAGE 2

5. Which picture shows the star <u>inside</u> the circle?

Ⓐ

Ⓑ

Ⓒ

6. Which picture shows the square <u>outside</u> the circle?

Ⓐ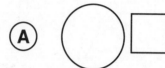

Ⓑ

Ⓒ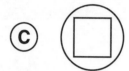

7. Which figure is open?

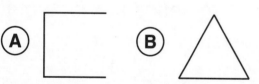

Ⓐ Ⓑ
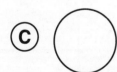 
Ⓒ Ⓓ

8. Which figure is closed?

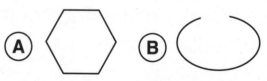

Ⓐ Ⓑ

Ⓒ Ⓓ

9. How many corners are there?

Ⓐ 4   Ⓑ 5
Ⓒ 6   Ⓓ 10

10. How many corners are there?

Ⓐ 3   Ⓑ 4
Ⓒ 5   Ⓓ 6

GO ON

Name _____

Posttest

CHAPTER 6
PAGE 3

11. How many sides are there?

- (A) 1
- (B) 2
- (C) 4
- (D) 5

12. How many sides are there?

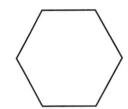

- (A) 3
- (B) 4
- (C) 5
- (D) 6

13. Which figure is the same size and shape as this one?

- (A)
- (B)
- (C)
- (D)

14. Which figure is the same size and shape as this one?

- (A)
- (B)
- (C)
- (D)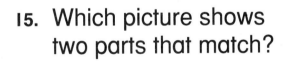

15. Which picture shows two parts that match?

- (A)
- (B)

16. Which picture shows two parts that match?

- (A)
- (B)

GO ON

Name _____

Posttest

CHAPTER 6
PAGE 4

17. What is the pattern?

Ⓐ ● ○

Ⓑ ● ●

Ⓒ ○ ●

Ⓓ ○ ○

18. What is the pattern?

Ⓐ △ □

Ⓑ □ △

Ⓒ □ □ △

Ⓓ △ △ □

19. Which shape comes next in the pattern?

Ⓐ △

Ⓑ ○

Ⓒ □

20. Which shape comes next in the pattern?

Ⓐ ▭

Ⓑ ⬭

Ⓒ △

GO ON

Name _____

Posttest

CHAPTER 6
PAGE 5

21. Which is the missing shape?

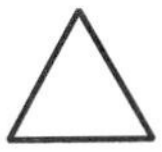

△ □ △ □ ? □

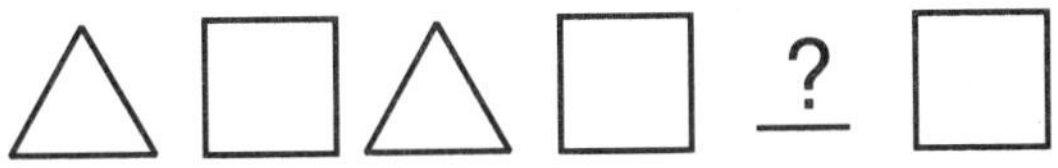

Ⓐ △   Ⓑ □

22. Which is the missing shape?

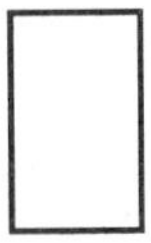

▯ ▯ ○ ▯ ? ○

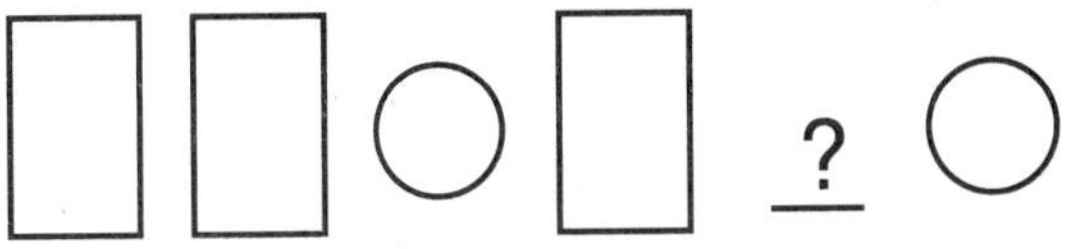

Ⓐ ▯   Ⓑ ○

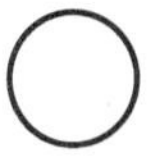

Use the graph to answer questions 23–24.

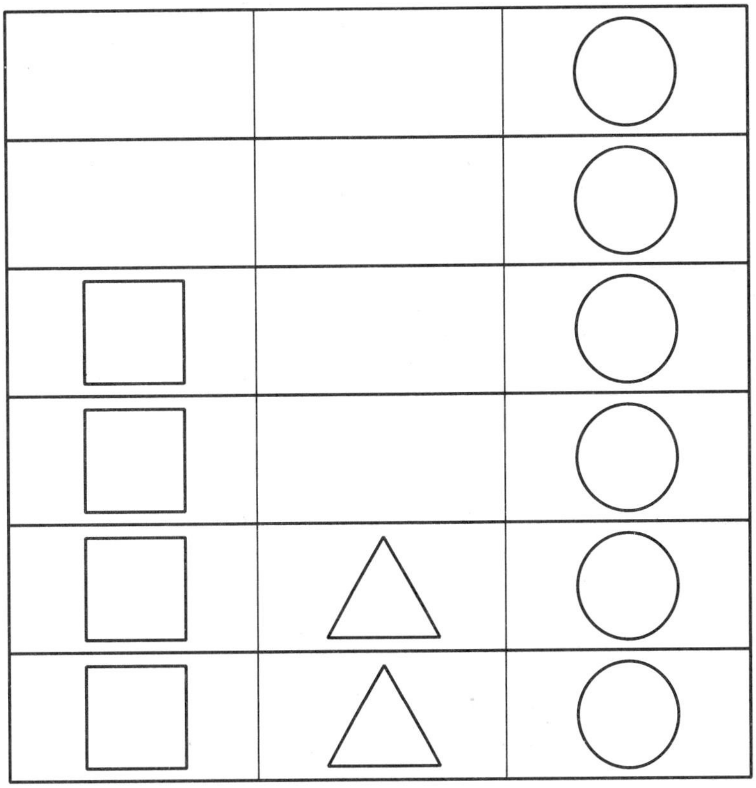

		○
		○
□		○
□		○
□	△	○
□	△	○

23. How many 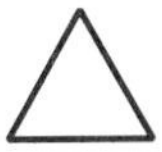 △ are there?

Ⓐ 1   Ⓑ 2
Ⓒ 3   Ⓓ 5

24. How many □ are there?

Ⓐ 2   Ⓑ 3
Ⓒ 4   Ⓓ 6

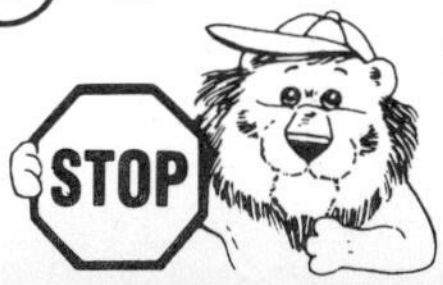

Name _____

Cumulative Test

CHAPTERS **1–6**
PAGE 1

Choose the correct answer.

1. How many ☐ are there?

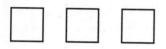

- (A) 5
- (B) 6
- (C) 7
- (D) 9

2. How many ☆ are there?

- (A) 4
- (B) 5
- (C) 6
- (D) 7

3. How many ○ are there?

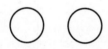
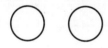

- (A) 2
- (B) 3
- (C) 4
- (D) 5

4. How many △ are there?

- (A) 6
- (B) 8
- (C) 9
- (D) not here

5. Which group has fewer than this?

- (A)
- (B)
- (C)

6. Which group has more than this?

- (A)
- (B)
- (C)

GO ON

Name _____

Cumulative Test

CHAPTERS 1–6
PAGE 2

7. What comes next?

2, 3, 4, __?__

- (A) 1
- (B) 5
- (C) 6
- (D) not here

8. What comes next?

6, 7, 8, __?__

- (A) 5
- (B) 6
- (C) 9
- (D) not here

9. Which addition sentence tells how many in all?

- (A) 2 + 2 = 4
- (B) 2 + 0 = 2
- (C) 4 + 4 = 8
- (D) 4 + 2 = 6

10. Which subtraction sentence tells the story?

- (A) 3 − 1 = 2
- (B) 2 − 1 = 1
- (C) 3 − 3 = 0
- (D) 4 − 1 = 3

11.

7 + 2 = __?__

- (A) 5
- (B) 8
- (C) 10
- (D) not here

12.

6 − 4 = __?__

- (A) 2
- (B) 3
- (C) 4
- (D) 10

13. 1 + 4 = __?__

- (A) 1
- (B) 3
- (C) 4
- (D) 5

14. 0 + 3 = __?__

- (A) 0
- (B) 1
- (C) 3
- (D) not here

Name _____

Cumulative Test

CHAPTERS 1-6
PAGE 3

15. 6 + 2 = ?

(A) 4 (B) 7
(C) 8 (D) 9

16. 5 + 5 = ?

(A) 0 (B) 1
(C) 5 (D) 10

17. 8
    +1

(A) 1 (B) 7
(C) 8 (D) 9

18. 3
    +4

(A) 1 (B) 6
(C) 7 (D) 8

19. Which is a name for 10?

(A) 3 + 4
(B) 6 + 1
(C) 9 + 1
(D) 9 + 0

20. Which is a name for 8?

(A) 3 + 3
(B) 4 + 5
(C) 6 + 1
(D) 7 + 1

21. 4
    −2

(A) 0 (B) 2
(C) 4 (D) 6

22. 10
    −6

(A) 4 (B) 5
(C) 10 (D) 16

23. 7 − 1 = ?

(A) 1 (B) 7
(C) 8 (D) not here

24. 2 − 0 = ?

(A) 0 (B) 1
(C) 2 (D) 3

GO ON

Name _____

Cumulative Test

CHAPTERS 1–6
PAGE 4

25. 8 − 3 = ?

(A) 5 (B) 6
(C) 11 (D) not here

26. 6 − 5 = ?

(A) 1 (B) 2
(C) 11 (D) 65

27. 5 dogs are playing.
I dog runs away.
How many are left?

(A) 1 dog (B) 4 dogs
(C) 5 dogs (D) 6 dogs

28. There are 9 buses.
Then 2 buses leave.
How many are left?

(A) 2 buses
(B) 7 buses
(C) 9 buses
(D) 11 buses

29. Which one is the same shape?

(A)

(B)

(C)

(D)

30. Which one is the same shape?

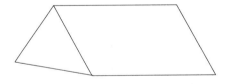

(A)

(B)

(C)

(D)

GO ON

Name _____

Cumulative Test

CHAPTERS 1–6
PAGE 5

31. Which plane shape matches the solid shape?

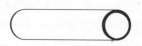

Ⓐ ☐   Ⓑ ◇

Ⓒ △   Ⓓ ◯

32. Which plane shape matches the face of the solid shape?

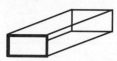

Ⓐ △   Ⓑ ◯

Ⓒ ☐   Ⓓ ▭

33. Which plane shape matches the face of the solid shape?

Ⓐ ⬭   Ⓑ △

Ⓒ ▭   Ⓓ ☐

34. Which one is the same size and shape?

Ⓐ ◯   Ⓑ

Ⓒ   Ⓓ

35. Which one is the same size and shape?

Ⓐ △   Ⓑ △

Ⓒ ◿   Ⓓ △

36. Which picture shows two parts that match?

Ⓐ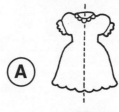

Ⓑ

Name _____

Cumulative Test

CHAPTERS 1–6
PAGE 6

37. Which picture shows two parts that match?

Ⓐ

Ⓑ

38. What is the pattern?

Ⓐ △ △
Ⓑ △ ◯
Ⓒ △ △ ◯
Ⓓ ◯ △ ◯

39. What is the pattern?

☆ ◯ ▢ ☆ ⬭ ▢

Ⓐ ⬭ ☆ ▢
Ⓑ ☆ ⬭ ▢
Ⓒ ▢ ⬭ ☆
Ⓓ ☆ ▢ ⬭

40. Which shape comes next in the pattern?

Ⓐ ▢
Ⓑ △
Ⓒ ☆

41. Which shape comes next in the pattern?

Ⓐ ◯
Ⓑ ▭
Ⓒ ▢

42. Which shape comes next in the pattern?

Ⓐ ▢
Ⓑ ◇
Ⓒ △

Cumulative Test

CHAPTERS 1–6
PAGE 7

43. Which is the missing shape?

   (A) ☆   (B) ◇

44. Which is the number pair? The sum is 5. The difference is 3.

(A) 2 and 3   (B) 3 and 6
(C) 1 and 4   (D) 3 and 5

45. What do you do to find how many 🐟 are left?

(A) Add   (B) Subtract

46. What do you do to find how many 🐶 in all?

(A) Add   (B) Subtract

47. How much did Rusty spend?

(A) 2¢   (B) 6¢
(C) 9¢   (D) 10¢

48.

There are 4 🌷.
There are 5 🌼.
How many more 🌼 than 🌷?

(A) 1 more
(B) 4 more
(C) 5 more
(D) 9 more

GO ON

Name _____

Cumulative Test

CHAPTERS 1-6
PAGE 8

Use the pictograph below to answer questions 49–50.

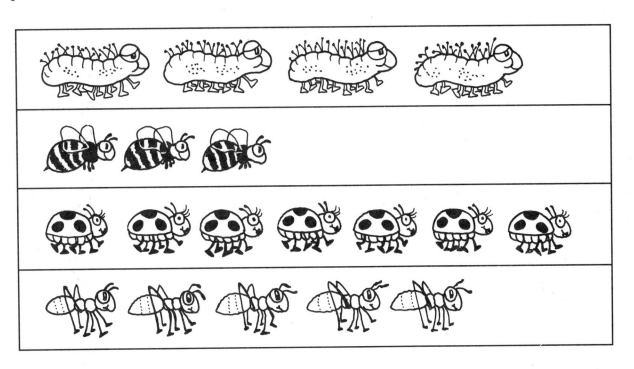

49. Which group has fewer,  ?

A

B

50. Which group has more than 6?

Name _____

Pretest

CHAPTER 7
PAGE 1

Choose the correct answer.

1. How many tens and ones are there?

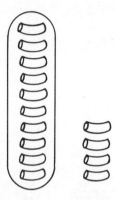

- (A) 1 ten 4 ones
- (B) 4 tens 1 one
- (C) 4 tens 10 ones
- (D) 10 tens 1 one

2. How many tens and ones are there?

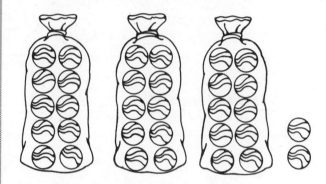

- (A) 30 tens 2 ones
- (B) 1 ten 3 ones
- (C) 3 tens 2 ones
- (D) 2 tens 3 ones

3. How many tens and ones are there?

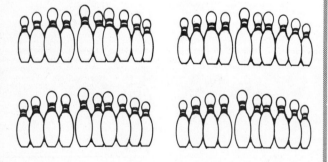

- (A) 4 tens 0 ones
- (B) 0 tens 4 ones
- (C) 1 ten 4 ones
- (D) 10 tens 1 one

4. How many tens and ones are there?

- (A) 3 tens 0 ones
- (B) 20 tens 1 one
- (C) 1 ten 2 ones
- (D) 2 tens 1 one

GO ON

Name _____

Pretest

CHAPTER **7**
PAGE 2

5. How many are there in all?

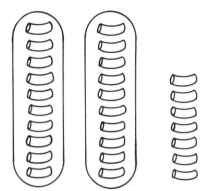

Ⓐ 9   Ⓑ 17
Ⓒ 27   Ⓓ 72

6. How many are there in all?

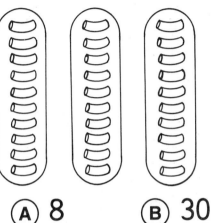

Ⓐ 8   Ⓑ 30
Ⓒ 53   Ⓓ not here

7. Which is the greater number?

Ⓐ 36   Ⓑ 55

8. Which number is less?

Ⓐ 29   Ⓑ 30

9. What number comes before?

 , 47

Ⓐ 36   Ⓑ 37
Ⓒ 46   Ⓓ 48

10. What number comes after?

70, ?

Ⓐ 69   Ⓑ 71
Ⓒ 77   Ⓓ 80

Name _____

Pretest

CHAPTER 7
PAGE 3

11. What number comes between?

86, ?, 88

Ⓐ 85   Ⓑ 87
Ⓒ 89   Ⓓ not here

12. What number comes between?

69, ?, 71

Ⓐ 60   Ⓑ 68
Ⓒ 70   Ⓓ 80

13. Which shoe is black?

first

Ⓐ second   Ⓑ third
Ⓒ fourth   Ⓓ sixth

14. Which one is the pig?

first

Ⓐ first   Ⓑ second
Ⓒ fourth   Ⓓ fifth

15. Which one is the  ?

Ⓐ first   Ⓑ second
Ⓒ third   Ⓓ fourth

16. Which one is the  ?

Ⓐ second   Ⓑ sixth
Ⓒ seventh Ⓓ eighth

GO ON

Name _____

Pretest

CHAPTER 7
PAGE 4

17. Count by twos. What number is missing?

| 2 | 4 | 6 | ? |

- (A) 5
- (B) 7
- (C) 8
- (D) 16

18. Count by tens. What number is missing?

| 10 | 20 | 30 | ? |

- (A) 4
- (B) 31
- (C) 35
- (D) 40

19. Count by fives. What number is missing?

| 20 | 25 | 30 | ? |

- (A) 31
- (B) 35
- (C) 40
- (D) not here

20. Count by twos. What number is missing?

| 44 | 46 | 48 | ? |

- (A) 49
- (B) 50
- (C) 58
- (D) 59

21. Which is the best estimate?

- (A) 40
- (B) 50
- (C) 60
- (D) 90

22. Which is the best estimate?

- (A) 30
- (B) 40
- (C) 50
- (D) 60

Name _____

Pretest

CHAPTER 7
PAGE 5

23.

How many 🏳 can you buy with 🪙🪙🪙 ?

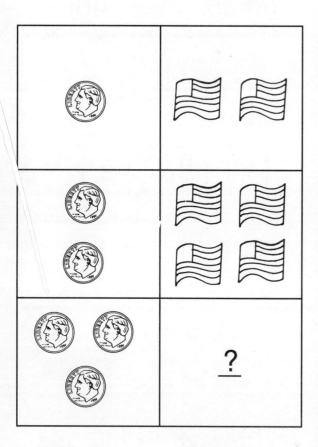

Ⓐ 3 Ⓑ 4
Ⓒ 6 Ⓓ 10

24.

How many ☺ can you buy with 🪙🪙🪙 ?

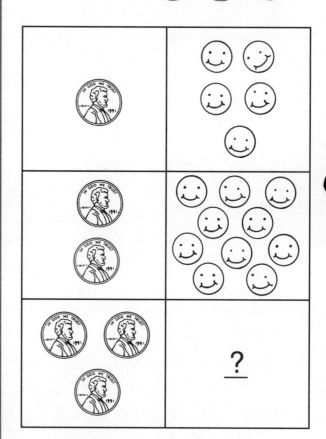

Ⓐ 11 Ⓑ 12
Ⓒ 15 Ⓓ 20

Name _____

Posttest

CHAPTER 7
PAGE 1

Choose the correct answer.

1. How many tens and ones are there?

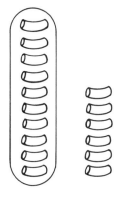

- (A) 1 ten 6 ones
- (B) 6 tens 1 one
- (C) 6 tens 10 ones
- (D) 10 tens 1 one

2. How many tens and ones are there?

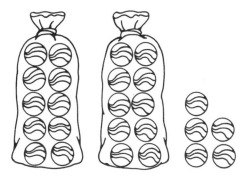

- (A) 20 tens 5 ones
- (B) 50 tens 2 ones
- (C) 5 tens 2 ones
- (D) 2 tens 5 ones

3. How many tens and ones are there?

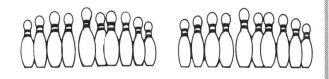

- (A) 3 tens 0 ones
- (B) 0 tens 3 ones
- (C) 1 ten 3 ones
- (D) 30 tens 0 ones

4. How many tens and ones are there?

- (A) 3 tens 0 ones
- (B) 10 tens 1 one
- (C) 1 ten 4 ones
- (D) 3 tens 4 ones

GO ON

Name _____

Posttest

CHAPTER 7
PAGE 2

5. How many are there in all?

- (A) 9
- (B) 26
- (C) 36
- (D) 63

6. How many are there in all?

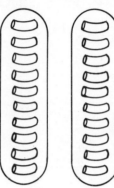

- (A) 10
- (B) 20
- (C) 82
- (D) not here

7. Which is the greater number?

- (A) 49
- (B) 50

8. Which number is less?

 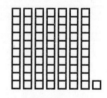

- (A) 38
- (B) 71

9. What number comes before?

? , 86

- (A) 75
- (B) 76
- (C) 85
- (D) 87

10. What number comes after?

50, ?

- (A) 49
- (B) 51
- (C) 60
- (D) 61

Name _____

Posttest

CHAPTER 7
PAGE 3

11. What number comes between?

44, ?, 46

- (A) 42
- (B) 45
- (C) 48
- (D) not here

12. What number comes between?

79, ?, 81

- (A) 70
- (B) 78
- (C) 80
- (D) 90

13. Which hat has a feather?

first

- (A) first
- (B) second
- (C) fourth
- (D) fifth

14. Which one is the horse?

first

- (A) first
- (B) second
- (C) fifth
- (D) sixth

15. Which one is the ?

first

- (A) first
- (B) second
- (C) third
- (D) fourth

16. Which one is the ?

- (A) first
- (B) second
- (C) fourth
- (D) fifth

Name _____

Posttest

CHAPTER 7
PAGE 4

17. Count by fives. What number is missing?

| 5 | 10 | 15 | ? |

Ⓐ 5 Ⓑ 16
Ⓒ 20 Ⓓ 25

18. Count by tens. What number is missing?

| 10 | 20 | 30 | ? |

Ⓐ 4 Ⓑ 31
Ⓒ 35 Ⓓ 40

19. Count by twos. What number is missing?

| 54 | 56 | 58 | ? |

Ⓐ 59 Ⓑ 60
Ⓒ 68 Ⓓ not here

20. Count by fives. What number is missing?

| 15 | 20 | 25 | ? |

Ⓐ 26 Ⓑ 30
Ⓒ 35 Ⓓ not here

21. Which is the best estimate?

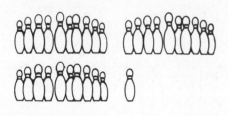

Ⓐ 3 Ⓑ 10
Ⓒ 30 Ⓓ 40

22. Which is the best estimate?

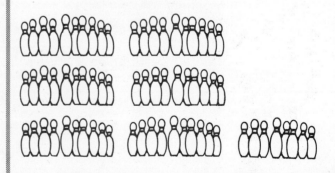

Ⓐ 10 Ⓑ 50
Ⓒ 60 Ⓓ 70

GO ON

Name _____

Posttest

CHAPTER 7
PAGE 5

23.

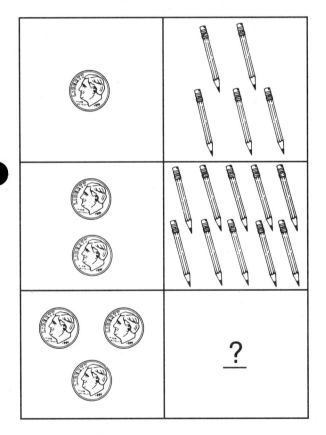

(A) 6 (B) 7
(C) 15 (D) 20

24.

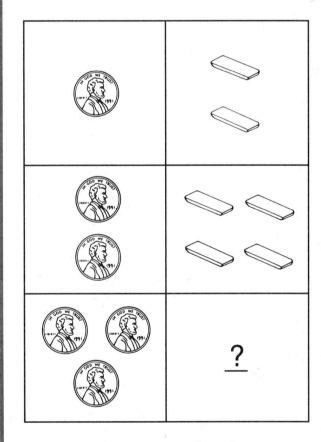

(A) 5 (B) 6
(C) 8 (D) 10

Name _____

Pretest

CHAPTER 8
PAGE 1

Choose the correct answer.

1. Which addition sentence tells how many in all?

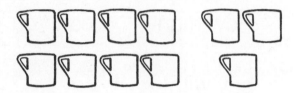

- Ⓐ 3 + 5 = 8
- Ⓑ 7 + 2 = 9
- Ⓒ 1 + 3 = 4
- Ⓓ 8 + 3 = 11

2. Which addition sentence tells how many in all?

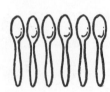 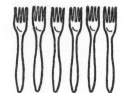

- Ⓐ 3 + 3 = 6
- Ⓑ 7 + 7 = 14
- Ⓒ 6 + 6 = 12
- Ⓓ 5 + 5 = 10

3. Which addition sentence tells how many in all?

- Ⓐ 4 + 1 = 5
- Ⓑ 4 + 5 = 9
- Ⓒ 5 + 5 = 10
- Ⓓ 3 + 4 = 7

4. Which addition sentence tells how many in all?

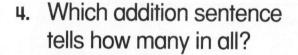
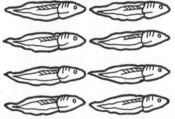

- Ⓐ 8 + 1 = 9
- Ⓑ 7 + 1 = 8
- Ⓒ 9 + 1 = 10
- Ⓓ 9 + 9 = 18

5. 2 + 9 = ?

- Ⓐ 7
- Ⓑ 10
- Ⓒ 11
- Ⓓ 12

6. 0 + 6 = ?

- Ⓐ 0
- Ⓑ 6
- Ⓒ 7
- Ⓓ not here

GO ON

Name _____

Pretest

CHAPTER 8
PAGE 2

7. $7 + 3 = \underline{?}$
   - (A) 1
   - (B) 4
   - (C) 10
   - (D) 11

8. $8 + 4 = \underline{?}$
   - (A) 4
   - (B) 10
   - (C) 11
   - (D) 12

9. Sam catches 5 fish. Dad catches 7 fish. How many fish do they catch in all?
   - (A) 6 fish
   - (B) 8 fish
   - (C) 12 fish
   - (D) 13 fish

10. Meg has 5 pet birds. She has 5 pet fish. How many pets does she have in all?
    - (A) 0 pets
    - (B) 5 pets
    - (C) 9 pets
    - (D) 10 pets

11. Which subtraction sentence tells the story?

- (A) $11 - 3 = 8$
- (B) $8 - 3 = 5$
- (C) $5 - 3 = 2$
- (D) $7 - 3 = 4$

12. Which subtraction sentence tells the story?

- (A) $9 - 3 = 6$
- (B) $6 - 3 = 3$
- (C) $12 - 9 = 3$
- (D) $11 - 8 = 3$

Name _____

Pretest

CHAPTER 8
PAGE 3

13. Which subtraction sentence tells the story?

- Ⓐ 8 − 2 = 6
- Ⓑ 6 − 2 = 4
- Ⓒ 12 − 2 = 10
- Ⓓ 10 − 2 = 8

14. Which subtraction sentence tells the story?

- Ⓐ 6 − 3 = 3
- Ⓑ 6 − 6 = 0
- Ⓒ 9 − 3 = 6
- Ⓓ 12 − 6 = 6

15. 11 − 6 = ?
- Ⓐ 5
- Ⓑ 6
- Ⓒ 15
- Ⓓ 17

16. 9 − 0 = ?
- Ⓐ 0
- Ⓑ 8
- Ⓒ 9
- Ⓓ not here

17. 10 − 1 = ?
- Ⓐ 1
- Ⓑ 9
- Ⓒ 10
- Ⓓ 11

18. 12 − 8 = ?
- Ⓐ 4
- Ⓑ 5
- Ⓒ 6
- Ⓓ 16

19. Mom buys 12 🍎. Pam eats 3 🍎. How many 🍎 are left?
- Ⓐ 9
- Ⓑ 10
- Ⓒ 11
- Ⓓ 15

20. There were 11 ☕. 5 were broken. How many ☕ are left?
- Ⓐ 5
- Ⓑ 6
- Ⓒ 7
- Ⓓ 14

GO ON

Name _____

Pretest

CHAPTER 8
PAGE 4

21. 7¢
    + 4¢
    Ⓐ 2¢   Ⓑ 3¢
    Ⓒ 10¢ Ⓓ 11¢

22. 9¢
    − 2¢
    Ⓐ 6¢   Ⓑ 7¢
    Ⓒ 10¢ Ⓓ 11¢

23. 8¢
    − 7¢
    Ⓐ 1¢   Ⓑ 3¢
    Ⓒ 4¢   Ⓓ 15¢

24. 6¢
    + 6¢
    Ⓐ 0¢   Ⓑ 6¢
    Ⓒ 11¢ Ⓓ 12¢

25. Which number sentence does not belong to the fact family?
    Ⓐ 7 + 4 = 11
    Ⓑ 11 − 7 = 4
    Ⓒ 3 + 4 = 7
    Ⓓ 4 + 7 = 11

26. Which number sentence does not belong to the fact family?
    Ⓐ 9 − 1 = 8
    Ⓑ 1 + 8 = 9
    Ⓒ 9 − 8 = 1
    Ⓓ 7 + 1 = 8

27. Which number sentence belongs to the same fact family as 10 − 6 = 4?
    Ⓐ 6 + 4 = 10
    Ⓑ 10 + 6 = 16
    Ⓒ 6 − 4 = 2
    Ⓓ 10 − 5 = 5

28. Which number sentence belongs to the same fact family as 12 − 7 = 5?
    Ⓐ 2 + 5 = 7
    Ⓑ 7 − 5 = 2
    Ⓒ 6 + 6 = 12
    Ⓓ 5 + 7 = 12

GO ON

Name _____

Pretest

CHAPTER 8
PAGE 5

Use the graph below to answer questions 29–30.

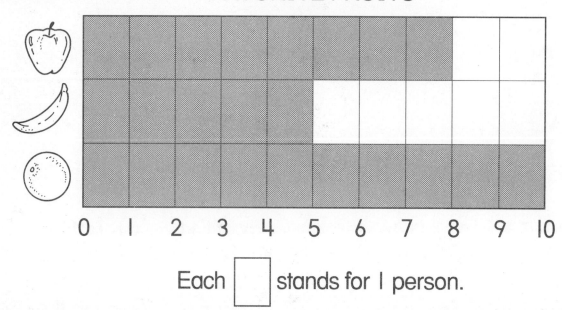

Each ▢ stands for 1 person.

29. How many people like best?

   Ⓐ 5 Ⓑ 6
   Ⓒ 8 Ⓓ 10

30. Which fruit is the favorite of the most people?

31. There are 8 black puppies. There are 4 white puppies. What do you do to find how many puppies there are in all?

   Ⓐ Add Ⓑ Subtract

32. Roy had 10 pennies. He lost 3 pennies. What do you do to find how many pennies are left?

   Ⓐ Add Ⓑ Subtract

Name _____

Posttest

CHAPTER **8**
PAGE 1

Choose the correct answer.

1. Which addition sentence tells how many in all?

- Ⓐ 8 + 3 = 11
- Ⓑ 6 + 3 = 9
- Ⓒ 7 + 3 = 10
- Ⓓ 9 + 3 = 12

2. Which addition sentence tells how many in all?

- Ⓐ 1 + 10 = 11
- Ⓑ 1 + 9 = 10
- Ⓒ 0 + 9 = 9
- Ⓓ 8 + 1 = 9

3. Which addition sentence tells how many in all?

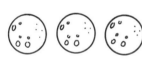

- Ⓐ 4 + 2 = 6
- Ⓑ 6 + 4 = 10
- Ⓒ 6 + 5 = 11
- Ⓓ 4 + 4 = 8

4. Which addition sentence tells how many in all?

- Ⓐ 2 + 5 = 7
- Ⓑ 7 + 1 = 8
- Ⓒ 9 + 1 = 10
- Ⓓ 7 + 2 = 9

5. 8 + 3 = ?

- Ⓐ 5
- Ⓑ 10
- Ⓒ 11
- Ⓓ 12

6. 9 + 0 = ?

- Ⓐ 0
- Ⓑ 6
- Ⓒ 9
- Ⓓ 10

Posttest

CHAPTER 8
PAGE 2

7. $5 + 5 = \underline{?}$
   - (A) 0
   - (B) 1
   - (C) 5
   - (D) 10

8. $2 + 8 = \underline{?}$
   - (A) 0
   - (B) 6
   - (C) 11
   - (D) not here

9. 2 ants are on a hill. 9 more ants come up the hill. How many ants in all?
   - (A) 7 ants
   - (B) 10 ants
   - (C) 11 ants
   - (D) 12 ants

10. There are 4 red balls. There are 8 green balls. How many balls in all?
    - (A) 2 balls
    - (B) 4 balls
    - (C) 10 balls
    - (D) 12 balls

11. Which subtraction sentence tells the story?

   - (A) $11 - 4 = 7$
   - (B) $7 - 4 = 3$
   - (C) $7 - 5 = 2$
   - (D) $11 - 3 = 8$

12. Which subtraction sentence tells the story?

   - (A) $5 - 4 = 1$
   - (B) $8 - 4 = 4$
   - (C) $10 - 4 = 6$
   - (D) $9 - 4 = 5$

GO ON

Name _____

Posttest

CHAPTER 8
PAGE 3

13. Which subtraction sentence tells the story?

- Ⓐ $6 - 6 = 0$
- Ⓑ $10 - 6 = 4$
- Ⓒ $12 - 6 = 6$
- Ⓓ $12 - 5 = 7$

14. Which subtraction sentence tells the story?

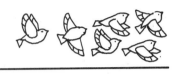

- Ⓐ $6 - 5 = 1$
- Ⓑ $11 - 6 = 5$
- Ⓒ $10 - 6 = 4$
- Ⓓ $12 - 7 = 5$

15. $11 - 3 = \underline{?}$

- Ⓐ 2
- Ⓑ 8
- Ⓒ 9
- Ⓓ not here

16. $10 - 8 = \underline{?}$

- Ⓐ 12
- Ⓑ 10
- Ⓒ 2
- Ⓓ 13

17. $9 - 3 = \underline{?}$

- Ⓐ 3
- Ⓑ 5
- Ⓒ 6
- Ⓓ 12

18. $12 - 7 = \underline{?}$

- Ⓐ 4
- Ⓑ 5
- Ⓒ 6
- Ⓓ not here

19. Mom buys 12 . The children eat 9 . How many are left?

- Ⓐ 3
- Ⓑ 6
- Ⓒ 9
- Ⓓ 12

20. There were 9 . 8 were broken. How many  are left?

- Ⓐ 1
- Ⓑ 2
- Ⓒ 10
- Ⓓ 17

GO ON

Name _____

Posttest

CHAPTER 8
PAGE 4

21. 5¢
    + 6¢
    ___
   (A) 1¢ (B) 2¢
   (C) 10¢ (D) 11¢

22. 12¢
    − 4¢
    ___
   (A) 2¢ (B) 8¢
   (C) 14¢ (D) 16¢

23. 11¢
    − 9¢
    ___
   (A) 2¢ (B) 8¢
   (C) 18¢ (D) not here

24. 1¢
    + 8¢
    ___
   (A) 0¢ (B) 7¢
   (C) 8¢ (D) 9¢

25. Which number sentence does not belong to the fact family for 3, 6, and 9?
   (A) 3 + 6 = 9
   (B) 9 − 3 = 6
   (C) 9 + 3 = 12
   (D) 9 − 6 = 3

26. Which number sentence does not belong to the fact family for 3, 7, and 10?
   (A) 3 + 7 = 10
   (B) 3 + 4 = 7
   (C) 10 − 3 = 7
   (D) 10 − 7 = 3

27. Which number sentence belongs to the same fact family as 7 + 4 = 11?
   (A) 11 − 7 = 4
   (B) 7 + 0 = 7
   (C) 3 + 4 = 7
   (D) 7 − 4 = 3

28. Which number sentence belongs to the same fact family as 12 − 5 = 7?
   (A) 2 + 5 = 7
   (B) 7 − 5 = 2
   (C) 7 + 5 = 12
   (D) 6 + 6 = 12

GO ON

Name _____

Posttest

CHAPTER 8
PAGE 5

Use the graph below to answer questions 29–30.

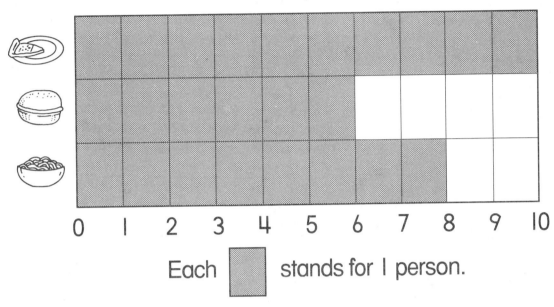

---

29. How many people like  best?

    (A) 5 (B) 6
    (C) 8 (D) 10

30. Which is the favorite lunch of the most people?

    (A) 🍕 (B) 🍔
    (C) 🍜

---

31. 12 cats were sleeping. 6 cats woke up. What do you do to find how many cats are still sleeping?

    (A) Add (B) Subtract

32. 8 chickens are sitting on nests. 8 chickens are eating. What do you do to find how many chickens there are in all?

    (A) Add (B) Subtract

Name _____

Pretest

CHAPTER 9
PAGE 1

Choose the correct answer.

1. Which is the shortest?
   - Ⓐ
   - Ⓑ
   - Ⓒ
   - Ⓓ

2. Which is the longest?
   - Ⓐ
   - Ⓑ
   - Ⓒ
   - Ⓓ

3. How long is the fork?

   - Ⓐ 1
   - Ⓑ 2
   - Ⓒ 7
   - Ⓓ 10

4. How long is the knife?

   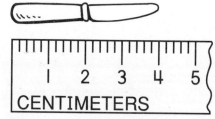

   - Ⓐ 1 centimeter
   - Ⓑ 2 centimeters
   - Ⓒ 4 centimeters
   - Ⓓ 5 centimeters

5. How long is the pin?

   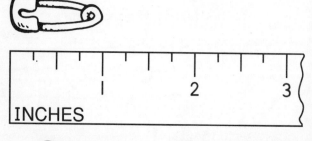

   - Ⓐ 1 inch
   - Ⓑ 2 inches
   - Ⓒ 3 inches
   - Ⓓ 4 inches

6. How long is the whistle?

   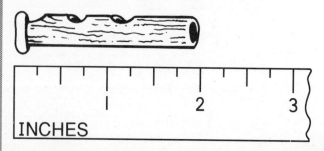

   - Ⓐ 1 inch
   - Ⓑ 3 inches
   - Ⓒ 2 inches
   - Ⓓ 4 inches

GO ON

Name _____

Pretest

CHAPTER 9
PAGE 2

7. How long is the ribbon?

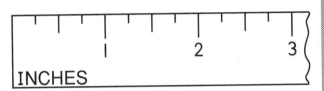

- (A) 1 inch
- (B) 2 inches
- (C) 3 inches
- (D) 4 inches

8. How long is the comb?

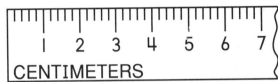

- (A) 2 centimeters
- (B) 4 centimeters
- (C) 6 centimeters
- (D) 10 centimeters

9. How long is the paper clip?

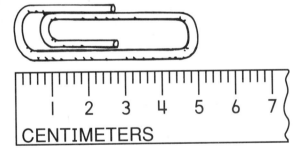

- (A) 2 centimeters
- (B) 3 centimeters
- (C) 4 centimeters
- (D) 5 centimeters

10. How long is the pin?

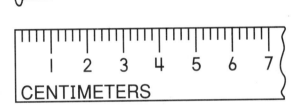

- (A) 1 centimeter
- (B) 3 centimeters
- (C) 5 centimeters
- (D) 6 centimeters

11. Which is heavier?

(A)

(B)

12. Which is heavier?

(A)

(B)

Name _____

Pretest

CHAPTER 9
PAGE 3

13. Which holds more?

Ⓐ

Ⓑ

14. Which is hot?

Ⓐ Ⓑ

15. Which holds about the same as a ?

Ⓐ   Ⓑ

Ⓒ

16. Which holds about the same as a ?

Ⓐ   Ⓑ

Ⓒ

17. Which is about as heavy as a ?

Ⓐ ⌬

Ⓑ ✏️

Ⓒ ✂️

18. Which is about as heavy as a ?

Ⓐ ✉️

Ⓑ

Ⓒ

 GO ON

Name _____

Pretest

CHAPTER 9
PAGE 4

19. Which shows fair shares?

Ⓐ   Ⓑ

Ⓒ   Ⓓ

20. Which shows fair shares?

Ⓐ   Ⓑ

Ⓒ   Ⓓ

21. Which shows one half?

Ⓐ   Ⓑ

Ⓒ   Ⓓ

22. Which shows one third?

Ⓐ   Ⓑ

Ⓒ   Ⓓ

23. Which shows one fourth?

Ⓐ

Ⓑ

Ⓒ

Ⓓ

24. What fraction does this picture show?

Ⓐ $\dfrac{1}{5}$ Ⓑ $\dfrac{1}{4}$

Ⓒ $\dfrac{1}{3}$ Ⓓ $\dfrac{1}{2}$

Name _____

Pretest

CHAPTER 9
PAGE 5

25. What fraction does this picture show?

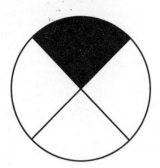

- Ⓐ $\frac{1}{5}$
- Ⓑ $\frac{1}{4}$
- Ⓒ $\frac{1}{3}$
- Ⓓ $\frac{1}{2}$

26. What fraction does this picture show?

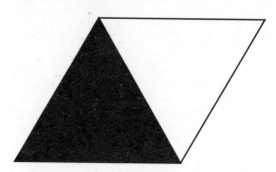

- Ⓐ $\frac{1}{5}$
- Ⓑ $\frac{1}{4}$
- Ⓒ $\frac{1}{3}$
- Ⓓ $\frac{1}{2}$

27. What fraction of the group is small?

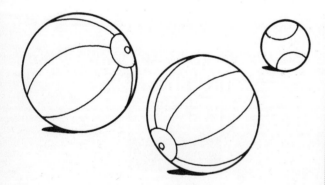

- Ⓐ $\frac{1}{5}$
- Ⓑ $\frac{1}{4}$
- Ⓒ $\frac{1}{3}$
- Ⓓ $\frac{1}{2}$

28. What fraction of the group are girls?

- Ⓐ $\frac{1}{5}$
- Ⓑ $\frac{1}{4}$
- Ⓒ $\frac{1}{3}$
- Ⓓ $\frac{2}{4}$

GO ON

Pretest

CHAPTER 9
PAGE 6

Use the table below to answer questions 29–32.

CHILDREN'S MEASUREMENTS (in inches)			
Child	Foot	Hand	Arm
Pat	6	5	12
Chris	5	4	10
Bobby	8	6	14

29. How long is Pat's arm?

   (A) 6 inches
   (B) 12 inches
   (C) 14 inches

30. How long is Bobby's hand?

   (A) 5 inches
   (B) 6 inches
   (C) 8 inches

31. Whose arm is shortest?

   (A) Pat's
   (B) Chris's
   (C) Bobby's

32. Whose foot is longest?

   (A) Pat's
   (B) Chris's
   (C) Bobby's

Name _____

Posttest

CHAPTER 9
PAGE 1

Choose the correct answer.

1. Which is the longest?

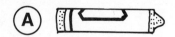

2. Which is the shortest?

3. How long is the worm?

Ⓐ 1
Ⓑ 2
Ⓒ 7
Ⓓ 8

4. How long is the frog?

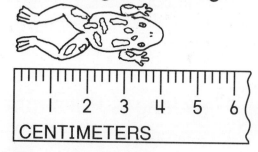

Ⓐ 1 centimeter
Ⓑ 2 centimeters
Ⓒ 4 centimeters
Ⓓ 5 centimeters

5. How long is the spoon?

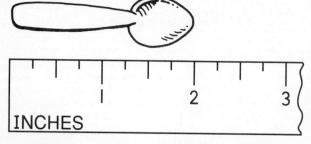

Ⓐ 1 inch   Ⓑ 2 inches
Ⓒ 3 inches Ⓓ 4 inches

6. How long is the paper clip?

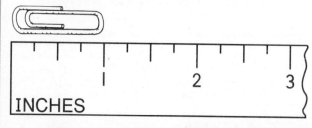

Ⓐ 1 inch   Ⓑ 2 inches
Ⓒ 3 inches Ⓓ 4 inches

GO ON

Name _____

Posttest

CHAPTER 9
PAGE 2

7. How long is the string?

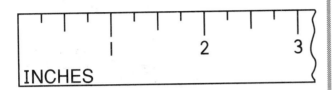

- (A) 1 inch
- (B) 2 inches
- (C) 3 inches
- (D) 4 inches

8. How long is the bean?

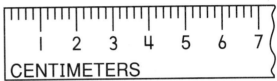

- (A) 1 centimeter
- (B) 2 centimeters
- (C) 3 centimeters
- (D) 5 centimeters

9. How long is the key?

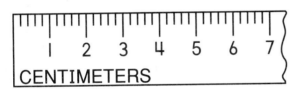

- (A) 1 centimeter
- (B) 3 centimeters
- (C) 4 centimeters
- (D) 5 centimeters

10. How long is the eraser?

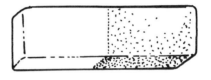

- (A) 3 centimeters
- (B) 5 centimeters
- (C) 6 centimeters
- (D) 7 centimeters

11. Which is heavier?

(A)

(B)

12. Which is heavier?

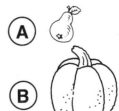

(A)

(B)

GO ON

Name _____

Posttest

CHAPTER 9
PAGE 3

13. Which holds more?

Ⓐ

Ⓑ

14. Which is hot?

Ⓐ

Ⓑ

15. Which holds about the same as a  ?

Ⓐ

Ⓑ

Ⓒ

16. Which holds about the same as a  ?

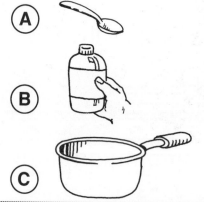

Ⓐ

Ⓑ

Ⓒ

17. Which is about as heavy as an 🍎 ?

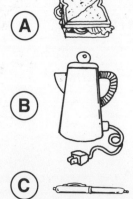

Ⓐ

Ⓑ

Ⓒ

18. Which is about as heavy as an ✏️ ?

Ⓐ

Ⓑ

Ⓒ

GO ON

Name _____

Posttest

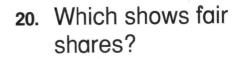

19. Which shows fair shares?

20. Which shows fair shares?

21. Which shows one half?

22. Which shows one fourth?

23. Which shows one third?

24. What fraction does this picture show?

Ⓐ $\dfrac{1}{5}$ Ⓑ $\dfrac{1}{4}$

Ⓒ $\dfrac{1}{3}$ Ⓓ $\dfrac{1}{2}$

GO ON

Posttest

CHAPTER 9
PAGE 5

25. What fraction does this picture show?

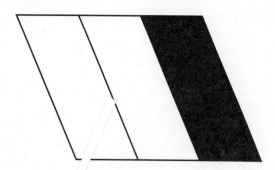

(A) $\frac{1}{5}$ (B) $\frac{1}{4}$
(C) $\frac{1}{3}$ (D) $\frac{1}{2}$

26. What fraction does this picture show?

(A) $\frac{1}{5}$ (B) $\frac{1}{4}$
(C) $\frac{1}{3}$ (D) $\frac{1}{2}$

27. What fraction of the group is large?

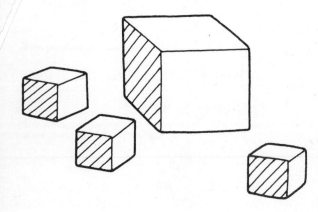

(A) $\frac{1}{5}$ (B) $\frac{1}{4}$
(C) $\frac{1}{3}$ (D) $\frac{1}{2}$

28. What fraction of the group are boys?

(A) $\frac{1}{5}$ (B) $\frac{1}{4}$
(C) $\frac{1}{3}$ (D) $\frac{1}{2}$

Name _____

Posttest

CHAPTER 9
PAGE 6

Use the table below to answer questions 29–32.

LENGTHS (in inches)			
Object	Red	Blue	Yellow
Crayon	2	5	3
Pencil	5	6	4
Chalk	1	6	3

29. Which crayon is shortest?

- (A) red
- (B) blue
- (C) yellow

30. Which chalk is longest?

- (A) red
- (B) blue
- (C) yellow

31. How long is the yellow pencil?

- (A) 3 inches
- (B) 4 inches
- (C) 5 inches

32. How long is the red chalk?

- (A) 1 inch
- (B) 2 inches
- (C) 3 inches

Name _____

Cumulative Test

CHAPTERS 1-9
PAGE 1

Choose the correct answer.

1.

    6 − 1 = ?

    Ⓐ 1   Ⓑ 5
    Ⓒ 6   Ⓓ 7

2.

    5 + 3 = ?

    Ⓐ 2   Ⓑ 3
    Ⓒ 7   Ⓓ 8

3. Which plane shape matches the face of the solid shape?

    Ⓐ   Ⓑ ○

    Ⓒ ▭   Ⓓ △

4. Which plane shape matches the face of the solid shape?

    Ⓐ   Ⓑ ▢

    Ⓒ ○   Ⓓ

5. Which figure is the same size and shape?

    Ⓐ △   Ⓑ △

    Ⓒ △   Ⓓ △

6. Which shows two parts that match?

    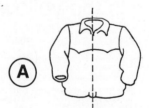
    Ⓐ

    Ⓑ

GO ON

Name _____

Cumulative Test

CHAPTERS 1–9
PAGE 2

7. Which shape comes next in the pattern?

- Ⓐ △
- Ⓑ □
- Ⓒ ○

8. Which shape comes next in the pattern?

- Ⓐ ○
- Ⓑ ○
- Ⓒ □

9. How many tens and ones are there?

- Ⓐ 2 tens and 7 ones
- Ⓑ 3 tens and 3 ones
- Ⓒ 2 tens and 4 ones
- Ⓓ 10 tens and 7 ones

10. How many tens and ones are there?

- Ⓐ 10 tens and 2 ones
- Ⓑ 3 tens and 0 ones
- Ⓒ 2 tens and 1 one
- Ⓓ 1 ten and 2 ones

11. How many are there in all?

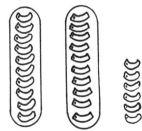

- Ⓐ 8
- Ⓑ 26
- Ⓒ 27
- Ⓓ not here

12. How many are there in all?

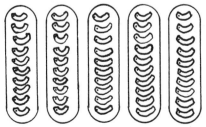

- Ⓐ 5
- Ⓑ 6
- Ⓒ 15
- Ⓓ 51

Name _____

Cumulative Test

CHAPTERS 1–9
PAGE 3

13. Which number is less?

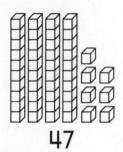

47   50

Ⓐ 47   Ⓑ 50

14. Which number is greater?

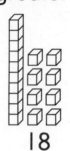 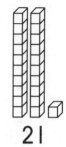

18   21

Ⓐ 18   Ⓑ 21

15. Which number comes after?

61, ?

Ⓐ 60   Ⓑ 62
Ⓒ 69   Ⓓ not here

16. Which number comes between?

49, ?, 51

Ⓐ 40   Ⓑ 48
Ⓒ 50   Ⓓ not here

17. Which one is the ?

first

Ⓐ second   Ⓑ third
Ⓒ first   Ⓓ fourth

18. Which one is the ?

Ⓐ fourth   Ⓑ fifth
Ⓒ second   Ⓓ third

Cumulative Test

CHAPTERS 1–9
PAGE 4

19. 4 + 5 = ?
   - Ⓐ 9
   - Ⓑ 8
   - Ⓒ 1
   - Ⓓ 10

20. 2 + 4 = ?
   - Ⓐ 2
   - Ⓑ 4
   - Ⓒ 6
   - Ⓓ 8

21. 9 + 3 = ?
   - Ⓐ 6
   - Ⓑ 11
   - Ⓒ 12
   - Ⓓ 93

22. 6 − 0 = ?
   - Ⓐ 0
   - Ⓑ 6
   - Ⓒ 5
   - Ⓓ 9

23. There are 3 red hats. There are 7 blue hats. How many hats are there in all?
   - Ⓐ 1 hat
   - Ⓑ 4 hats
   - Ⓒ 9 hats
   - Ⓓ 10 hats

24. Roy has 6 books. Nancy has 5 books. How many books do they have in all?
   - Ⓐ 1 book
   - Ⓑ 12 books
   - Ⓒ 11 books
   - Ⓓ 65 books

25. 11 − 7 = ?
   - Ⓐ 4
   - Ⓑ 6
   - Ⓒ 16
   - Ⓓ 18

26. There were 12 eggs. 4 were broken. How many are left?
   - Ⓐ 2 eggs
   - Ⓑ 7 eggs
   - Ⓒ 8 eggs
   - Ⓓ 16 eggs

27. There were 9 birds. Then 3 flew away. How many birds are left?
   - Ⓐ 5 birds
   - Ⓑ 6 birds
   - Ⓒ 7 birds
   - Ⓓ 12 birds

28. 8¢ − 3¢
   - Ⓐ 4¢
   - Ⓑ 5¢
   - Ⓒ 6¢
   - Ⓓ 11¢

GO ON

Name _____

Cumulative Test

CHAPTERS 1–9
PAGE 5

29. 6¢
    +3¢

  (A) 1¢   (B) 8¢
  (C) 9¢   (D) not here

30. 7¢
    +1¢

  (A) 1¢   (B) 6¢
  (C) 7¢   (D) not here

31. Which one is longest?

(A)

(B)

(C)

(D)

32. How long is the 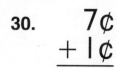?

(A) 1 ⌒ long
(B) 4 ⌒ long
(C) 6 ⌒ long
(D) 8 ⌒ long

33. Use your inch ruler. How long is the worm?

(A) 1 inch   (B) 2 inches
(C) 3 inches (D) 4 inches

34. Use your centimeter ruler. How long is the stick?

(A) 1 centimeter
(B) 3 centimeters
(C) 5 centimeters
(D) 7 centimeters

GO ON

Name _____

Cumulative Test

CHAPTERS 1–9
PAGE 6

35. Which is heavier, 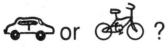 or ?

- Ⓐ
- Ⓑ

36. Which is cold?

- Ⓐ
- Ⓑ

37. Which holds <u>about</u> the same as a ?

- Ⓐ
- Ⓑ
- Ⓒ

38. Which is <u>about</u> as heavy as an ?

- Ⓐ
- Ⓑ
- Ⓒ

39. Which shows fair shares?

- Ⓐ
- Ⓑ
- Ⓒ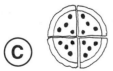
- Ⓓ

40. Which shows one third?

- Ⓐ
- Ⓑ
- Ⓒ
- Ⓓ

GO ON

Cumulative Test

CHAPTERS 1-9
PAGE 7

41. What fraction does this picture show?

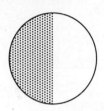

- A) $\frac{1}{5}$
- B) $\frac{1}{4}$
- C) $\frac{1}{3}$
- D) $\frac{1}{2}$

42. What fraction of the group is small?

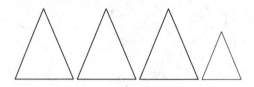

- A) $\frac{1}{5}$
- B) $\frac{1}{4}$
- C) $\frac{1}{3}$
- D) $\frac{1}{2}$

43. 2 pencils for a dime. How many pencils can you buy with 3 dimes?

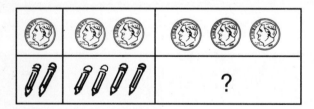

- A) 2
- B) 3
- C) 5
- D) 6

44. Which is the number pair? The sum is 7. The difference is 5.

- A) 2 and 7
- B) 3 and 4
- C) 5 and 2
- D) 1 and 6

GO ON

Name _____

Cumulative Test

CHAPTERS 1–9
PAGE 8

45. How much did the girl spend?

- (A) 2¢
- (B) 5¢
- (C) 7¢
- (D) 12¢

46. There are 3 .
There are 5 .
How many more  are there than ?

- (A) 2 more
- (B) 3 more
- (C) 5 more
- (D) 8 more

Use the table below to answer questions 47–48.

Money to Spend

Children	Dimes	Pennies
Ali	3	2
Barb	1	5
Connie	4	0
Dave	2	3

47. Who has the most dimes?

- (A) Ali
- (B) Barb
- (C) Connie
- (D) Dave

48. What do you do to find how much money Ali and Barb have in all?

- (A) Add
- (B) Subtract

Cumulative Test

CHAPTERS 1–9
PAGE 9

Use the graph below to answer question 49.

Favorite Colors

Red	👧 👧 👧 👧
Blue	👧 👧 👧 👧 👧 👧
Yellow	👧 👧 👧
Green	👧 👧 👧 👧 👧

Each 👧 stands for 1 person.

49. Which color is the favorite of the most people?

(A) Blue   (B) Yellow   (C) Green   (D) Red

Use the graph below to answer question 50.

Favorite Toys

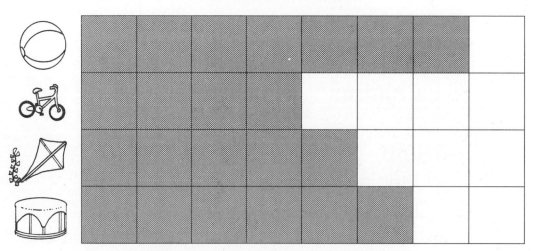

Each ▨ stands for 1 person.

50. Which toy is the favorite of the most people?

(A)   (B)   (C)   (D)

Name _____

Pretest

CHAPTER 10
PAGE 1

● Choose the correct answer.

1. Which happened last?

Ⓐ

Ⓑ

Ⓒ

Ⓓ

2. Which happened first?

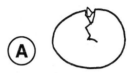
Ⓐ

Ⓑ

Ⓒ

Ⓓ

3. Which takes more time?

Ⓐ

Ⓑ

4. Which takes less time?

Ⓐ

Ⓑ

GO ON

Name _____

Pretest

CHAPTER 10
PAGE 2

5. Which one shows 2 o'clock?

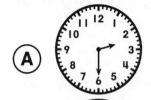
A

B

C

D

6. Which one shows 7:30?

A

B

C

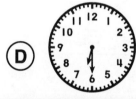
D

7. What time is it?

A) 3:00   B) 6:00
C) 9:00   D) 12:00

8. What time is it?

A) 8:30   B) 9:30
C) 7:30   D) 6:30

9. What time is it?

A) 11:00   B) 12:00
C) 1:00   D) 7:00

10. What time is it?

A) 6:00   B) 12:30
C) 1:30   D) not here

Name _____

Pretest

CHAPTER 10
PAGE 3

• Use the calendar below to answer questions 11–14.

April						
Sunday	Monday	Tuesday	Wednesday	Thursday	Friday	Saturday
			1	2	3	4
5	6	7	8	9	10	11

11. On what day does this month begin?

   (A) Friday
   (B) Wednesday
   (C) Monday
   (D) Thursday

12. What day of the week is the third day of this month?

   (A) Tuesday
   (B) Friday
   (C) Saturday
   (D) Monday

13. What day comes just after Wednesday?

   (A) Tuesday
   (B) Thursday
   (C) Friday
   (D) Saturday

14. What day comes just before Saturday?

   (A) Friday
   (B) Wednesday
   (C) Monday
   (D) Sunday

GO ON

Name _____

Pretest

CHAPTER 10
PAGE 4

15. How much money?

- Ⓐ 5¢
- Ⓑ 6¢
- Ⓒ 30¢
- Ⓓ 60¢

16. How much money?

- Ⓐ 3¢
- Ⓑ 5¢
- Ⓒ 15¢
- Ⓓ 30¢

17. How much money?

- Ⓐ 3¢
- Ⓑ 20¢
- Ⓒ 25¢
- Ⓓ 30¢

18. How much money?

- Ⓐ 25¢
- Ⓑ 27¢
- Ⓒ 35¢
- Ⓓ 46¢

Name _____

Pretest

CHAPTER 10
PAGE 5

19. How much money?

Ⓐ 5¢   Ⓑ 14¢
Ⓒ 24¢ Ⓓ 29¢

20. How much money?

Ⓐ 17¢   Ⓑ 28¢
Ⓒ 32¢   Ⓓ 37¢

Use coins to answer questions 21–24.

21. Rita has 1 nickel and 2 pennies. How much money does she have?

Ⓐ 2¢   Ⓑ 3¢
Ⓒ 7¢   Ⓓ 12¢

22. Kit has 2 dimes and 2 nickels. How much money does he have?

Ⓐ 4¢   Ⓑ 10¢
Ⓒ 20¢   Ⓓ 30¢

23. Lois has 5 dimes. How much money does she have?

Ⓐ 5¢   Ⓑ 10¢
Ⓒ 25¢   Ⓓ 50¢

24. Ned has 1 quarter, 1 dime, and 3 pennies. How much money does he have?

Ⓐ 5¢   Ⓑ 33¢
Ⓒ 38¢   Ⓓ 50¢

Name _____

Pretest

CHAPTER 10
PAGE 6

25. Which group of coins is the same amount as ?

A

B

C

D

26. Which group of coins is the same amount as ?

A

B

C

D

GO ON

Name _____

Pretest

CHAPTER 10
PAGE 7

27. Which group of coins is the same amount as   ?

Ⓐ

Ⓑ

Ⓒ

Ⓓ

28. Which group of coins is the same amount as

 ?

Ⓐ

Ⓑ

Ⓒ

Ⓓ

GO ON

Pretest

CHAPTER 10
PAGE 8

29. The movie will be over in 1 hour. At what time will the movie be over?

Ⓐ 7:00 Ⓑ 6:00
Ⓒ 8:30 Ⓓ 9:00

30. The plane will take off in 30 minutes. At what time will the plane take off?

Ⓐ 3:00 Ⓑ 2:00
Ⓒ 3:30 Ⓓ 5:30

31. Karen is going to play for 30 minutes. At what time will she finish?

Ⓐ 5:00 Ⓑ 4:30
Ⓒ 7:30 Ⓓ 7:00

32. Art class will be over in 1 hour. At what time will art class be over?

Ⓐ 11:30 Ⓑ 1:00
Ⓒ 9:30   Ⓓ 11:00

Name _____

Posttest

CHAPTER 10
PAGE 1

Choose the correct answer.

1. Which happened last?

Ⓐ

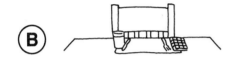
Ⓑ

Ⓒ

Ⓓ

2. Which happened first?

Ⓐ

Ⓑ

Ⓒ

Ⓓ

3. Which takes more time?

Ⓐ

Ⓑ

4. Which takes less time?

Ⓐ

Ⓑ

Name _____

Posttest

CHAPTER 10
PAGE 2

5. Which one shows 1 o'clock?

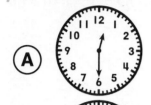
Ⓐ

Ⓑ

Ⓒ

Ⓓ

6. Which one shows 5:30?

Ⓐ

Ⓑ

Ⓒ

Ⓓ

7. What time is it?

Ⓐ 2:00   Ⓑ 4:00
Ⓒ 3:00   Ⓓ 12:00

8. What time is it?

Ⓐ 4:30   Ⓑ 3:30
Ⓒ 8:30   Ⓓ 6:30

9. What time is it?

Ⓐ 1:00   Ⓑ 12:00
Ⓒ 11:00 Ⓓ 2:00

10. What time is it?

Ⓐ 5:30   Ⓑ 6:30
Ⓒ 7:30   Ⓓ not here

Name _____

Posttest

CHAPTER 10
PAGE 3

Use the calendar below to answer questions 11–14.

May						
Sunday	Monday	Tuesday	Wednesday	Thursday	Friday	Saturday
					1	2
3	4	5	6	7	8	9

11. On what day does this month begin?

   (A) Sunday
   (B) Friday
   (C) Saturday
   (D) Monday

12. What day of the week is the third day of this month?

   (A) Tuesday
   (B) Sunday
   (C) Saturday
   (D) Thursday

13. What day comes just after Tuesday?

   (A) Monday
   (B) Friday
   (C) Wednesday
   (D) Saturday

14. What day comes just before Friday?

   (A) Thursday
   (B) Saturday
   (C) Sunday
   (D) Tuesday

GO ON

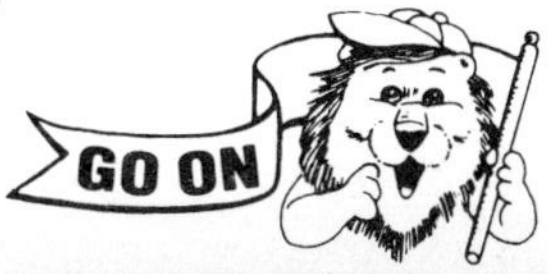

Name _____

Posttest

CHAPTER 10
PAGE 4

15. How much money?

Ⓐ 4¢   Ⓑ 5¢
Ⓒ 20¢ Ⓓ 40¢

16. How much money?

Ⓐ 6¢   Ⓑ 16¢
Ⓒ 30¢ Ⓓ 60¢

17. How much money?

Ⓐ 5¢   Ⓑ 21¢
Ⓒ 30¢ Ⓓ 50¢

18. How much money?

Ⓐ 5¢   Ⓑ 17¢
Ⓒ 22¢ Ⓓ 35¢

Name _____

Posttest

CHAPTER 10
PAGE 5

19. How much money?

Ⓐ 4¢ Ⓑ 28¢
Ⓒ 31¢ Ⓓ 40¢

20. How much money?

Ⓐ 6¢ Ⓑ 26¢
Ⓒ 41¢ Ⓓ 51¢

Use coins to answer questions 21–24.

21. Rae has 2 nickels and 1 penny. How much money does she have?

Ⓐ 3¢ Ⓑ 11¢
Ⓒ 15¢ Ⓓ not here

22. Kim has 1 dime and 3 nickels. How much money does she have?

Ⓐ 4¢ Ⓑ 20¢
Ⓒ 25¢ Ⓓ 40¢

23. Lola has 6 dimes. How much money does she have?

Ⓐ 6¢ Ⓑ 10¢
Ⓒ 30¢ Ⓓ 60¢

24. Nick has 1 quarter, 1 dime, and 1 nickel. How much money does he have?

Ⓐ 3¢ Ⓑ 31¢
Ⓒ 40¢ Ⓓ 45¢

Name _____

Posttest

CHAPTER 10
PAGE 6

25. Which group of coins is the same amount as ?

A

B

C

D

26. Which group of coins is the same amount as  ?

A

B

C

D

Name _____

Posttest

CHAPTER 10
PAGE 7

27. Which group of coins is the same amount as

   ?

Ⓐ

Ⓑ

Ⓒ

Ⓓ

28. Which group of coins is the same amount as

   ?

Ⓐ

Ⓑ

Ⓒ

Ⓓ

Posttest

CHAPTER 10
PAGE 8

29. The movie will be over in 1 hour. At what time will the movie be over?

- (A) 9:30
- (B) 8:00
- (C) 10:30
- (D) 10:00

30. The school day will end in 30 minutes. At what time will the school day end?

- (A) 2:00
- (B) 4:30
- (C) 3:30
- (D) 3:00

31. Terry is going to play for 30 minutes. At what time will he finish?

- (A) 4:30
- (B) 5:30
- (C) 8:00
- (D) 6:00

32. The show will start in 1 hour. At what time will the show start?

- (A) 7:30
- (B) 5:30
- (C) 7:00
- (D) 8:30

Name _____

Pretest

CHAPTER 11
PAGE 1

Choose the letter of the correct answer.

1. $3 + 8 = \underline{?}$

   Ⓐ 5 Ⓑ 10
   Ⓒ 11 Ⓓ 38

2. $6 + 0 = \underline{?}$

   Ⓐ 0 Ⓑ 1
   Ⓒ 6 Ⓓ 60

3. $7 - 5 = \underline{?}$

   Ⓐ 2 Ⓑ 3
   Ⓒ 12 Ⓓ 75

4. $9 - 1 = \underline{?}$

   Ⓐ 7 Ⓑ 8
   Ⓒ 10 Ⓓ 91

5. How many cubes in all?

   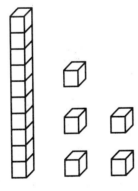

   Ⓐ 6 cubes
   Ⓑ 13 cubes
   Ⓒ 14 cubes
   Ⓓ 15 cubes

6. How many cubes in all?

   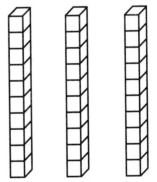

   Ⓐ 3 cubes
   Ⓑ 20 cubes
   Ⓒ 30 cubes
   Ⓓ 33 cubes

GO ON

Name _____

Pretest

CHAPTER 11
PAGE 2

7. How many cubes in all?

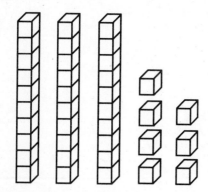

- (A) 10 cubes
- (B) 30 cubes
- (C) 36 cubes
- (D) 37 cubes

8. How many cubes in all?

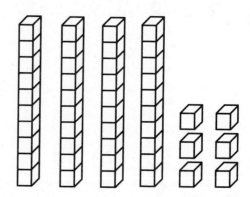

- (A) 46 cubes
- (B) 60 cubes
- (C) 64 cubes
- (D) 66 cubes

9. Which addition sentence tells how many cubes in all?

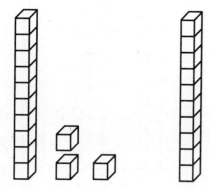

- (A) $4 + 1 = 5$
- (B) $10 + 3 = 13$
- (C) $10 + 10 = 20$
- (D) $13 + 10 = 23$

10. Which addition sentence tells how many cubes in all?

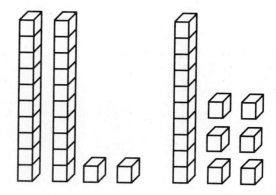

- (A) $2 + 6 = 8$
- (B) $3 + 7 = 10$
- (C) $20 + 10 = 30$
- (D) $22 + 16 = 38$

GO ON

Pretest

**CHAPTER 11
PAGE 3**

11. 10
    + 60

    (A) 6 (B) 7
    (C) 70 (D) 71

12. 43
    + 21

    (A) 22 (B) 61
    (C) 63 (D) 64

13. 32
    + 54

    (A) 86 (B) 87
    (C) 88 (D) not here

14. 33
    + 33

    (A) 0 (B) 33
    (C) 66 (D) 99

15. Chris has 10 bats. Pat has 12 bats. How many bats in all?

    (A) 2 bats
    (B) 10 bats
    (C) 22 bats
    (D) 32 bats

16. There are 23 girls and 24 boys playing. How many children are playing?

    (A) 7 children
    (B) 27 children
    (C) 41 children
    (D) 47 children

17. Tod found 35 nuts. Bob found 14 nuts. How many nuts are there?

    (A) 9 nuts
    (B) 19 nuts
    (C) 37 nuts
    (D) 49 nuts

18. Sara had 20 berries. She picked 16 more. How many does she have?

    (A) 4 berries
    (B) 18 berries
    (C) 36 berries
    (D) 81 berries

GO ON

Name _____

Pretest

CHAPTER 11
PAGE 4

19. Which subtraction sentence tells the story?

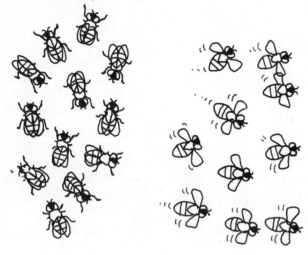

- Ⓐ 22 − 10 = 12
- Ⓑ 12 − 10 = 2
- Ⓒ 12 − 2 = 10
- Ⓓ 22 − 11 = 11

20. Which subtraction sentence tells the story?

- Ⓐ 23 − 13 = 10
- Ⓑ 13 − 10 = 3
- Ⓒ 23 − 10 = 13
- Ⓓ 10 − 3 = 7

21. 40
    −20
    ─────
- Ⓐ 0   Ⓑ 2
- Ⓒ 20   Ⓓ 60

22. 68
    −41
    ─────
- Ⓐ 7   Ⓑ 27
- Ⓒ 28   Ⓓ 72

23. 54
    −20
    ─────
- Ⓐ 3   Ⓑ 30
- Ⓒ 32   Ⓓ not here

24. 49
    −39
    ─────
- Ⓐ 10   Ⓑ 11
- Ⓒ 19   Ⓓ not here

GO ON

Pretest

CHAPTER 11
PAGE 5

25. There were 19 .
Then 10 were sold.
How many are left?

Ⓐ 1

Ⓑ 9

Ⓒ 10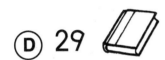

Ⓓ 29

26. Kate had 53 .
She gave away 22.
How many are left?

Ⓐ 1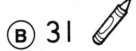

Ⓑ 31

Ⓒ 32

Ⓓ 51

27. There were 24 .
Then 11 ran away.
How many are left?

Ⓐ 11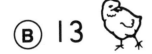

Ⓑ 13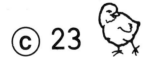

Ⓒ 23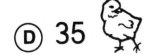

Ⓓ 35

28. There were 17 .
Then 12 flew away.
How many are left?

Ⓐ 5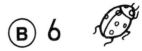

Ⓑ 6

Ⓒ 15

Ⓓ 29

GO ON

Pretest

CHAPTER 11
PAGE 6

29. Mona had 63¢. Then she spent 21¢. <u>About</u> how much money does she have now?

- Ⓐ about 40¢
- Ⓑ about 50¢

30. Sasha spent 48¢. Then she spent 31¢. <u>About</u> how much did she spend in all?

- Ⓐ about 70¢
- Ⓑ about 80¢

31. Ned had 25 . He gave away 15. How many does he have now?

- Ⓐ 10
- Ⓑ 40
- Ⓒ 45
- Ⓓ 100

32. Mike is 4 years old. His sister is 3 years older. How old is Mike's sister?

- Ⓐ 1 year old
- Ⓑ 3 years old
- Ⓒ 7 years old
- Ⓓ 40 years old

Name _____

Posttest

CHAPTER 11
PAGE 1

Choose the letter of the correct answer.

1. $4 + 7 = \underline{?}$

   (A) 1   (B) 10
   (C) 11 (D) 47

2. $0 + 7 = \underline{?}$

   (A) 0   (B) 1
   (C) 7   (D) 8

3. $6 - 6 = \underline{?}$

   (A) 0   (B) 1
   (C) 6   (D) 12

4. $11 - 8 = \underline{?}$

   (A) 2   (B) 3
   (C) 4   (D) 19

5. How many cubes in all?

   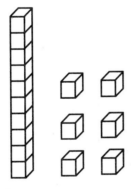

   (A) 7 cubes
   (B) 14 cubes
   (C) 15 cubes
   (D) 16 cubes

6. How many cubes in all?

   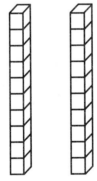

   (A) 2 cubes
   (B) 10 cubes
   (C) 20 cubes
   (D) 22 cubes

GO ON

Name _____

Posttest

CHAPTER 11
PAGE 2

7. How many cubes in all?

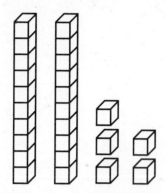

- (A) 25 cubes
- (B) 26 cubes
- (C) 52 cubes
- (D) 62 cubes

8. How many cubes in all?

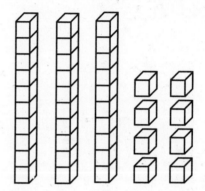

- (A) 38 cubes
- (B) 80 cubes
- (C) 83 cubes
- (D) 88 cubes

9. Which addition sentence tells how many cubes in all?

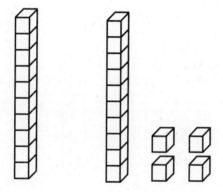

- (A) 1 + 5 = 6
- (B) 10 + 14 = 24
- (C) 9 + 13 = 22
- (D) 10 + 5 = 15

10. Which addition sentence tells how many cubes in all?

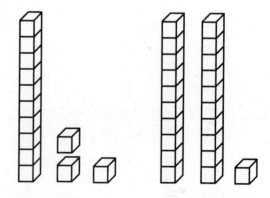

- (A) 10 + 3 = 13
- (B) 13 + 21 = 34
- (C) 4 + 3 = 7
- (D) 13 + 20 = 33

GO ON

Name _____

Posttest

CHAPTER 11
PAGE 3

11. 70
    +10

    (A) 7 (B) 8
    (C) 80 (D) 81

12. 54
    +21

    (A) 33 (B) 71
    (C) 74 (D) 75

13. 36
    +12

    (A) 48 (B) 83
    (C) 84 (D) not here

14. 22
    +22

    (A) 0 (B) 22
    (C) 44 (D) not here

15. Carl has 14 tops.
    Pam has 10 tops.
    How many tops in all?

    (A) 4 tops
    (B) 10 tops
    (C) 14 tops
    (D) 24 tops

16. There are 42 boys and 32 girls playing. How many children are playing?

    (A) 10 children
    (B) 70 children
    (C) 72 children
    (D) 74 children

17. Toni found 40 rocks.
    Lee found 13 rocks.
    How many rocks are there?

    (A) 13 rocks
    (B) 17 rocks
    (C) 43 rocks
    (D) 53 rocks

18. Walt had 36 berries.
    He picked 23 more.
    How many does he have?

    (A) 13 berries
    (B) 19 berries
    (C) 59 berries
    (D) 69 berries

GO ON

Name _____

Posttest

CHAPTER 11
PAGE 4

19. Which subtraction sentence tells the story?

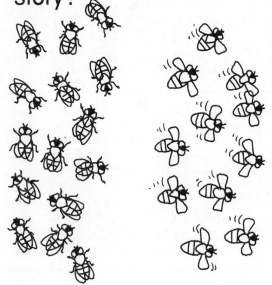

- Ⓐ $15 - 11 = 4$
- Ⓑ $15 - 2 = 13$
- Ⓒ $26 - 10 = 16$
- Ⓓ $26 - 11 = 15$

20. Which subtraction sentence tells the story?

- Ⓐ $13 - 10 = 3$
- Ⓑ $23 - 10 = 13$
- Ⓒ $23 - 1 = 22$
- Ⓓ $13 - 3 = 10$

21.
```
  60
- 30
```
- Ⓐ 0
- Ⓑ 3
- Ⓒ 30
- Ⓓ 90

22.
```
  57
- 31
```
- Ⓐ 6
- Ⓑ 26
- Ⓒ 31
- Ⓓ 62

23.
```
  45
- 20
```
- Ⓐ 5
- Ⓑ 20
- Ⓒ 24
- Ⓓ not here

24.
```
  68
- 58
```
- Ⓐ 10
- Ⓑ 11
- Ⓒ 18
- Ⓓ not here

Name _____

Posttest

CHAPTER 11
PAGE 5

25. There were 18 .
Then 10 were sold.
How many are left?

 Ⓐ 1
 Ⓑ 8
 Ⓒ 10
 Ⓓ 28

26. Tom had 45 .
He gave away 33.
How many are left?

 Ⓐ 2
 Ⓑ 12
 Ⓒ 32
 Ⓓ 78

27. There were 35 .
Then 14 ran away.
How many are left?

 Ⓐ 21
 Ⓑ 31
 Ⓒ 39
 Ⓓ 49

28. There were 18 .
Then 11 ran away.
How many are left?

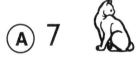 Ⓐ 7
 Ⓑ 8
 Ⓒ 17
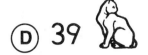 Ⓓ 39

GO ON

Posttest

CHAPTER 11
PAGE 6

29. Erin had 52¢. Then she spent 30¢. <u>About</u> how much money does she have now?

Ⓐ about 20¢
Ⓑ about 30¢

30. Mark spent 52¢. Then he spent 21¢. <u>About</u> how much did he spend in all?

Ⓐ about 70¢
Ⓑ about 80¢

31. Eric had 32 . He gave away 11. How many does he have now?

Ⓐ 21   Ⓑ 43
Ⓒ 53 Ⓓ 320

32. Marie is 15 years old. Her brother is 3 years older. How old is Marie's brother?

Ⓐ 8 years old
Ⓑ 18 years old
Ⓒ 45 years old

Name _____

Pretest

CHAPTER **12**
PAGE 1

Choose the letter of the correct answer.

1. Which addition sentence tells how many in all?

  Ⓐ 6 + 8 = 14
  Ⓑ 6 + 2 = 8
  Ⓒ 2 + 4 = 6
  Ⓓ 8 + 2 = 10

2. Which addition sentence tells how many in all?

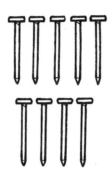 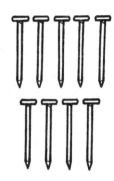

  Ⓐ 0 + 0 = 0
  Ⓑ 9 + 9 = 18
  Ⓒ 9 + 8 = 17
  Ⓓ 9 + 0 = 9

3. Which addition sentence tells how many in all?

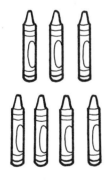 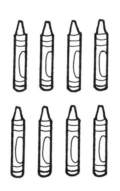

  Ⓐ 1 + 7 = 8
  Ⓑ 8 + 1 = 9
  Ⓒ 8 + 9 = 17
  Ⓓ 7 + 8 = 15

4. Which addition sentence tells how many in all?

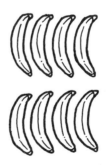 

  Ⓐ 8 + 9 = 17
  Ⓑ 7 + 1 = 8
  Ⓒ 8 + 1 = 9
  Ⓓ 8 + 8 = 16

GO ON

Name _____

Pretest

CHAPTER 12
PAGE 2

5. $5 + 8 =$?
   - A) 12
   - B) 13
   - C) 14
   - D) 15

6. $8 + 8 =$?
   - A) 0
   - B) 8
   - C) 15
   - D) 16

7. $9 + 5 =$?
   - A) 4
   - B) 13
   - C) 14
   - D) 15

8. $6 + 7 =$?
   - A) 1
   - B) 11
   - C) 12
   - D) not here

9. $\begin{array}{r} 9 \\ +7 \end{array}$
   - A) 2
   - B) 15
   - C) 16
   - D) 17

10. $8 + 6 =$?
    - A) 2
    - B) 4
    - C) 13
    - D) 14

Use counters to answer questions 11–14.

11. Mary has 9 🎈. Karl has 4 🎈. How many 🎈 are there in all?
    - A) 5
    - B) 9
    - C) 13
    - D) 14

12. There are 7 red 🎩 on the rack. There are 7 blue 🎩 on the rack. How many 🎩 are there in all?
    - A) 0
    - B) 2
    - C) 13
    - D) 14

GO ON

Name _____
Pretest

CHAPTER 12
PAGE 3

13. Mark has 9 🌼.
    Pam has 7 🌼.
    How many 🌼 in all?

    Ⓐ 2 🌼   Ⓑ 9 🌼
    Ⓒ 15 🌼 Ⓓ 16 🌼

14. There are 7 red 🧤 on a rack. There are 6 blue 🧤 on the rack. How many 🧤 in all?

    Ⓐ 1 🧤   Ⓑ 12 🧤
    Ⓒ 13 🧤 Ⓓ 14 🧤

15. Which subtraction sentence tells the story?

    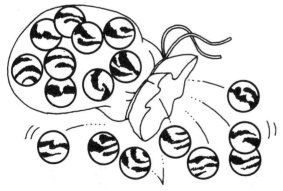

    Ⓐ 9 − 1 = 8
    Ⓑ 17 − 8 = 9
    Ⓒ 16 − 8 = 8
    Ⓓ 18 − 9 = 9

16. Which subtraction sentence tells the story?

    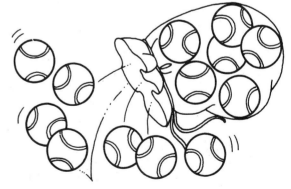

    Ⓐ 7 − 1 = 6
    Ⓑ 14 − 7 = 7
    Ⓒ 6 − 6 = 0
    Ⓓ 13 − 7 = 6

GO ON

Name _____

Pretest

CHAPTER 12
PAGE 4

17. Which subtraction sentence tells the story?

- Ⓐ 9 − 7 = 2
- Ⓑ 16 − 7 = 9
- Ⓒ 9 − 8 = 1
- Ⓓ 15 − 9 = 6

18. Which subtraction sentence tells the story?

- Ⓐ 15 − 8 = 7
- Ⓑ 8 − 7 = 1
- Ⓒ 8 − 1 = 7
- Ⓓ 14 − 8 = 6

19.   15
         − 6
         ___

- Ⓐ 8
- Ⓑ 9
- Ⓒ 11
- Ⓓ not here

20.   14
         − 7
         ___

- Ⓐ 3
- Ⓑ 6
- Ⓒ 13
- Ⓓ not here

21. 13 − 7 = ?

- Ⓐ 4
- Ⓑ 6
- Ⓒ 7
- Ⓓ 14

22. 16 − 8 = ?

- Ⓐ 2
- Ⓑ 7
- Ⓒ 8
- Ⓓ 12

GO ON

Pretest

CHAPTER 12
PAGE 5

● Use counters to answer questions 23 and 24.

23. Sally baked 18 .
    Her friends ate 9 .
    How many are left?

    Ⓐ 0    Ⓑ 8

    Ⓒ 9 🍪   Ⓓ 11 🍪

24. There were 15 🎈.
    Then 7 🎈 popped.
    How many are left?

    Ⓐ 2 🎈   Ⓑ 8 🎈

    Ⓒ 9 🎈   Ⓓ not here

25. Which number sentence does <u>not</u> belong to the fact family for 5, 9, and 14?

    Ⓐ 5 + 4 = 9
    Ⓑ 9 + 5 = 14
    Ⓒ 14 − 9 = 5
    Ⓓ 14 − 5 = 9

26. Which number sentence does <u>not</u> belong to the fact family for 8, 7, and 15?

    Ⓐ 15 − 7 = 8
    Ⓑ 8 + 7 = 15
    Ⓒ 15 − 8 = 7
    Ⓓ 8 − 7 = 1

27. Which number sentence belongs to the same fact family as 13 − 8 = 5?

    Ⓐ 5 + 3 = 8
    Ⓑ 8 + 5 = 13
    Ⓒ 5 − 3 = 2
    Ⓓ 8 − 5 = 3

28. Which number sentence belongs to the same fact family as 16 − 7 = 9?

    Ⓐ 7 + 2 = 9
    Ⓑ 9 − 7 = 2
    Ⓒ 14 − 7 = 7
    Ⓓ 9 + 7 = 16

Name _____

Pretest

CHAPTER 12
PAGE 6

Use the table below to answer questions 29 – 30.

FAVORITE SNACKS	Boys	Girls
Popcorn	||||	||||-|||
Fruit	||||-	||||- ||||
Cheese	||||- ||||	||||-	

29. How many boys and girls like cheese best?

    Ⓐ 15   Ⓑ 9
    Ⓒ 12   Ⓓ 3

30. How many boys and girls like fruit best?

    Ⓐ 14   Ⓑ 5
    Ⓒ 4   Ⓓ 15

31.

There are 9 balloons.
There are 3 children.
How many balloons does each child get?

Ⓐ 2 balloons
Ⓑ 3 balloons
Ⓒ 4 balloons
Ⓓ 6 balloons

32.

There are 16 stars.
There are 2 children.
How many stars does each child get?

Ⓐ 2 stars
Ⓑ 4 stars
Ⓒ 8 stars
Ⓓ 14 stars

Name _____

Posttest

CHAPTER 12
PAGE 1

● Choose the letter of the correct answer.

1. Which addition sentence tells how many in all?

Ⓐ 8 + 7 = 15
Ⓑ 8 + 2 = 10
Ⓒ 7 + 1 = 8
Ⓓ 5 + 2 = 7

2. Which addition sentence tells how many in all?

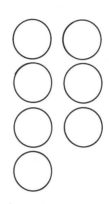

Ⓐ 0 + 0 = 0
Ⓑ 7 + 0 = 7
Ⓒ 6 + 7 = 13
Ⓓ 7 + 7 = 14

3. Which addition sentence tells how many in all?

Ⓐ 4 + 5 = 9
Ⓑ 9 + 6 = 15
Ⓒ 8 + 5 = 13
Ⓓ 9 + 5 = 14

4. Which addition sentence tells how many in all?

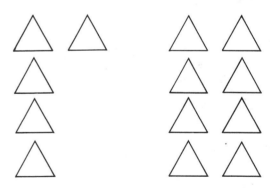

Ⓐ 5 + 3 = 8
Ⓑ 5 + 8 = 13
Ⓒ 3 + 8 = 11
Ⓓ 5 + 9 = 14

GO ON

Name _____

Posttest

CHAPTER 12
PAGE 2

5. $6 + 7 = \underline{?}$

   (A) 12 (B) 13
   (C) 14 (D) 15

6. $9 + 9 = \underline{?}$

   (A) 0   (B) 9
   (C) 18 (D) 19

7. $8 + 6 = \underline{?}$

   (A) 2   (B) 12
   (C) 13 (D) 14

8. $9 + 8 = \underline{?}$

   (A) 1   (B) 16
   (C) 17 (D) not here

9. $\begin{array}{r} 6 \\ +9 \\ \hline \end{array}$

   (A) 3   (B) 15
   (C) 16 (D) 17

10. $8 + 8 = \underline{?}$

    (A) 0   (B) 8
    (C) 14 (D) 16

Use counters to answer questions 11–14.

11. Mark has 7 .
    Pam has 9 .
    How many in all?

    (A) 2   (B) 9
    (C) 15 (D) 16

12. There are 6 red  on a rack. There are 7 blue on the rack. How many  in all?

    (A) 1   (B) 12
    (C) 13 (D) 14

GO ON

Name _____

Posttest

CHAPTER 12
PAGE 3

13. Mary has 9 🎈.
Karl has 5 🎈.
How many 🎈 are
there in all?

Ⓐ 5 🎈   Ⓑ 9 🎈

Ⓒ 13 🎈   Ⓓ 14 🎈

14. There are 7 red 🎩 on
the rack. There are
7 blue 🎩 on the rack.
How many 🎩 are there
in all?

Ⓐ 0 🎩   Ⓑ 2 🎩

Ⓒ 14 🎩   Ⓓ 13 🎩

15. Which subtraction
sentence tells the story?

Ⓐ 8 − 8 = 0
Ⓑ 13 − 5 = 8
Ⓒ 14 − 5 = 98
Ⓓ 8 − 5 = 3

16. Which subtraction
sentence tells the story?

Ⓐ 9 − 6 = 3
Ⓑ 15 − 9 = 6
Ⓒ 10 − 4 = 6
Ⓓ 15 − 6 = 9

GO ON

Name _____

Posttest

CHAPTER 12
PAGE 4

17. Which subtraction sentence tells the story?

- (A) $8 - 6 = 2$
- (B) $14 - 6 = 8$
- (C) $15 - 7 = 8$
- (D) $13 - 7 = 6$

18. Which subtraction sentence tells the story?

- (A) $14 - 5 = 9$
- (B) $9 - 5 = 4$
- (C) $15 - 9 = 6$
- (D) $5 - 5 = 0$

19. 16
    $-\ 8$

- (A) 2
- (B) 8
- (C) 12
- (D) 16

20. 15
    $-\ 7$

- (A) 2
- (B) 6
- (C) 12
- (D) 8

21. $17 - 8 = \underline{?}$

- (A) 1
- (B) 9
- (C) 8
- (D) 11

22. $13 - 7 = \underline{?}$

- (A) 2
- (B) 8
- (C) 6
- (D) not here

GO ON

Posttest

CHAPTER 12
PAGE 5

● Use counters to answer questions 23 and 24.

23. Mac picked 16 🍑. He gave 9 to his friend. How many 🍑 are left?

   Ⓐ 3 🍑 Ⓑ 5 🍑
   Ⓒ 7 🍑 Ⓓ 13 🍑

24. There were 18 🥚. Then 9 🥚 were broken. How many 🥚 are left?

   Ⓐ 2 🥚 Ⓑ 8 🥚
   Ⓒ 9 🥚 Ⓓ 11 🥚

25. Which number sentence does <u>not</u> belong to the fact family for 4, 9, and 13?

   Ⓐ 9 − 4 = 5
   Ⓑ 13 − 9 = 4
   Ⓒ 4 + 9 = 13
   Ⓓ 9 + 4 = 13

26. Which number sentence does <u>not</u> belong to the fact family for 7, 8, and 15?

   Ⓐ 8 − 7 = 1
   Ⓑ 15 − 8 = 7
   Ⓒ 7 + 8 = 15
   Ⓓ 8 + 7 = 15

27. Which number sentence belongs to the same fact family as 14 − 6 = 8?

   Ⓐ 8 − 6 = 2
   Ⓑ 14 − 8 = 6
   Ⓒ 4 + 4 = 8
   Ⓓ 2 + 4 = 6

28. Which number sentence belongs to the same fact family as 17 − 8 = 9?

   Ⓐ 9 − 1 = 8
   Ⓑ 8 + 1 = 9
   Ⓒ 8 + 9 = 17
   Ⓓ 18 − 9 = 9

GO ON

Name _____

Posttest

CHAPTER 12
PAGE 6

Use the table below to answer questions 29–30.

THINGS CHILDREN LIKE TO DO	Girls	Boys
Jumping	IIII	HHT IIII
Running	HHT	HHT IIII
Swinging	HHT III	HHT I

29. How many girls and boys like running best?

- (A) 14
- (B) 7
- (C) 13
- (D) 0

30. How many girls and boys like jumping best?

- (A) 14
- (B) 13
- (C) 15
- (D) 17

31.

There are 8 books.
There are 2 children.
How many books does each child get?

- (A) 2 books
- (B) 4 books
- (C) 6 books
- (D) 8 books

32.

There are 15 boats.
There are 5 children.
How many boats does each child get?

- (A) 1 boat
- (B) 2 boats
- (C) 3 boats
- (D) 5 boats

Name _____

Cumulative Test

CHAPTERS 1–12
PAGE 1

Choose the correct answer.

1.

    $7 + 1 = \underline{\ ?\ }$

    Ⓐ 1   Ⓑ 7
    Ⓒ 8   Ⓓ 9

2.

    $6 - 4 = \underline{\ ?\ }$

    Ⓐ 2   Ⓑ 3
    Ⓒ 9   Ⓓ 10

3. Which plane shape matches the face of the solid shape?

    Ⓐ ○   Ⓑ △
    Ⓒ □   Ⓓ △

4. Which plane shape matches the face of the solid shape?

    Ⓐ ○   Ⓑ □
    Ⓒ △   Ⓓ ▭

5. Which figure is the same size and shape?

    Ⓒ △   Ⓓ □

6. Which shows two parts that match?

    Ⓐ

    Ⓑ

Name _____

Cumulative Test

CHAPTERS 1–12
PAGE 2

7. Which shape comes next in the pattern?

 ?

Ⓐ ▢

Ⓑ ☆

Ⓒ ▭

8. Which shape comes next in the pattern?

△ ○ ◇ △ ○ ◇ △ ?

Ⓐ △

Ⓑ ○

Ⓒ ◇

9. Which number comes after?

39, ?

Ⓐ 30 Ⓑ 38
Ⓒ 40 Ⓓ 50

10. Which number comes before?

?, 72

Ⓐ 61 Ⓑ 62
Ⓒ 70 Ⓓ 71

11. Which number comes between?

9, ?, 11

Ⓐ 8 Ⓑ 10
Ⓒ 12 Ⓓ not here

12. Which one is 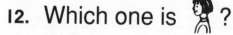 ?

Ⓐ fourth Ⓑ seventh
Ⓒ third   Ⓓ fifth

Name _____

Cumulative Test

CHAPTERS 1-12
PAGE 3

13. Which one is the  ?

    (A) fourth (B) second
    (C) sixth   (D) third

14. Use your inch ruler. How long is the needle?

    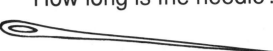

    (A) 1 inch   (B) 2 inches
    (C) 3 inches (D) 4 inches

15. Use your centimeter ruler. How long is the feather?

    (A) 1 centimeter
    (B) 2 centimeters
    (C) 4 centimeters
    (D) 6 centimeters

16. Which holds about the same as a ?

    (A)

    (B)

    (C)

17. Which is about as heavy as a  ?

    (A)

    (B)

    (C)

18. Which is about the same temperature as a ?

    (A)   (B)

    (C)

Name _____

Cumulative Test

CHAPTERS 1–12
PAGE 4

19. Which shows fair shares?

Ⓐ

Ⓑ

Ⓒ

Ⓓ

20. Which shows one fourth?

Ⓐ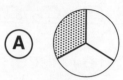

Ⓑ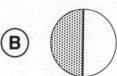

Ⓒ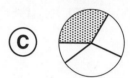

Ⓓ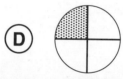

21. What fraction does this picture show?

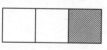

Ⓐ $\dfrac{1}{5}$ Ⓑ $\dfrac{1}{4}$
Ⓒ $\dfrac{1}{3}$ Ⓓ $\dfrac{1}{2}$

22. What time is it?

Ⓐ 7:00   Ⓑ 8:00
Ⓒ 9:00   Ⓓ 12:00

23. What time is it?

Ⓐ 1:30   Ⓑ 2:30
Ⓒ 6:30   Ⓓ not here

24. How much money is there?

Ⓐ 5¢   Ⓑ 15¢
Ⓒ 17¢   Ⓓ 32¢

Name _____

Cumulative Test

CHAPTERS 1–12
PAGE 5

25. How much money is there?

- (A) 6¢
- (B) 23¢
- (C) 41¢
- (D) 63¢

26. How much money is there?

- (A) 7¢
- (B) 25¢
- (C) 40¢
- (D) not here

27. How many cubes are there in all?

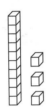

- (A) 3 cubes
- (B) 4 cubes
- (C) 10 cubes
- (D) 13 cubes

28. How many cubes are there in all?

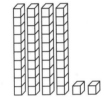

- (A) 4 cubes
- (B) 6 cubes
- (C) 24 cubes
- (D) 42 cubes

29. Which addition sentence tells how many cubes in all?

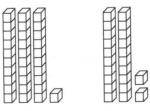

- (A) 31 + 24 = 55
- (B) 4 + 6 = 10
- (C) 30 + 1 = 31
- (D) 30 + 20 = 50

30. 70
     + 14
     ─────

- (A) 70
- (B) 74
- (C) 80
- (D) 84

Name _____

Cumulative Test

CHAPTERS 1-12
PAGE 6

31. 22
    +22

(A) 0   (B) 4
(C) 22   (D) 44

32. Ella has 46 beads. Gus has 21 beads. How many beads are there in all?

(A) 25 beads
(B) 26 beads
(C) 66 beads
(D) 67 beads

33. Which subtraction sentence tells the story?

(A) 23 − 10 = 13
(B) 13 − 10 = 3
(C) 13 − 3 = 10
(D) 23 − 0 = 23

34. 73
    −21

(A) 51   (B) 52
(C) 53   (D) 94

35. 54
    −34

(A) 20   (B) 24
(C) 28   (D) 88

36. There were 41 balloons. Then 30 popped. How many are left?

(A) 1 balloon
(B) 10 balloons
(C) 11 balloons
(D) 71 balloons

GO ON

Cumulative Test

CHAPTERS 1–12
PAGE 7

37. $9 + 8 = \underline{?}$

    A) 1   B) 16
    C) 17 D) not here

38. $7 + 7 = \underline{?}$

    A) 0   B) 7
    C) 14 D) 77

39. There are 5 red balls. There are 7 blue balls. How many balls are there in all?

    A) 2 balls
    B) 12 balls
    C) 13 balls
    D) 14 balls

40. $16 - 9 = \underline{?}$

    A) 3   B) 7
    C) 13 D) not here

41. $14 - 6 = \underline{?}$

    A) 2   B) 8
    C) 9   D) 12

42. There were 14 grapes. Then Tim ate 5. How many are left?

    A) 9 grapes
    B) 10 grapes
    C) 11 grapes
    D) 19 grapes

Cumulative Test

CHAPTERS 1-12
PAGE 8

43.

Mother will be home in 1 hour.
At what time will she be home?

(A) 4:00 (B) 5:00
(C) 6:00 (D) 7:00

44.

There are 20 🍪.
There are 10 children.
How many 🍪 will each child get?

(A) 2 (B) 4
(C) 5 (D) 20

45. 3 🔔 for 🪙
How many 🔔 can you buy with 🪙🪙🪙?

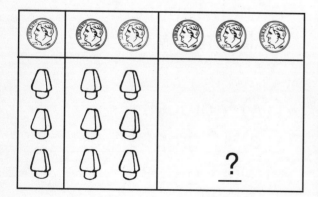

(A) 3 (B) 4
(C) 7 (D) 9

46. Which is the number pair?
The sum is 8.
The difference is 0.

(A) 1 and 7
(B) 0 and 8
(C) 8 and 8
(D) 4 and 4

Name _____

Cumulative Test

CHAPTERS 1-12
PAGE 9

Use the graph below to answer question 47.

GLASSES OF MILK IN ONE DAY

Hal	🥛 🥛 🥛
Terri	🥛 🥛 🥛 🥛 🥛
Ira	🥛 🥛
Nan	🥛 🥛 🥛 🥛

Each  🥛 stands for 1 glass.

47. How many glasses of milk did Ira drink?

   Ⓐ 2 glasses   Ⓑ 3 glasses   Ⓒ 4 glasses   Ⓓ 5 glasses

Use the graph below to answer question 48.

FAVORITE JUICES

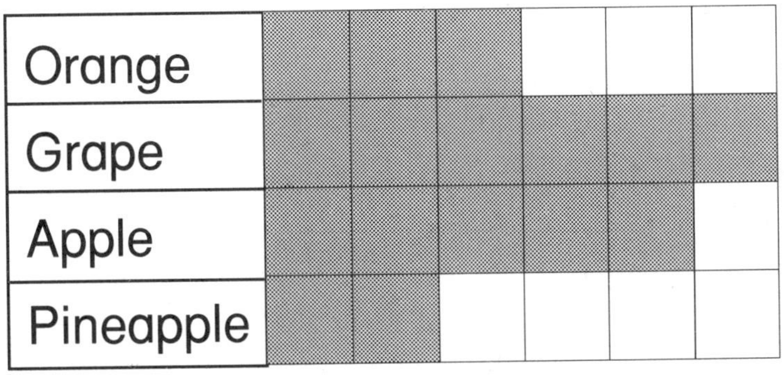

Orange	▓ ▓ ▓ ☐ ☐ ☐
Grape	▓ ▓ ▓ ▓ ▓ ▓
Apple	▓ ▓ ▓ ▓ ▓ ☐
Pineapple	▓ ▓ ☐ ☐ ☐ ☐

Each ▓ stands for 1 person.

48. Which juice is the favorite of the most people?

   Ⓐ orange   Ⓑ grape   Ⓒ apple   Ⓓ pineapple

GO ON

Name _____

Cumulative Test

CHAPTERS 1-12
PAGE 10

Use the table below to answer questions 49–50.

EATING TIMES (in minutes)			
	Breakfast	Lunch	Dinner
Amy	10	15	25
Bob	10	20	30

49. How long did it take Bob to eat lunch?

(A) 10 minutes
(B) 15 minutes
(C) 20 minutes
(D) 30 minutes

50. What do you do to find how much longer it took Amy to eat dinner than to eat lunch?

(A) Add   (B) Subtract

STOP

Name _____

End-of-Book Test

PAGE 1

Choose the correct answer.

1.

    7 − 1 = ?

    (A) 1   (B) 6
    (C) 7   (D) 8

2.

    5 + 2 = ?

    (A) 3   (B) 7
    (C) 8   (D) 9

3. Which plane shape matches the face of the solid shape?

    (A) ▭   (B) ○
    (C) ▫   (D) △

4. Which plane shape matches the face of the solid shape?

    (A) △   (B) ▭
    (C) ▫   (D) ○

5. Which figure is the same size and shape?

    (A) △   (B) ▫
    (C) ▭   (D) ▭

6. Which shows two parts that match?

    (A)

    (B)

GO ON

Name _____

End-of-Book Test

PAGE 2

7. Which shape comes next in the pattern?

   ?

Ⓐ ◇   Ⓑ △   Ⓒ ○

8. Which shape comes next in the pattern?

   ?

Ⓐ ◇   Ⓑ ○   Ⓒ ▭

9. Which number comes just before?

? , 31

Ⓐ 20   Ⓑ 21
Ⓒ 30   Ⓓ 32

10. Which number comes just after?

68, ?

Ⓐ 67   Ⓑ 69
Ⓒ 70   Ⓓ 79

11. Which number comes between?

44, ? , 46

Ⓐ 43   Ⓑ 45
Ⓒ 47   Ⓓ not here

12. Which one is 👤 ?

Ⓐ second   Ⓑ third
Ⓒ fifth   Ⓓ sixth

13. Which one is the 🏎 ?

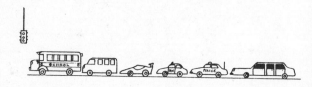

Ⓐ second   Ⓑ third
Ⓒ fourth   Ⓓ sixth

14. Use your centimeter ruler. How long is the key?

Ⓐ 1 centimeter
Ⓑ 2 centimeters
Ⓒ 3 centimeters
Ⓓ 4 centimeters

GO ON

Name _____
End-of-Book Test

PAGE 3

15. Use your inch ruler. How long is the leaf?

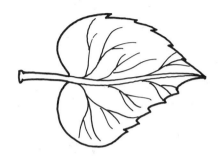

Ⓐ 1 inch Ⓑ 2 inches
Ⓒ 3 inches Ⓓ 4 inches

16. Which holds about the same as ?

Ⓐ
Ⓑ
Ⓒ

17. Which is about as heavy as a ?

Ⓐ   Ⓑ   Ⓒ

18. Which is about the same temperature as ?

Ⓐ

Ⓑ

Ⓒ

19. Which shows fair shares?

Ⓐ Ⓑ
Ⓒ Ⓓ

20. Which shows one half?

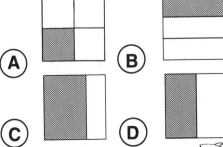

Ⓐ Ⓑ
Ⓒ Ⓓ

GO ON

End-of-Book Test

PAGE 4

21. What fraction does this picture show?

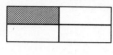

- Ⓐ $\frac{1}{5}$
- Ⓑ $\frac{1}{4}$
- Ⓒ $\frac{1}{3}$
- Ⓓ $\frac{1}{2}$

22. What time is it?

- Ⓐ 3:00
- Ⓑ 4:00
- Ⓒ 5:00
- Ⓓ 12:00

23. What time is it?

- Ⓐ 6:30
- Ⓑ 10:30
- Ⓒ 11:30
- Ⓓ not here

24. How much money is there?

- Ⓐ 5¢
- Ⓑ 35¢
- Ⓒ 40¢
- Ⓓ 50¢

25. How much money is there?

- Ⓐ 5¢
- Ⓑ 17¢
- Ⓒ 27¢
- Ⓓ 32¢

26. How much money is there?

- Ⓐ 10¢
- Ⓑ 20¢
- Ⓒ 25¢
- Ⓓ not here

27. How many cubes are there in all?

- Ⓐ 6 cubes
- Ⓑ 7 cubes
- Ⓒ 10 cubes
- Ⓓ 16 cubes

28. How many cubes are there in all?

- Ⓐ 10 cubes
- Ⓑ 17 cubes
- Ⓒ 37 cubes
- Ⓓ 73 cubes

Name _____

End-of-Book Test

PAGE 5

29. Which addition sentence tells how many cubes there are in all?

- Ⓐ 23 + 14 = 37
- Ⓑ 3 + 4 = 7
- Ⓒ 5 + 5 = 10
- Ⓓ 20 + 10 = 30

30. 45
      + 30
      ____

- Ⓐ 15
- Ⓑ 65
- Ⓒ 70
- Ⓓ 75

31. 44
      + 44
      ____

- Ⓐ 0
- Ⓑ 4
- Ⓒ 44
- Ⓓ 88

32. Tina has 25 beans. Bert has 14 beans. How many beans are there in all?

- Ⓐ 11 beans
- Ⓑ 29 beans
- Ⓒ 31 beans
- Ⓓ 39 beans

33. Which subtraction sentence tells the story?

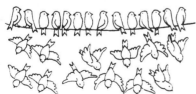
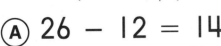

- Ⓐ 26 − 12 = 14
- Ⓑ 14 − 10 = 4
- Ⓒ 14 − 2 = 12
- Ⓓ 12 − 2 = 10

34. 56
      − 31
      ____

- Ⓐ 21
- Ⓑ 25
- Ⓒ 27
- Ⓓ 87

GO ON

End-of-Book Test

35. 63
    − 23

- Ⓐ 40
- Ⓑ 43
- Ⓒ 46
- Ⓓ 86

36. Fred won 24 stars. Amy won 10 stars. How many more stars did Fred win?

- Ⓐ 4 stars
- Ⓑ 10 stars
- Ⓒ 14 stars
- Ⓓ 34 stars

37. 8 + 5 = ?

- Ⓐ 12
- Ⓑ 13
- Ⓒ 14
- Ⓓ not here

38. 9 + 0 = ?

- Ⓐ 0
- Ⓑ 9
- Ⓒ 18
- Ⓓ 90

39. There are 4 white 🌼. There are 7 yellow 🌼. How many 🌼 in all?

- Ⓐ 3
- Ⓑ 10
- Ⓒ 11
- Ⓓ 12

40. 17 − 8 = ?

- Ⓐ 1
- Ⓑ 9
- Ⓒ 11
- Ⓓ not here

41. 12 − 5 = ?

- Ⓐ 3
- Ⓑ 7
- Ⓒ 8
- Ⓓ not here

42. Use pennies to solve. Pam has 15 pennies. She spends 8 of them. How many pennies does Pam have left?

- Ⓐ 3 pennies
- Ⓑ 7 pennies
- Ⓒ 8 pennies
- Ⓓ 22 pennies

GO ON

Name _____
End-of-Book Test

PAGE 7

43. Story time will start in 1 hour. What time will it start?

Ⓐ 1:00   Ⓑ 2:00
Ⓒ 3:00   Ⓓ 12:00

44. There are 12 eggs. There are 3 children. How many eggs does each child get?

Ⓐ 3 eggs   Ⓑ 4 eggs
Ⓒ 6 eggs   Ⓓ 12 eggs

45. 4 🧶 for 🪙   How many 🧶 can you buy with 🪙🪙🪙?

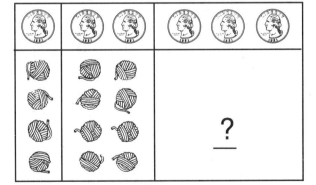

Ⓐ 4   Ⓑ 9
Ⓒ 10   Ⓓ 12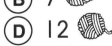

46. Which is the number pair? The sum is 6. The difference is 2.

Ⓐ 4 and 6
Ⓑ 3 and 3
Ⓒ 2 and 2
Ⓓ 2 and 4

End-of-Book Test

PAGE 8

Use the graph below to answer question 47.

PINE CONES FOUND

Tom	🌰 🌰 🌰 🌰
Sara	🌰 🌰 🌰
Dan	🌰
Alice	🌰 🌰 🌰 🌰 🌰 🌰

Each 🌰 stands for one pine cone.

47. How many 🌰 did Dan find?

(A) 1 pine cone   (B) 3 pine cones
(C) 4 pine cones   (D) 6 pine cones

Use the graph below to answer question 48.

FLOWERS PICKED

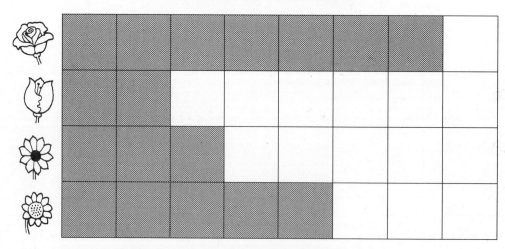

Each ▨ stands for one flower.

48. How many were picked?

   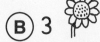   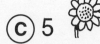

Name _____

End-of-Book Test

PAGE 9

Use the table below to answer questions 49–50.

Number of Children

	School Bus	City Bus
Boys	14	4
Girls	12	5

49. How many girls are on the school bus?

   (A) 4 girls
   (B) 5 girls
   (C) 12 girls
   (D) 17 girls

50. What do you do to find how many children are on the city bus?

   (A) Add   (B) Subtract

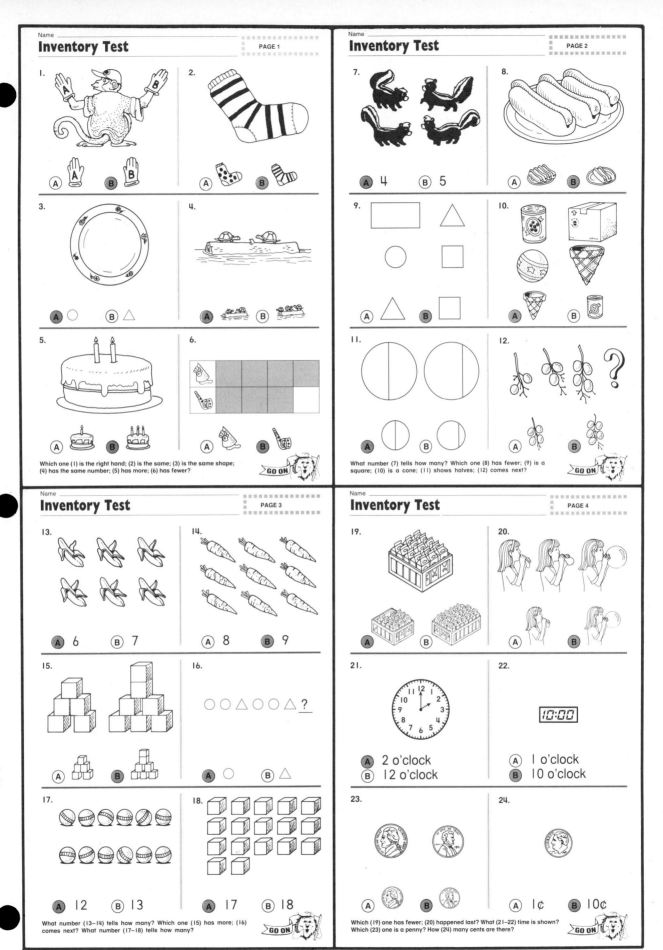

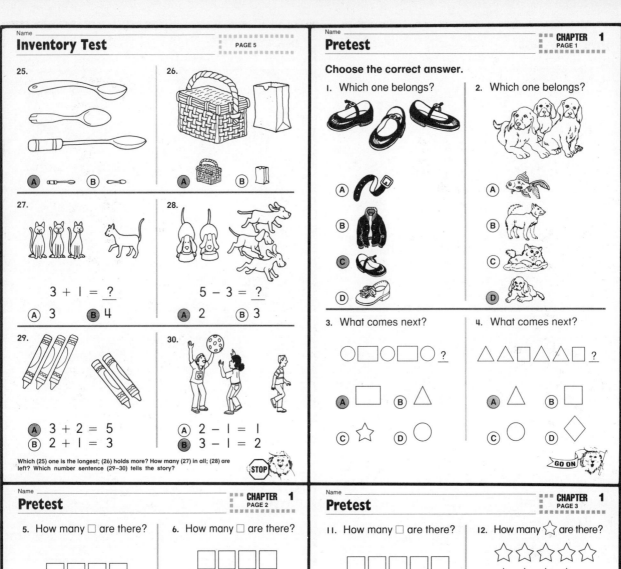

Pretest

CHAPTER 1
PAGE 4

Use the pictograph below to answer questions 17–20.

Number of Boats

17. Which group has more, ⛵ or 🚣?
 - (A)
 - (B)

18. Which group has fewer, 🚣 or ⛵?
 - (A)
 - (B)

19. Which group has more than 4?
 - (A)
 - (B)
 - (C)
 - (D)

20. Which group has fewer than 3?
 - (A)
 - (B)
 - (C)
 - (D)

Posttest

CHAPTER 1
PAGE 1

Choose the correct answer.

1. Which one belongs?
 - (A)
 - (B)
 - (C)
 - (D)

2. Which one belongs?
 - (A)
 - (B)
 - (C)
 - (D)

3. What comes next? △○△○△ ?
 - (A) □
 - (B) ○
 - (C) △
 - (D) □

4. What comes next? □○○□○○ ?
 - (A) △
 - (B) □
 - (C) ○
 - (D) ☆

Posttest

CHAPTER 1
PAGE 2

5. How many □ are there?
 - (A) 3
 - (B) 4
 - (C) 5
 - (D) 6

6. How many □ are there?
 - (A) 1
 - (B) 2
 - (C) 4
 - (D) not here

7. How many □ are there?
 - (A) 6
 - (B) 8
 - (C) 9
 - (D) not here

8. How many □ are there?
 - (A) 2
 - (B) 8
 - (C) 9
 - (D) 10

9. How many ○ are there?
 - (A) 6
 - (B) 7
 - (C) 8
 - (D) 9

10. How many □ are there?
 - (A) 0
 - (B) 1
 - (C) 2
 - (D) 10

Posttest

CHAPTER 1
PAGE 3

11. How many ☆ are there?
 - (A) 0
 - (B) 6
 - (C) 7
 - (D) 8

12. How many ○ are there?
 - (A) 4
 - (B) 5
 - (C) 6
 - (D) not here

13. Which group has fewer than this?
 - (A)
 - (B)
 - (C)

14. Which group has more than this?
 - (A)
 - (B)
 - (C)

15. What comes next? 2, 3, 4, ?
 - (A) 5
 - (B) 6
 - (C) 8
 - (D) 9

16. What comes next? 6, 7, 8, ?
 - (A) 4
 - (B) 5
 - (C) 9
 - (D) 10

Standardized Format • Test Answers

Posttest — CHAPTER 1, PAGE 4

Use the pictograph below to answer questions 17–20.

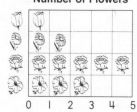

Number of Flowers

17. Which group has fewer, or ?

 (A)   (B)

18. Which group has more,  or ?

(A) (B)

19. Which group has more than 4?

 (A)   (B)
(C)   (D)

20. Which group has fewer than 2?

(A)   (B)
(C)   (D)

Pretest — CHAPTER 2, PAGE 1

Choose the correct answer.

1. Which addition sentence tells how many are there?

(A) $4 + 2 = 6$
●(B) $4 + 1 = 5$
(C) $1 + 3 = 4$
(D) $2 + 1 = 3$

2. Which addition sentence tells how many are there?

(A) $2 + 1 = 3$
(B) $3 + 3 = 6$
●(C) $2 + 3 = 5$
(D) $3 + 1 = 4$

3. Which addition sentence tells how many are there?

●(A) $2 + 4 = 6$
(B) $2 + 2 = 4$
(C) $2 + 3 = 5$
(D) $1 + 1 = 2$

4. Which addition sentence tells how many are there?

(A) $3 + 3 = 6$
(B) $1 + 3 = 4$
(C) $1 + 1 = 2$
●(D) $1 + 2 = 3$

Pretest — CHAPTER 2, PAGE 2

5. Which addition sentence tells how many are there?

●(A) $2 + 2 = 4$
(B) $1 + 2 = 3$
(C) $1 + 1 = 2$
(D) $2 + 3 = 5$

6. Which addition sentence tells how many are there?

(A) $1 + 3 = 4$
●(B) $2 + 1 = 3$
(C) $2 + 3 = 5$
(D) $1 + 1 = 2$

7. $3 + 1 = \underline{?}$

(A) 3   ●(B) 4
(C) 5   (D) not here

8. $3 + 3 = \underline{?}$

(A) 1   (B) 3
(C) 5   ●(D) not here

9. $0 + 2 = \underline{?}$

(A) 0   ●(B) 2
(C) 3   (D) 4

10. $3 + 2 = \underline{?}$

(A) 2   (B) 3
●(C) 5   (D) 6

Pretest — CHAPTER 2, PAGE 3

11. $\begin{array}{r}1\\+5\\\hline\end{array}$

(A) 1   (B) 4
(C) 5   ●(D) not here

12. $\begin{array}{r}2\\+2\\\hline\end{array}$

(A) 0   (B) 2
(C) 3   ●(D) 4

13. $\begin{array}{r}5\\+0\\\hline\end{array}$

(A) 0   (B) 1
●(C) 5   (D) 6

14. $\begin{array}{r}1\\+1\\\hline\end{array}$

(A) 0   (B) 1
●(C) 2   (D) not here

15. Which is a name for 3?

●(A) $3 + 0$
(B) $1 + 3$
(C) $3 + 3$
(D) $0 + 6$

16. Which is a name for 5?

(A) $1 + 5$
●(B) $2 + 3$
(C) $3 + 3$
(D) $3 + 1$

17. Which is a name for 6?

(A) $1 + 3$
(B) $3 + 2$
●(C) $3 + 3$
(D) $5 + 0$

18. Which is a name for 2?

(A) $1 + 1$
(B) $2 + 1$
(C) $1 + 3$
(D) $2 + 2$

Standardized Format • Test Answers

Pretest — Chapter 2, Page 4

19. Which is a name for 1?
- (A) 1 + 1
- (B) 1 + 2
- (C) 2 + 2
- (D) 1 + 0

20. Which is a name for 4?
- (A) 3 + 1
- (B) 4 + 1
- (C) 3 + 3
- (D) 0 + 3

21.
4 cats sleep.
2 cats play.
How many cats are there?
- (A) 2 cats
- (B) 4 cats
- (C) 5 cats
- (D) 6 cats

22.
1 girl plays.
3 more girls play.
How many girls play?
- (A) 1 girl
- (B) 3 girls
- (C) 4 girls
- (D) 5 girls

23.
1 dog sits.
1 dog runs.
How many dogs are there?
- (A) 1 dog
- (B) 2 dogs
- (C) 3 dogs
- (D) 4 dogs

24.
3 boys read.
3 boys write.
How many boys are there?
- (A) 3 boys
- (B) 4 boys
- (C) 5 boys
- (D) 6 boys

Posttest — Chapter 2, Page 1

Choose the correct answer.

1. Which addition sentence tells how many are there?
- (A) 4 + 2 = 6
- (B) 3 + 2 = 5
- (C) 2 + 2 = 4
- (D) 2 + 1 = 3

2. Which addition sentence tells how many are there?
- (A) 1 + 1 = 2
- (B) 2 + 3 = 5
- (C) 2 + 2 = 4
- (D) 4 + 2 = 6

3. Which addition sentence tells how many are there?
- (A) 1 + 4 = 5
- (B) 2 + 4 = 6
- (C) 2 + 2 = 4
- (D) 1 + 3 = 4

4. Which addition sentence tells how many are there?
- (A) 2 + 1 = 3
- (B) 3 + 1 = 4
- (C) 3 + 3 = 6
- (D) 3 + 2 = 5

Posttest — Chapter 2, Page 2

5. Which addition sentence tells how many are there?
- (A) 3 + 3 = 6
- (B) 1 + 2 = 3
- (C) 2 + 2 = 4
- (D) 3 + 2 = 5

6. Which addition sentence tells how many are there?
- (A) 1 + 3 = 4
- (B) 1 + 2 = 3
- (C) 2 + 3 = 5
- (D) 1 + 1 = 2

7. 2 + 3 = ?
- (A) 3
- (B) 4
- (C) 5
- (D) 6

8. 0 + 4 = ?
- (A) 0
- (B) 3
- (C) 4
- (D) 5

9. 1 + 1 = ?
- (A) 1
- (B) 2
- (C) 3
- (D) not here

10. 3 + 3 = ?
- (A) 3
- (B) 4
- (C) 5
- (D) not here

Posttest — Chapter 2, Page 3

11. 2 + 2
- (A) 2
- (B) 3
- (C) 4
- (D) 5

12. 5 + 1
- (A) 1
- (B) 4
- (C) 5
- (D) 6

13. 3 + 0
- (A) 0
- (B) 3
- (C) 4
- (D) not here

14. 2 + 4
- (A) 2
- (B) 3
- (C) 5
- (D) not here

15. Which is a name for 4?
- (A) 3 + 1
- (B) 2 + 3
- (C) 3 + 3
- (D) 0 + 3

16. Which is a name for 1?
- (A) 0 + 1
- (B) 1 + 1
- (C) 1 + 2
- (D) 2 + 2

17. Which is a name for 5?
- (A) 1 + 3
- (B) 3 + 2
- (C) 3 + 3
- (D) 4 + 0

18. Which is a name for 6?
- (A) 1 + 5
- (B) 5 + 0
- (C) 3 + 2
- (D) 2 + 2

Standardized Format • Test Answers

Posttest — Chapter 2, Page 4

19. Which is a name for 3?
    - (A) 1 + 1
    - (B) 0 + 2
    - (C) 2 + 2
    - (D) 3 + 0

20. Which is a name for 2?
    - (A) 3 + 1
    - (B) 0 + 2
    - (C) 1 + 0
    - (D) 2 + 2

21.
    2 frogs jump.
    3 frogs sit.
    How many frogs are there?
    - (A) 2 frogs
    - (B) 4 frogs
    - (C) 5 frogs
    - (D) 6 frogs

22.
    1 girl plays.
    1 girl reads.
    How many girls are there?
    - (A) 0 girls
    - (B) 1 girl
    - (C) 2 girls
    - (D) 3 girls

23. 4 boys smile.
    2 more boys smile.
    How many boys smile in all?
    - (A) 2 boys
    - (B) 4 boys
    - (C) 5 boys
    - (D) 6 boys

24. 1 cat sleeps.
    5 cats drink.
    How many cats are there?
    - (A) 1 cat
    - (B) 4 cats
    - (C) 5 cats
    - (D) 6 cats

Pretest — Chapter 3, Page 1

Choose the correct answer.

1. Which subtraction sentence tells the story?

   - (A) 6 − 1 = 5
   - (B) 6 − 6 = 0
   - (C) 6 − 3 = 3
   - (D) 6 − 2 = 4

2. Which subtraction sentence tells the story?

   - (A) 4 − 1 = 3
   - (B) 4 − 0 = 4
   - (C) 4 − 2 = 2
   - (D) 4 − 1 = 3

3. Which subtraction sentence tells the story?

   - (A) 4 − 0 = 4
   - (B) 4 − 1 = 3
   - (C) 5 − 1 = 4
   - (D) 5 − 5 = 0

4. Which subtraction sentence tells the story?

   - (A) 3 − 3 = 0
   - (B) 3 − 0 = 3
   - (C) 6 − 5 = 1
   - (D) 6 − 3 = 3

Pretest — Chapter 3, Page 2

5.
   4 − 2 = ?
   - (A) 1
   - (B) 2
   - (C) 3
   - (D) 6

6.
   5 − 4 = ?
   - (A) 0
   - (B) 2
   - (C) 5
   - (D) not here

7. 3 − 3 = ?
   - (A) 0
   - (B) 1
   - (C) 3
   - (D) 5

8. 2 − 1 = ?
   - (A) 0
   - (B) 1
   - (C) 2
   - (D) 3

9. 4 − 3 = ?
   - (A) 1
   - (B) 2
   - (C) 3
   - (D) not here

10. 3 − 0 = ?
    - (A) 0
    - (B) 1
    - (C) 2
    - (D) 3

11. 3 − 1 = ?
    - (A) 0
    - (B) 1
    - (C) 2
    - (D) 3

12. 6 − 5 = ?
    - (A) 1
    - (B) 2
    - (C) 3
    - (D) 5

Pretest — Chapter 3, Page 3

13. 6
    − 4
    - (A) 0
    - (B) 1
    - (C) 2
    - (D) 10

14. 1
    − 1
    - (A) 0
    - (B) 1
    - (C) 2
    - (D) not here

15. 5
    − 3
    - (A) 2
    - (B) 3
    - (C) 5
    - (D) 8

16. 4
    − 2
    - (A) 0
    - (B) 3
    - (C) 4
    - (D) not here

17. Which subtraction sentence goes with the addition sentence?
    ○ ○   ○ ○
      ○
    3 + 2 = 5
    - (A) 3 − 3 = 0
    - (B) 3 − 2 = 1
    - (C) 5 − 1 = 4
    - (D) 5 − 2 = 3

18. Which subtraction sentence goes with the addition sentence?
    ○ ○ ○ ○   ○
       ○ ○
    5 + 1 = 6
    - (A) 5 − 1 = 4
    - (B) 6 − 1 = 5
    - (C) 6 − 3 = 4
    - (D) 3 − 3 = 0

Standardized Format • Test Answers

Pretest — Chapter 3, Page 4

19. Which number sentence does <u>not</u> belong?

- (A) $1 + 3 = 4$
- (B) $4 - 3 = 1$
- (C) $4 - 1 = 3$
- (D) $3 - 1 = 2$

20. Which number sentence does <u>not</u> belong?

- (A) $3 - 2 = 1$
- (B) $5 - 2 = 3$
- (C) $5 - 3 = 2$
- (D) $3 + 2 = 5$

21.

What do you do to find how many in all?

- (A) Add
- (B) Subtract

22.

What do you do to find how many are left?

- (A) Add
- (B) Subtract

Pretest — Chapter 3, Page 5

23. Sara sees 4 . Then 1 swims away.

Which is the correct question?

- (A) How many are left?
- (B) How many are there in all?

24. Tom sees 1 . Then he sees 1 more .

Which is the correct question?

- (A) How many are left?
- (B) How many are there in all?

Posttest — Chapter 3, Page 1

Choose the correct answer.

1. Which subtraction sentence tells the story?

- (A) $3 - 2 = 1$
- (B) $3 - 1 = 2$
- (C) $4 - 1 = 3$
- (D) $4 - 2 = 2$

2. Which subtraction sentence tells the story?

- (A) $5 - 1 = 4$
- (B) $5 - 2 = 3$
- (C) $3 - 2 = 1$
- (D) $3 - 3 = 0$

3. Which subtraction sentence tells the story?

- (A) $6 - 3 = 3$
- (B) $3 - 1 = 2$
- (C) $3 - 3 = 0$
- (D) $4 - 1 = 3$

4. Which subtraction sentence tells the story?

- (A) $4 - 1 = 3$
- (B) $5 - 4 = 1$
- (C) $5 - 1 = 4$
- (D) $6 - 1 = 5$

Posttest — Chapter 3, Page 2

5.

$6 - 4 = \underline{?}$

- (A) 1
- (B) 2
- (C) 4
- (D) not here

6.

$5 - 3 = \underline{?}$

- (A) 2
- (B) 3
- (C) 5
- (D) 8

7.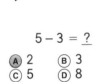

$4 - 3 = \underline{?}$

- (A) 1
- (B) 3
- (C) 4
- (D) not here

8.

$2 - 2 = \underline{?}$

- (A) 0
- (B) 1
- (C) 2
- (D) 4

9. $5 - 4 = \underline{?}$

- (A) 1
- (B) 2
- (C) 3
- (D) 4

10. $6 - 2 = \underline{?}$

- (A) 0
- (B) 2
- (C) 4
- (D) 6

11. $2 - 1 = \underline{?}$

- (A) 0
- (B) 1
- (C) 2
- (D) 3

12. $4 - 0 = \underline{?}$

- (A) 0
- (B) 2
- (C) 3
- (D) not here

Standardized Format • Test Answers

Posttest — Chapter 3, Page 3

13. 4
    −4
    (A) 0 (B) 1
    (C) 2 (D) 4

14. 6
    −3
    (A) 0 (B) 2
    (C) 3 (D) 6

15. 5
    −1
    (A) 2 (B) 3
    (C) 5 (D) not here

16. 6
    −5
    (A) 0 (B) 1
    (C) 2 (D) 3

17. Which subtraction sentence goes with the addition sentence?

○ ○ ○

2 + 1 = 3

(A) 2 − 1 = 1
(B) 3 − 3 = 0
(C) 2 − 2 = 0
(D) 3 − 1 = 2

18. Which subtraction sentence goes with the addition sentence?

○ ○ ○ ○

2 + 2 = 4

(A) 2 − 0 = 2
(B) 4 − 2 = 2
(C) 3 − 3 = 0
(D) 3 − 2 = 1

Posttest — Chapter 3, Page 4

19. Which number sentence does not belong?

○ ○   ○ ○
  ○ ○

(A) 4 + 2 = 6
(B) 6 − 4 = 2
(C) 4 − 2 = 2
(D) 6 − 2 = 4

20. Which number sentence does not belong?

○ ○   ○ ○
○ ○

(A) 5 − 1 = 4
(B) 4 + 1 = 5
(C) 5 − 4 = 1
(D) 4 − 1 = 3

21.

What do you do to find how many  in all?

(A) Add   (B) Subtract

22.

What do you do to find how many are left?

(A) Add   (B) Subtract

Posttest — Chapter 3, Page 5

23. There are 5 in this . Then 2 fly away.
Which is the correct question?

(A) How many 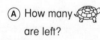 are left?

(B) How many are there in all?

24. Sara sees 5  in a pond. Then she sees 1 more .
Which is the correct question?

(A) How many  are left?

(B) How many  are there in all?

Cumulative Test — Chapters 1–3, Page 1

Choose the letter of the correct answer.

1. How many ○ are there?

○ ○ ○ ○
○ ○ ○

(A) 3 (B) 6
(C) 7 (D) 8

2. How many □ are there?

□ □ □
□ □ □

(A) 5 (B) 6
(C) 7 (D) 9

3. How many △ are there?

△ △ △ △
△ △ △

(A) 5 (B) 6
(C) 7 (D) 8

4. How many ☆ are there?

☆ ☆ ☆ ☆ ☆
☆ ☆ ☆ ☆ ☆

(A) 0 (B) 1
(C) 9 (D) not here

5. How many ◇ are there?

◇ ◇ ◇ ◇
◇ ◇ ◇

(A) 6 (B) 8
(C) 9 (D) not here

6. How many ▭ are there?

▭ ▭
▭

(A) 3 (B) 4
(C) 5 (D) 6

Standardized Format • Test Answers

Cumulative Test — CHAPTERS 1–3, PAGE 2

7. Which group has fewer than this?

- Ⓐ
- ⬤B
- Ⓒ

8. Which group has fewer than this?

- ⬤A
- Ⓑ
- Ⓒ

9. Which group has more than this?

- Ⓐ
- Ⓑ
- Ⓒ

10. Which group has more than this?

- Ⓐ
- Ⓑ
- Ⓒ

Cumulative Test — CHAPTERS 1–3, PAGE 3

11. Which comes next?
3, 4, 5, _?_
- Ⓐ 1
- Ⓑ 2
- ⬤C 6
- Ⓓ 9

12. Which comes next?
7, 8, 9, _?_
- Ⓐ 0
- Ⓑ 1
- Ⓒ 6
- ⬤D 10

13. Which comes next?
0, 1, 2, _?_
- Ⓐ 4
- Ⓑ 5
- Ⓒ 6
- ⬤D not here

14. Which comes next?
4, 5, 6, _?_
- Ⓐ 3
- Ⓑ 5
- ⬤C 7
- Ⓓ 9

15. Which addition sentence tells how many in all?

- Ⓐ 2 + 1 = 3
- Ⓑ 3 + 2 = 5
- Ⓒ 4 + 2 = 6
- ⬤D 3 + 1 = 4

16. Which addition sentence tells how many in all?

- Ⓐ 2 + 2 = 4
- ⬤B 2 + 4 = 6
- Ⓒ 4 + 4 = 8
- Ⓓ 2 + 3 = 5

Cumulative Test — CHAPTERS 1–3, PAGE 4

17. Which addition sentence tells how many in all?

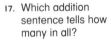

- ⬤A 2 + 2 = 4
- Ⓑ 1 + 1 = 2
- Ⓒ 3 + 0 = 3
- Ⓓ 4 + 4 = 8

18. Which addition sentence tells how many in all?

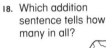

- Ⓐ 1 + 2 = 3
- Ⓑ 1 + 1 = 2
- ⬤C 2 + 3 = 5
- Ⓓ 2 + 2 = 4

19. Which addition sentence tells how many in all?

- Ⓐ 2 + 3 = 5
- ⬤B 1 + 5 = 6
- Ⓒ 1 + 0 = 1
- Ⓓ 2 + 5 = 7

20. Which addition sentence tells how many in all?

- Ⓐ 2 + 2 = 4
- Ⓑ 4 + 1 = 5
- Ⓒ 1 + 1 = 2
- ⬤D 3 + 3 = 6

Cumulative Test — CHAPTERS 1–3, PAGE 5

21. 5 + 1 = _?_
- Ⓐ 1
- Ⓑ 4
- Ⓒ 5
- ⬤D 6

22. 2 + 0 = _?_
- Ⓐ 0
- Ⓑ 1
- ⬤C 2
- Ⓓ 3

23. 2 + 2 = _?_
- Ⓐ 0
- Ⓑ 1
- Ⓒ 2
- ⬤D 4

24. 1 + 4 = _?_
- Ⓐ 1
- Ⓑ 4
- ⬤C 5
- Ⓓ 6

25. 0
      +1
- Ⓐ 0
- ⬤B 1
- Ⓒ 2
- Ⓓ not here

26. 3
      +3
- Ⓐ 0
- Ⓑ 1
- Ⓒ 3
- ⬤D not here

27. 2
      +3
- ⬤A 5
- Ⓑ 6
- Ⓒ 7
- Ⓓ 8

28. Which subtraction sentence tells the story?

- Ⓐ 4 − 2 = 2
- ⬤B 6 − 2 = 4
- Ⓒ 6 − 1 = 5
- Ⓓ 4 − 3 = 1

Standardized Format • Test Answers

Cumulative Test — CHAPTERS 1–3, PAGE 6

29. Which subtraction sentence tells the story?
- (A) 5 − 2 = 3 ●
- (B) 5 − 1 = 4
- (C) 3 − 2 = 1
- (D) 3 − 3 = 0

30. Which subtraction sentence tells the story?
- (A) 3 − 1 = 2
- (B) 4 − 3 = 1 ●
- (C) 4 − 2 = 2
- (D) 3 − 3 = 0

31. Which subtraction sentence tells the story?
- (A) 2 − 1 = 1
- (B) 1 − 1 = 0
- (C) 4 − 1 = 3
- (D) 3 − 1 = 2 ●

32. 4 − 1 = ?
- (A) 1
- (B) 2
- (C) 3 ●
- (D) 4

33. 6 − 3 = ?
- (A) 0
- (B) 3 ●
- (C) 6
- (D) not here

34. 5 − 0 = ?
- (A) 0
- (B) 1
- (C) 5 ●
- (D) 6

Cumulative Test — CHAPTERS 1–3, PAGE 7

35. 4 − 4 = ?
- (A) 0 ●
- (B) 1
- (C) 2
- (D) 4

36. 5 − 3
- (A) 2 ●
- (B) 3
- (C) 5
- (D) 8

37. 6 − 4 = ?
- (A) 0
- (B) 1
- (C) 2 ●
- (D) 3

38. 2 − 1 = ?
- (A) 1 ●
- (B) 2
- (C) 3
- (D) 4

39. 4 − 0 = ?
- (A) 0
- (B) 1
- (C) 4 ●
- (D) 5

40. 6 − 5 = ?
- (A) 0
- (B) 1 ●
- (C) 5
- (D) 6

41. 3 − 3 = ?
- (A) 0 ●
- (B) 1
- (C) 3
- (D) 6

42. 4 − 1 = ?
- (A) 0
- (B) 1
- (C) 2
- (D) not here ●

43. Jack sees 3 fish. Then he sees 2 more fish. Which is the correct question?
- (A) How many fish are there in all? ●
- (B) How many fish are left?

44. Jane sees 5 birds. Then 4 birds fly away. Which is the correct question?
- (A) How many birds are there in all?
- (B) How many birds are left? ●

Cumulative Test — CHAPTERS 1–3, PAGE 8

Use the pictograph below to answer questions 45–50.

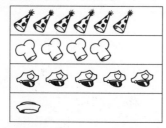

45. Which group has fewer, hats or chef hats?
- (A) hat
- (B) chef hat ●

46. Which group has more, bells or boats?
- (A) bell ●
- (B) boat

47. Which group has more than 5?
- (A) hat ●
- (B) chef hat
- (C) bell
- (D) boat

48. Which group has fewer than 4?
- (A) hat
- (B) chef hat
- (C) bell
- (D) boat ●

49. Which group has more than 5?
- (A) bell
- (B) hat ●

50. Which group has fewer, chef hat or boat?
- (A) chef hat
- (B) boat ●

Pretest — CHAPTER 4, PAGE 1

Choose the correct answer.

1. 5 + 3 = ?
- (A) 5
- (B) 7
- (C) 8 ●
- (D) 9

2. 6 + 4 = ?
- (A) 2
- (B) 6
- (C) 9
- (D) 10 ●

3. 8 + 1 = ?
- (A) 7
- (B) 8
- (C) 9 ●
- (D) 10

4. 5 + 2 = ?
- (A) 6
- (B) 7 ●
- (C) 8
- (D) 9

Standardized Format • Test Answers

Pretest — Chapter 4, Page 2

5.

$3 + 3 = \underline{?}$

(A) 0 (B) 5 **(C) 6** (D) 9

6.

$5 + 5 = \underline{?}$

(A) 5 (B) 6 (C) 9 **(D) 10**

7. $0 + 2 = \underline{?}$

(A) 0 **(B) 2** (C) 3 (D) not here

8. $5 + 1 = \underline{?}$

(A) 5 **(B) 6** (C) 7 (D) 9

9. $4 + 4 = \underline{?}$

(A) 0 (B) 4 **(C) 8** (D) 9

10. $2 + 7 = \underline{?}$

(A) 5 (B) 7 (C) 8 **(D) 9**

11. $4 + 3 = \underline{?}$

(A) 7 (B) 8 (C) 9 (D) not here

12. $3 + 2 = \underline{?}$

(A) 1 (B) 3 (C) 4 (D) not here

13. $8 + 2 = \underline{?}$

(A) 6 (B) 8 (C) 9 **(D) 10**

14. $2 + 2 = \underline{?}$

(A) 0 (B) 2 (C) 4 (D) 5

Pretest — Chapter 4, Page 3

15. Which is a name for 9?

(A) $1 + 6$ (B) $4 + 3$ (C) $5 + 2$ **(D) $9 + 0$**

16. Which is a name for 10?

(A) $4 + 4$ (B) $6 + 3$ (C) $7 + 0$ **(D) $8 + 2$**

17. Which addition sentence tells how many in all?

(A) $5 + 1 + 1 = 7$ **(B) $4 + 2 + 1 = 7$** (C) $4 + 1 = 5$ (D) $2 + 1 = 3$

18. Which addition sentence tells how many in all?

(A) $5 + 2 = 7$ (B) $2 + 2 = 4$ **(C) $5 + 2 + 2 = 9$** (D) $4 + 2 + 2 = 8$

19. $3 + 0 + 3 = \underline{?}$

(A) 0 (B) 3 **(C) 6** (D) 9

20. $4 + 1 + 3 = \underline{?}$

(A) 4 (B) 5 **(C) 8** (D) not here

Pretest — Chapter 4, Page 4

21. How much did Kara spend?

(A) 3¢ (B) 4¢ (C) 7¢ **(D) 10¢**

22. How much did Karl spend?

(A) 2¢ (B) 5¢ **(C) 6¢** (D) 7¢

23. Which sentence is **not** needed?

Lin sees 5 birds.
Then she sees 3 more birds.
Then she sees 2 cats.
How many birds does she see in all?

(A) Lin sees 5 birds.
(B) Then she sees 3 more birds.
(C) Then she sees 2 cats.

24. Which sentence is **not** needed?

Tom feeds 4 dogs.
Tom feeds 3 cats.
Then he feeds 2 more cats.
How many cats does he feed in all?

(A) Tom feeds 4 dogs.
(B) Tom feeds 3 cats.
(C) Then he feeds 2 more cats.

Posttest — Chapter 4, Page 1

Choose the correct answer.

1.

$7 + 1 = \underline{?}$

(A) 6 (B) 7 **(C) 8** (D) 9

2.

$8 + 2 = \underline{?}$

(A) 6 (B) 8 (C) 9 **(D) 10**

3.

$6 + 3 = \underline{?}$

(A) 3 (B) 6 (C) 7 **(D) 9**

4.

$5 + 2 = \underline{?}$

(A) 3 **(B) 7** (C) 8 (D) 9

Name

Posttest

CHAPTER 4
PAGE 2

5.

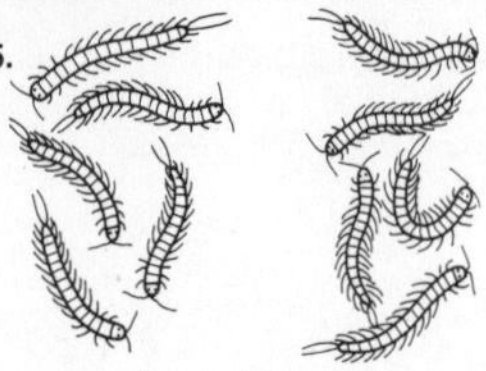

5 + 5 = ?

(A) 0 (B) 5
(C) 9 **(D) 10**

6.

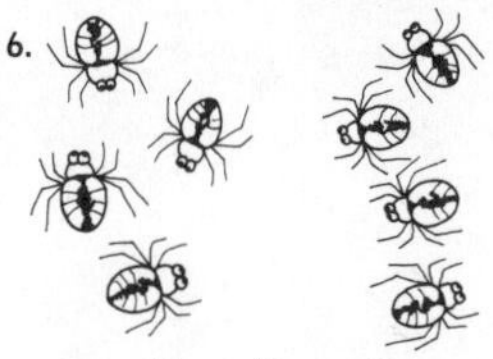

4 + 4 = ?

(A) 4 (B) 7
(C) 8 (D) 9

7. 0 + 3 = ?

(A) 0 (B) 1
(C) 3 (D) not here

8. 1 + 4 = ?

(A) 1 (B) 4
(C) 5 (D) 6

9. 6 + 2 = ?

(A) 2 (B) 4
(C) 7 **(D) 8**

10. 5 + 3 = ?

(A) 7 **(B) 8**
(C) 9 (D) not here

11. 6 + 4 = ?

(A) 2 (B) 8
(C) 9 **(D) not here**

12. 7 + 0 = ?

(A) 0 (B) 1
(C) 7 (D) 8

13. 4 + 5 = ?

(A) 0 (B) 1
(C) 9 (D) 10

14. 3 + 3 = ?

(A) 0 (B) 3
(C) 6 (D) not here

GO ON

Name

Posttest

CHAPTER 4
PAGE 3

15. Which is a name for 8?

(A) 8 + 1 (B) 3 + 3
(C) 6 + 3 **(D) 4 + 4**

16. Which is a name for 7?

(A) 1 + 8 (B) 3 + 2
(C) 4 + 1 **(D) 5 + 2**

17. Which addition sentence tells how many in all?

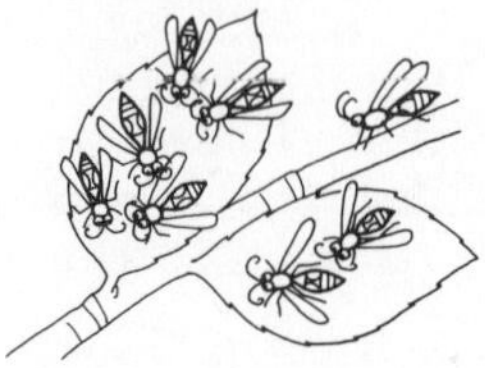

(A) 5 + 1 = 6
(B) 5 + 1 + 2 = 8
(C) 4 + 2 + 2 = 8
(D) 1 + 2 = 3

18. Which addition sentence tells how many in all?

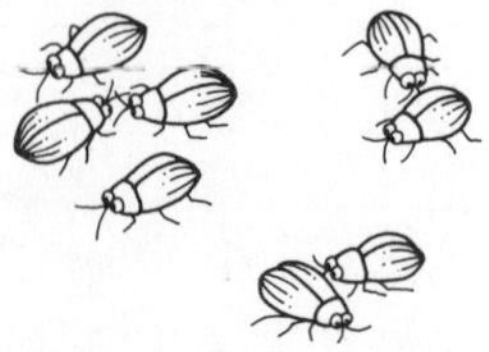

(A) 4 + 2 + 2 = 8
(B) 2 + 2 + 2 = 6
(C) 2 + 0 = 2
(D) 2 + 2 = 4

19. 4 + 0 + 4 = ?

(A) 0 (B) 4
(C) 8 (D) 9

20. 3 + 2 + 5 = ?

(A) 5 (B) 7
(C) 9 **(D) 10**

GO ON

Name

Posttest

CHAPTER 4
PAGE 4

21. How much did Kathy spend?

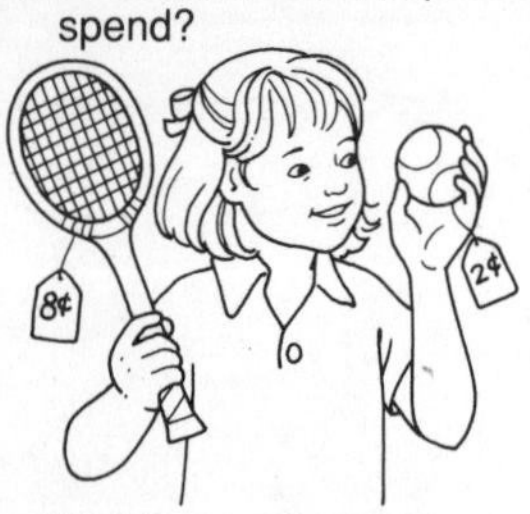

(A) 2¢ (B) 6¢
(C) 9¢ **(D) 10¢**

22. How much did Carlos spend?

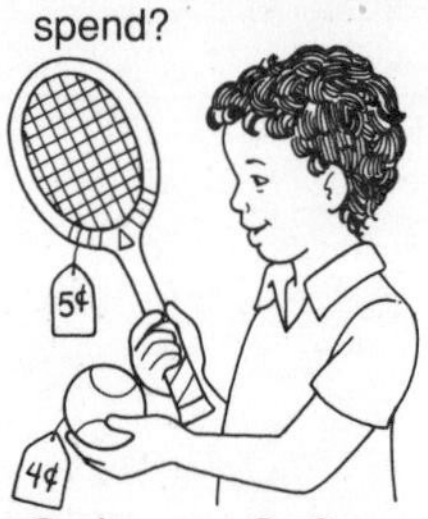

(A) 1¢ (B) 8¢
(C) 9¢ (D) 10¢

23. Which sentence is not needed?

Nan sees 4 cows.
Then she sees 3 more cows.
Then she sees 2 dogs.
How many cows does she see in all?

(A) Nan sees 4 cows.
(B) Then she sees 3 more cows.
(C) Then she sees 2 dogs.

24. Which sentence is not needed?

Sam feeds 6 fish.
Sam feeds 4 birds.
Then he feeds 2 more birds.
How many birds does he feed in all?

(A) Sam feeds 6 fish.
(B) Sam feeds 4 birds.
(C) Then he feeds 2 more birds.

STOP

Name

Pretest

CHAPTER 5
PAGE 1

Choose the correct answer.

1.

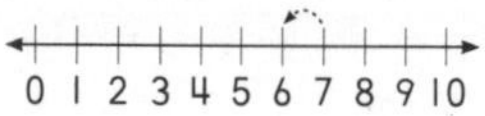

7 − 1 = ?

(A) 5 **(B) 6**
(C) 7 (D) 8

2.

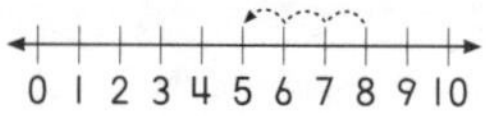

8 − 3 = ?

(A) 4 **(B) 5**
(C) 6 (D) 7

3.

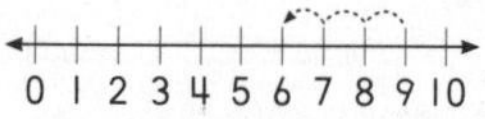

9 − 3 = ?

(A) 6 (B) 7
(C) 8 (D) 9

4.

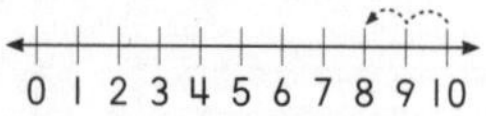

10 − 2 = ?

(A) 5 (B) 6
(C) 7 **(D) 8**

5.

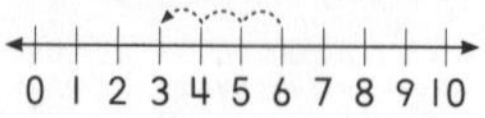

6 − 3 = ?

(A) 3 (B) 4
(C) 5 (D) 9

6.

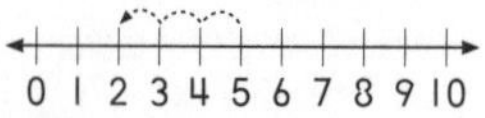

5 − 3 = ?

(A) 2 (B) 3
(C) 4 (D) 8

GO ON

HBJ material copyrighted under notice appearing earlier in this work.

Name _____ CHAPTER 5
Pretest PAGE 2

7. 7
   −2

   (A) 5   (B) 6
   (C) 8   (D) 9

8. 10
   − 7

   (A) 2   (B) 3
   (C) 4   (D) not here

9. 8
   −4

   (A) 0   (B) 3
   (C) 5   (D) not here

10. 6
    −6

   (A) 0   (B) 1
   (C) 6   (D) 9

11. 9 − 2 = ?

   (A) 7   (B) 8
   (C) 12 (D) not here

12. 8 − 6 = ?

   (A) 1   (B) 3
   (C) 4   (D) not here

13. 7 − 0 = ?

   (A) 0   (B) 1
   (C) 7   (D) 8

14. 9 − 8 = ?

   (A) 1   (B) 2
   (C) 4   (D) 9

GO ON

Name _____ CHAPTER 5
Pretest PAGE 3

15. Use counters to solve.
    Dale has 6 🐱.
    Then 4 run away.
    How many are left?

    (A) 2 🐱
    (B) 4 🐱
    (C) 6 🐱
    (D) 10 🐱

16. Use counters to solve.
    There are 8 🚗.
    Then 1 leaves.
    How many are left?

    (A) 6 🚗
    (B) 7 🚗
    (C) 8 🚗
    (D) 9 🚗

17.
    3 + 4 = 7
    7 − 4 = ?

    (A) 1   (B) 2
    (C) 3   (D) 4

18.
    4 + 5 = 9
    9 − 5 = ?

    (A) 1   (B) 2
    (C) 3   (D) 4

GO ON

Name _____ CHAPTER 5
Pretest PAGE 4

19. Which subtraction fact goes with this addition fact?

    2 + 7 = 9

    (A) 7 − 5 = 2
    (B) 7 − 2 = 5
    (C) 9 − 5 = 4
    (D) 9 − 7 = 2

20. Which subtraction fact goes with this addition fact?

    6 + 4 = 10

    (A) 6 − 4 = 2
    (B) 4 − 2 = 2
    (C) 10 − 6 = 4
    (D) 10 − 2 = 8

21. Which is the number pair?

    The sum is 6.
    The difference is 4.

    (A) 4 and 2
    (B) 3 and 3
    (C) 5 and 1
    (D) 6 and 2

22. Which is the number pair?

    The sum is 7.
    The difference is 7.

    (A) 4 and 3
    (B) 7 and 0
    (C) 7 and 7
    (D) 6 and 1

GO ON

Name _____ CHAPTER 5
Pretest PAGE 5

23.
    There are 5 👧.
    There is 1 👦.
    How many more 👧 than 👦 are there?

    (A) 3 more
    (B) 4 more
    (C) 5 more
    (D) 6 more

24.
    There are 8 🥄.
    There are 4 ☕.
    How many more 🥄 than ☕ are there?

    (A) 4 more
    (B) 6 more
    (C) 8 more
    (D) 10 more

STOP

Posttest — Chapter 5, Page 1

Choose the correct answer.

1.
   0 1 2 3 4 5 6 7 8 9 10

   $8 - 1 = \underline{?}$

   (A) 6 (B) 7 ●
   (C) 8 (D) 9

2.
   0 1 2 3 4 5 6 7 8 9 10

   $7 - 4 = \underline{?}$

   (A) 3 ● (B) 4
   (C) 5   (D) 7

3.
   0 1 2 3 4 5 6 7 8 9 10

   $10 - 4 = \underline{?}$

   (A) 4   (B) 5
   (C) 6 ● (D) 7

4.
   0 1 2 3 4 5 6 7 8 9 10

   $6 - 2 = \underline{?}$

   (A) 4 ● (B) 5
   (C) 6   (D) 8

5.
   0 1 2 3 4 5 6 7 8 9 10

   $9 - 1 = \underline{?}$

   (A) 1   (B) 7
   (C) 8 ● (D) 10

6.
   0 1 2 3 4 5 6 7 8 9 10

   $8 - 3 = \underline{?}$

   (A) 5 ● (B) 6
   (C) 7   (D) 8

Posttest — Chapter 5, Page 2

7. 9
   −2

   (A) 7 ● (B) 8
   (C) 9   (D) not here

8. 10
   −8

   (A) 0   (B) 1
   (C) 8   (D) not here ●

9. 7
   −5

   (A) 1   (B) 2 ●
   (C) 3   (D) 7

10. 5
    −5

    (A) 0 ● (B) 1
    (C) 5   (D) 10

11. $10 - 3 = \underline{?}$

    (A) 6   (B) 7 ●
    (C) 8   (D) not here

12. $8 - 5 = \underline{?}$

    (A) 2   (B) 3 ●
    (C) 4   (D) not here

13. $6 - 0 = \underline{?}$

    (A) 0   (B) 1
    (C) 6 ● (D) 7

14. $9 - 6 = \underline{?}$

    (A) 3 ● (B) 4
    (C) 5   (D) 6

Posttest — Chapter 5, Page 3

15. Use counters to solve.

    There are 5 🚗.
    Then 1 leaves.
    How many are left?

    (A) 1 🚗
    (B) 4 🚗 ●
    (C) 5 🚗
    (D) 6 🚗

16. Use counters to solve.

    There are 9 🐦.
    Then 4 fly away.
    How many are left?

    (A) 4 🐦
    (B) 5 🐦 ●
    (C) 6 🐦
    (D) 7 🐦

17.
    △△ ▲▲
    △△ ▲

    $4 + 3 = 7$
    $7 - 3 = \underline{?}$

    (A) 1   (B) 4 ●
    (C) 5   (D) 10

18.
    ●● ◯◯
    ●●◯◯◯

    $3 + 5 = 8$
    $8 - 5 = \underline{?}$

    (A) 1   (B) 2
    (C) 3 ● (D) 4

Posttest — Chapter 5, Page 4

19. Which subtraction fact goes with this addition fact?

    $6 + 1 = 7$

    (A) $6 - 1 = 5$
    (B) $1 - 1 = 0$
    (C) $8 - 1 = 7$
    (D) $7 - 1 = 6$ ●

20. Which subtraction fact goes with this addition fact?

    $5 + 5 = 10$

    (A) $5 - 0 = 5$
    (B) $5 - 5 = 0$
    (C) $10 - 5 = 5$ ●
    (D) $10 - 0 = 10$

21. Which is the number pair?

    The sum is 5.
    The difference is 3.

    (A) 3 and 3
    (B) 5 and 0
    (C) 3 and 2
    (D) 4 and 1 ●

22. Which is the number pair?

    The sum is 9.
    The difference is 9.

    (A) 8 and 1
    (B) 6 and 3
    (C) 3 and 3
    (D) 9 and 0 ●

Standardized Format • Test Answers

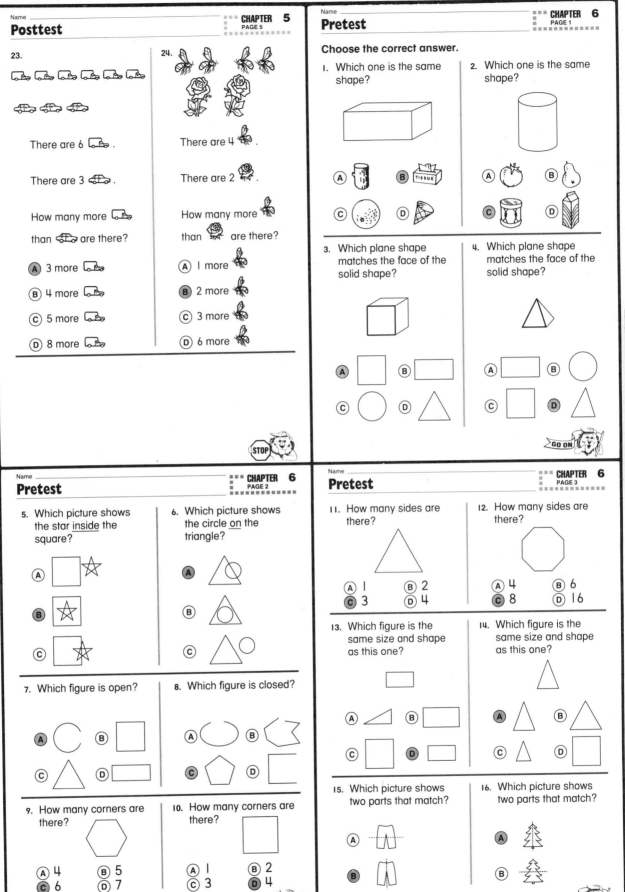

Pretest — Chapter 6, Page 4

17. What is the pattern?
▲△△▲△△
- (A) △▲
- (B) ▲▲
- (C) ▲△
- (D) △△

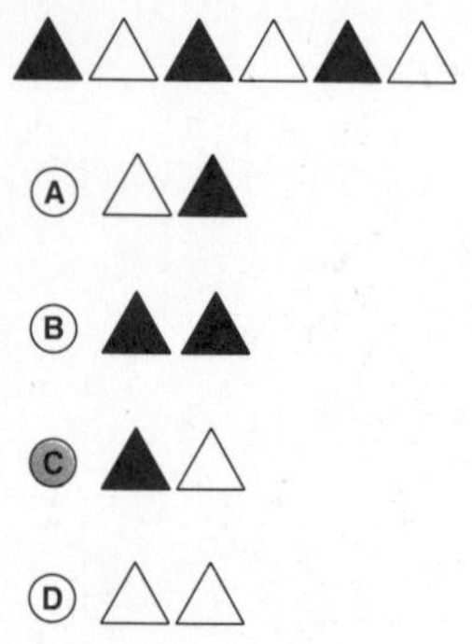

18. What is the pattern?
△○○△○○
- (A) ○△
- (B) △○△
- (C) ○○
- (D) △○○

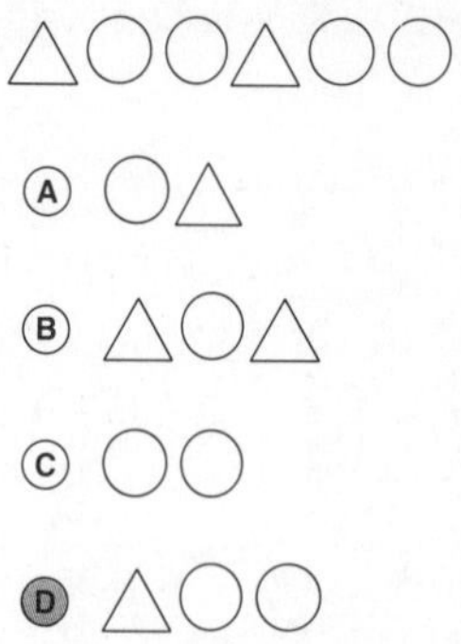

19. Which shape comes next in the pattern?
○□□○□□○□ ?
- (A) △
- (B) □
- (C) ○

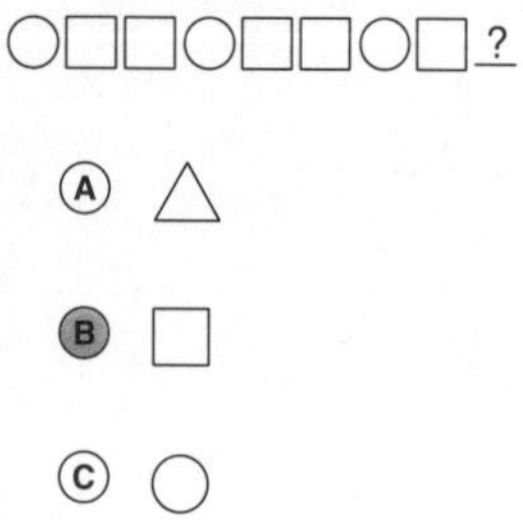

20. Which shape comes next in the pattern?
□○△□○△□ ?
- (A) △
- (B) ○
- (C) □

Pretest — Chapter 6, Page 5

21. Which is the missing shape?
□ □ ○ ? ○
- (A) □
- (B) ○

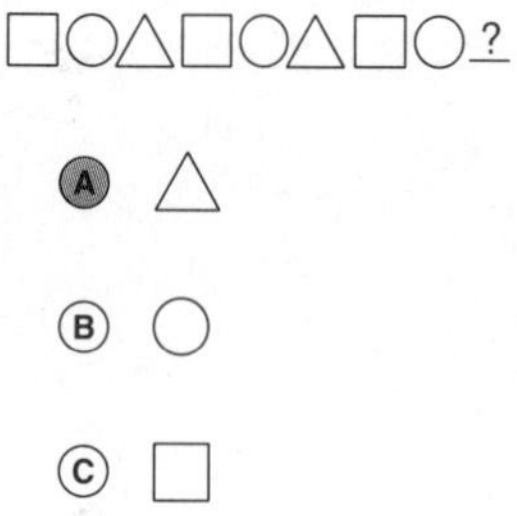

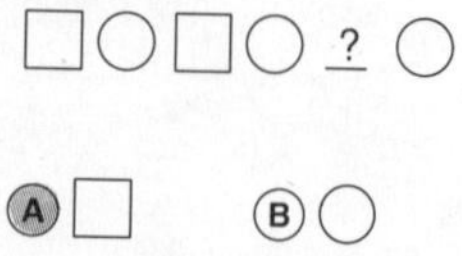

22. Which is the missing shape?
△ △ ○ △ ? ○
- (A) ○
- (B) △

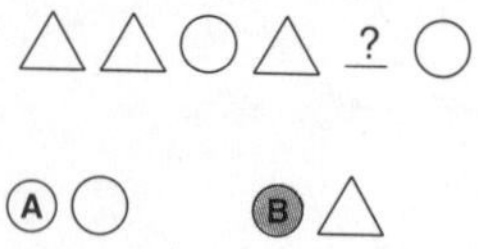

Use the graph to answer questions 23–24.

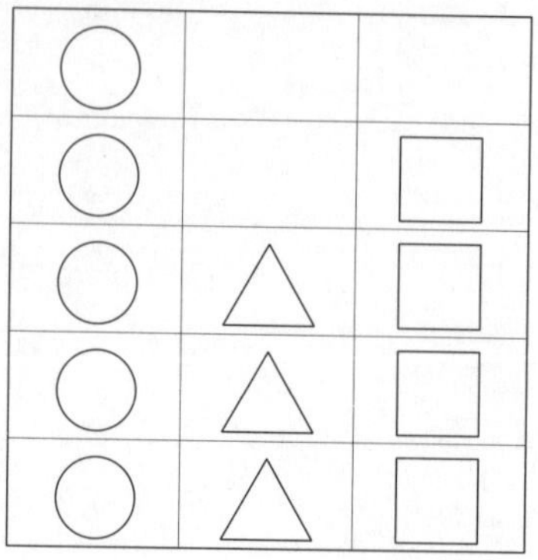

23. How many ○ are there?
- (A) 2
- (B) 3
- (C) 4
- (D) 5

24. How many □ are there?
- (A) 3
- (B) 4
- (C) 5
- (D) 6

Posttest — Chapter 6, Page 1

Choose the correct answer.

1. Which one is the same shape?
- (A) die
- (B) lemon
- (C) pencil
- (D) party hat

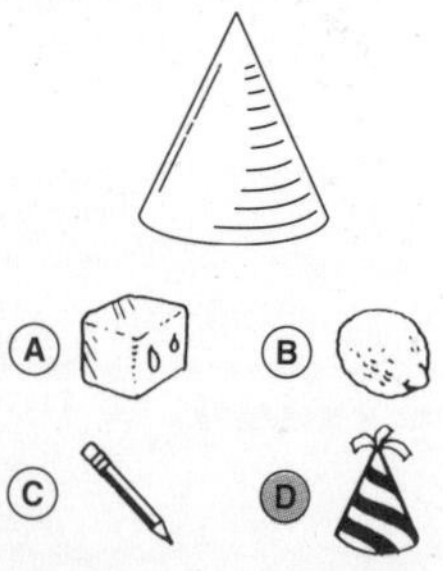

2. Which one is the same shape?
- (A) box
- (B) bat
- (C) wheel
- (D) beehive

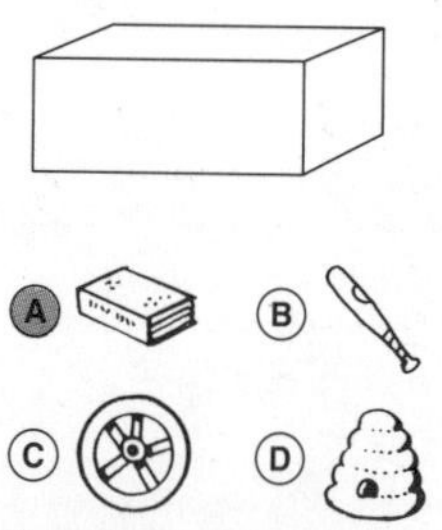

3. Which plane shape matches the face of the solid shape? (cylinder)
- (A) square
- (B) rectangle
- (C) circle
- (D) triangle

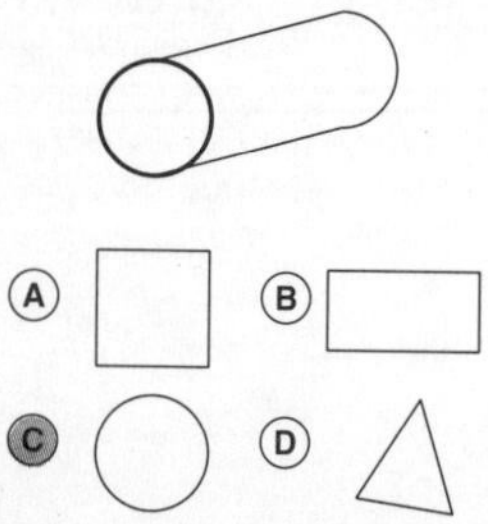

4. Which plane shape matches the face of the solid shape? (cube)
- (A) circle
- (B) square
- (C) triangle
- (D) triangle

Posttest — Chapter 6, Page 2

5. Which picture shows the star <u>inside</u> the circle?
- (A) star beside circle
- (B) star inside circle
- (C) star partially inside circle

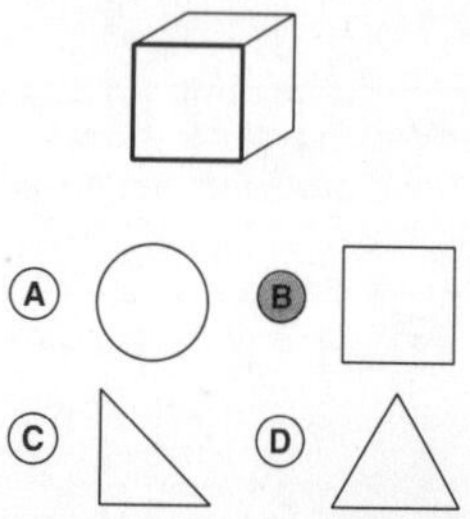

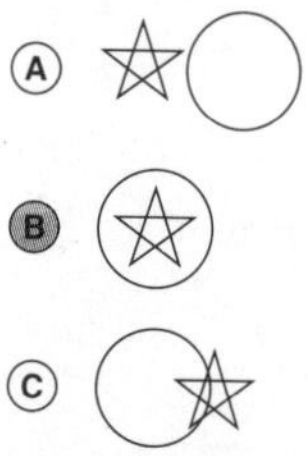

6. Which picture shows the square <u>outside</u> the circle?
- (A) circle and square separate
- (B) square overlapping circle
- (C) square inside circle

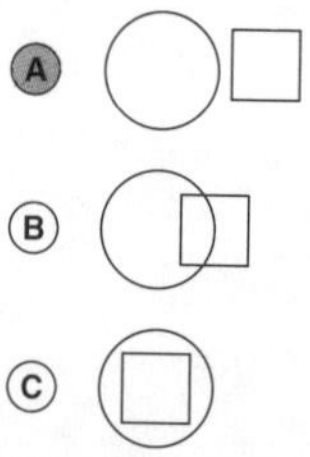

7. Which figure is open?
- (A) open rectangle shape
- (B) triangle
- (C) circle
- (D) rectangle

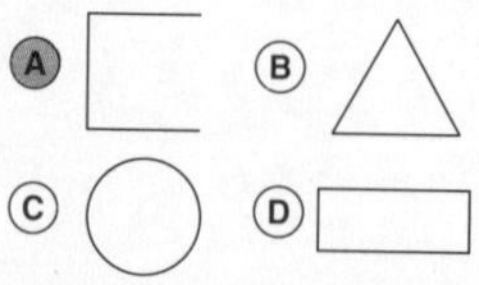

8. Which figure is closed?
- (A) hexagon
- (B) oval
- (C) triangle
- (D) rectangle

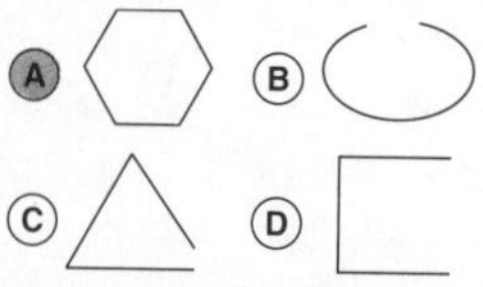

9. How many corners are there? (pentagon)
- (A) 4
- (B) 5
- (C) 6
- (D) 10

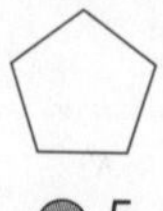

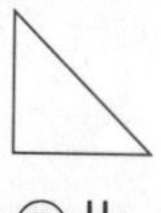

10. How many corners are there? (triangle)
- (A) 3
- (B) 4
- (C) 5
- (D) 6

Standardized Format • Test Answers

Posttest — Chapter 6, Page 3

11. How many sides are there?
- (A) 1
- (B) 2
- (C) 4 ●
- (D) 5

12. How many sides are there?
- (A) 3
- (B) 4
- (C) 5
- (D) 6

13. Which figure is the same size and shape as this one?
- (A)
- (B)
- (C)
- (D) ●

14. Which figure is the same size and shape as this one?
- (A) ●
- (B)
- (C)
- (D)

15. Which picture shows two parts that match?
- (A) ●
- (B)

16. Which picture shows two parts that match?
- (A) ●
- (B)

Posttest — Chapter 6, Page 4

17. What is the pattern?
- (A) ●
- (B) ●●
- (C) ○● ●
- (D) ○○ ●

18. What is the pattern?
- (A) △□
- (B) □△
- (C) □□△
- (D) △△□ ●

19. Which shape comes next in the pattern?
- (A) △
- (B) ○
- (C) □

20. Which shape comes next in the pattern?
- (A) □
- (B) ○
- (C) △ ●

Posttest — Chapter 6, Page 5

21. Which is the missing shape?
- (A) △ ●
- (B) □

22. Which is the missing shape?
- (A) □ ●
- (B) ○

Use the graph to answer questions 23–24.

23. How many △ are there?
- (A) 1
- (B) 2 ●
- (C) 3
- (D) 5

24. How many □ are there?
- (A) 2
- (B) 3
- (C) 4 ●
- (D) 6

Cumulative Test — Chapters 1–6, Page 1

Choose the correct answer.

1. How many □ are there?
- (A) 5
- (B) 6
- (C) 7
- (D) 9

2. How many ☆ are there?
- (A) 4
- (B) 5
- (C) 6
- (D) 7 ●

3. How many ○ are there?
- (A) 2
- (B) 3
- (C) 4 ●
- (D) 5

4. How many △ are there?
- (A) 6
- (B) 8
- (C) 9 ●
- (D) not here

5. Which group has fewer than this?
- (A) ●
- (B)
- (C)

6. Which group has more than this?
- (A) ●
- (B)
- (C)

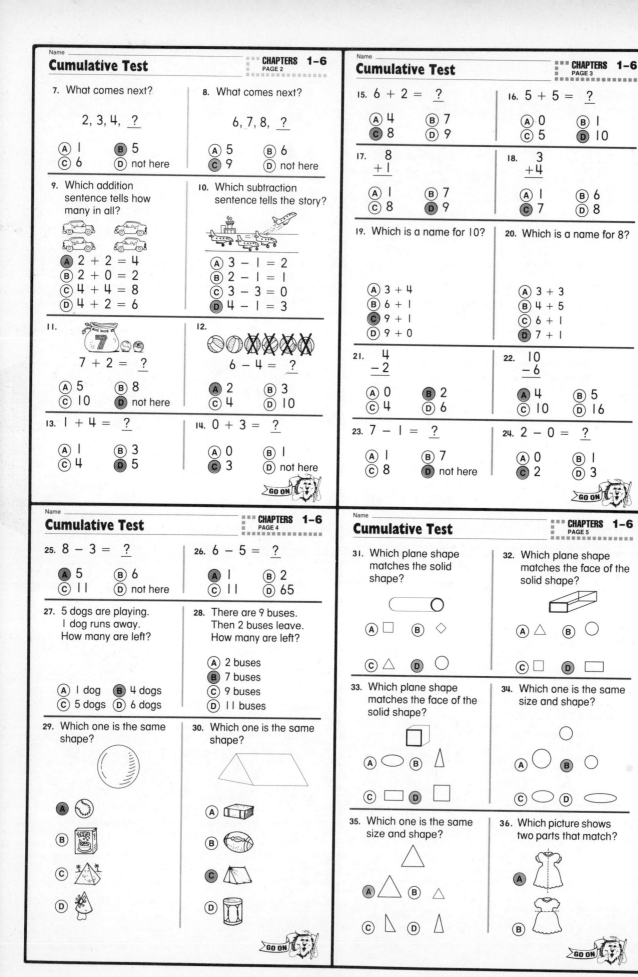

Cumulative Test — Chapters 1–6, Page 6

37. Which picture shows two parts that match?
- (A) tie (whole)
- (B) tie (split) ●

38. What is the pattern? △ △ ○ △ △ ○
- (A) △ △
- (B) △ ○
- (C) △ △ ○ ●
- (D) ○ △ ○

39. What is the pattern? ☆ ○ ☐ ☆ ○ ☐
- (A) ○ ☆ ☐
- (B) ☆ ○ ☐ ●
- (C) ☐ ○ ☆
- (D) ☆ ☐ ○

40. Which shape comes next in the pattern? ☆ △ ☐ ☆ △ ☐ ?
- (A) ☐
- (B) △
- (C) ☆ ●

41. Which shape comes next in the pattern? ☐ ○ ☐ ○ ?
- (A) ○
- (B) ▬
- (C) ☐ ●

42. Which shape comes next in the pattern? ◇ △ ☐ ◇ △ ☐ ◇ ?
- (A) ☐
- (B) ◇
- (C) △ ●

Cumulative Test — Chapters 1–6, Page 7

43. Which is the missing shape? ☆ ◇ ☆ ◇ ? ◇
- (A) ☆ ●
- (B) ◇

44. Which is the number pair? The sum is 5. The difference is 3.
- (A) 2 and 3
- (B) 3 and 6
- (C) 1 and 4
- (D) 3 and 5

45. What do you do to find how many fish are left?
- (A) Add
- (B) Subtract ●

46. What do you do to find how many dogs in all?
- (A) Add ●
- (B) Subtract

47. How much did Rusty spend? (4¢ car and 6¢ boat)
- (A) 2¢
- (B) 6¢
- (C) 9¢
- (D) 10¢

48. There are 4 tulips. There are 5 flowers. How many more flowers than tulips?
- (A) 1 more
- (B) 4 more
- (C) 5 more
- (D) 9 more

Cumulative Test — Chapters 1–6, Page 8

Use the pictograph below to answer questions 49–50.

49. Which group has fewer, caterpillars or ants?
- (A) caterpillar ●
- (B) ant

50. Which group has more than 6?
- (A) caterpillar
- (B) bee
- (C) ant
- (D) ladybug

Pretest — Chapter 7, Page 1

Choose the correct answer.

1. How many tens and ones are there?
- (A) 1 ten 4 ones ●
- (B) 4 tens 1 one
- (C) 4 tens 10 ones
- (D) 10 tens 1 one

2. How many tens and ones are there?
- (A) 30 tens 2 ones
- (B) 1 ten 3 ones
- (C) 3 tens 2 ones
- (D) 2 tens 3 ones

3. How many tens and ones are there?
- (A) 4 tens 0 ones ●
- (B) 0 tens 4 ones
- (C) 1 ten 4 ones
- (D) 10 tens 1 one

4. How many tens and ones are there?
- (A) 3 tens 0 ones
- (B) 20 tens 1 one
- (C) 1 ten 2 ones
- (D) 2 tens 1 one

Standardized Format • Test Answers

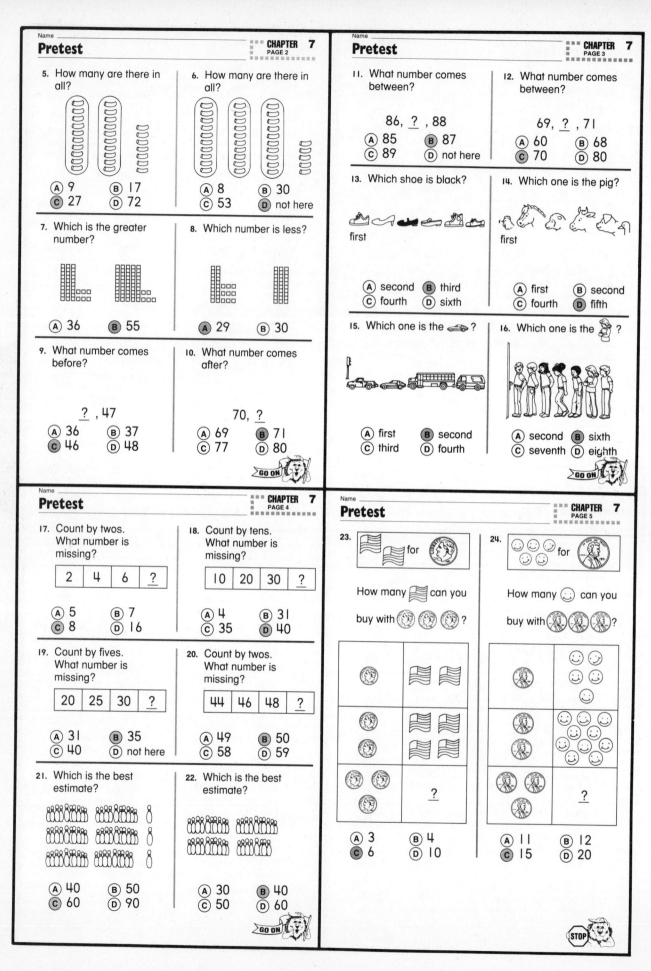

Posttest — Chapter 7, Page 1

Choose the correct answer.

1. How many tens and ones are there?
   - (A) 1 ten 6 ones
   - (B) 6 tens 1 one
   - (C) 6 tens 10 ones
   - (D) 10 tens 1 one

2. How many tens and ones are there?
   - (A) 20 tens 5 ones
   - (B) 50 tens 2 ones
   - (C) 5 tens 2 ones
   - (D) 2 tens 5 ones

3. How many tens and ones are there?
   - (A) 3 tens 0 ones
   - (B) 0 tens 3 ones
   - (C) 1 ten 3 ones
   - (D) 30 tens 0 ones

4. How many tens and ones are there?
   - (A) 3 tens 0 ones
   - (B) 10 tens 1 one
   - (C) 1 ten 4 ones
   - (D) 3 tens 4 ones

Posttest — Chapter 7, Page 2

5. How many are there in all?
   - (A) 9
   - (B) 26
   - (C) 36
   - (D) 63

6. How many are there in all?
   - (A) 10
   - (B) 20
   - (C) 82
   - (D) not here

7. Which is the greater number?
   - (A) 49
   - (B) 50

8. Which number is less?
   - (A) 38
   - (B) 71

9. What number comes before?

   ?, 86
   - (A) 75
   - (B) 76
   - (C) 85
   - (D) 87

10. What number comes after?

    50, ?
    - (A) 49
    - (B) 51
    - (C) 60
    - (D) 61

Posttest — Chapter 7, Page 3

11. What number comes between?

    44, ?, 46
    - (A) 42
    - (B) 45
    - (C) 48
    - (D) not here

12. What number comes between?

    79, ?, 81
    - (A) 70
    - (B) 78
    - (C) 80
    - (D) 90

13. Which hat has a feather?

    first
    - (A) first
    - (B) second
    - (C) fourth
    - (D) fifth

14. Which one is the horse?

    first
    - (A) first
    - (B) second
    - (C) fifth
    - (D) sixth

15. Which one is the ⛵?

    first
    - (A) first
    - (B) second
    - (C) third
    - (D) fourth

16. Which one is the 👤?
    - (A) first
    - (B) second
    - (C) fourth
    - (D) fifth

Posttest — Chapter 7, Page 4

17. Count by fives. What number is missing?

    | 5 | 10 | 15 | ? |
    - (A) 5
    - (B) 16
    - (C) 20
    - (D) 25

18. Count by tens. What number is missing?

    | 10 | 20 | 30 | ? |
    - (A) 4
    - (B) 31
    - (C) 35
    - (D) 40

19. Count by twos. What number is missing?

    | 54 | 56 | 58 | ? |
    - (A) 59
    - (B) 60
    - (C) 68
    - (D) not here

20. Count by fives. What number is missing?

    | 15 | 20 | 25 | ? |
    - (A) 26
    - (B) 30
    - (C) 35
    - (D) not here

21. Which is the best estimate?
    - (A) 3
    - (B) 10
    - (C) 30
    - (D) 40

22. Which is the best estimate?
    - (A) 10
    - (B) 50
    - (C) 60
    - (D) 70

Standardized Format • Test Answers

Posttest — Chapter 7, Page 5

23.  for

How many ✏ can you buy with 🪙🪙🪙?

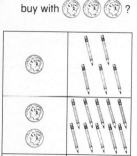

- (A) 6
- (B) 7
- ● (C) 15
- (D) 20

24.  for

How many ▭ can you buy with 🪙🪙🪙?

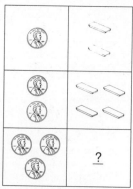

- (A) 5
- ● (B) 6
- (C) 8
- (D) 10

Pretest — Chapter 8, Page 1

Choose the correct answer.

1. Which addition sentence tells how many in all?

   - (A) 3 + 5 = 8
   - (B) 7 + 2 = 9
   - (C) 1 + 3 = 4
   - ● (D) 8 + 3 = 11

2. Which addition sentence tells how many in all?

   - (A) 3 + 3 = 6
   - (B) 7 + 7 = 14
   - ● (C) 6 + 6 = 12
   - (D) 5 + 5 = 10

3. Which addition sentence tells how many in all?

   - (A) 4 + 1 = 5
   - ● (B) 4 + 5 = 9
   - (C) 5 + 5 = 10
   - (D) 3 + 4 = 7

4. Which addition sentence tells how many in all?

   - ● (A) 8 + 1 = 9
   - (B) 7 + 1 = 8
   - (C) 9 + 1 = 10
   - (D) 9 + 9 = 18

5. 2 + 9 = ?
   - (A) 7
   - (B) 10
   - ● (C) 11
   - (D) 12

6. 0 + 6 = ?
   - (A) 0
   - ● (B) 6
   - (C) 7
   - (D) not here

Pretest — Chapter 8, Page 2

7. 7 + 3 = ?
   - (A) 1
   - (B) 4
   - ● (C) 10
   - (D) 11

8. 8 + 4 = ?
   - (A) 4
   - (B) 10
   - (C) 11
   - ● (D) 12

9. Sam catches 5 fish. Dad catches 7 fish. How many fish do they catch in all?
   - (A) 6 fish
   - (B) 8 fish
   - ● (C) 12 fish
   - (D) 13 fish

10. Meg has 5 pet birds. She has 5 pet fish. How many pets does she have in all?
    - (A) 0 pets
    - (B) 5 pets
    - (C) 9 pets
    - ● (D) 10 pets

11. Which subtraction sentence tells the story?

    - ● (A) 11 − 3 = 8
    - (B) 8 − 3 = 5
    - (C) 5 − 3 = 2
    - (D) 7 − 3 = 4

12. Which subtraction sentence tells the story?

    - (A) 9 − 3 = 6
    - (B) 6 − 3 = 3
    - ● (C) 12 − 9 = 3
    - (D) 11 − 8 = 3

Pretest — Chapter 8, Page 3

13. Which subtraction sentence tells the story?

    - (A) 8 − 2 = 6
    - (B) 6 − 2 = 4
    - (C) 12 − 2 = 10
    - ● (D) 10 − 2 = 8

14. Which subtraction sentence tells the story?

    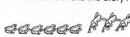

    - (A) 6 − 3 = 3
    - (B) 6 − 6 = 0
    - ● (C) 9 − 3 = 6
    - (D) 12 − 6 = 6

15. 11 − 6 = ?
    - ● (A) 5
    - (B) 6
    - (C) 15
    - (D) 17

16. 9 − 0 = ?
    - (A) 0
    - (B) 8
    - ● (C) 9
    - (D) not here

17. 10 − 1 = ?
    - (A) 1
    - ● (B) 9
    - (C) 10
    - (D) 11

18. 12 − 8 = ?
    - ● (A) 4
    - (B) 5
    - (C) 6
    - (D) 16

19. Mom buys 12 🍎. Pam eats 3 🍎. How many 🍎 are left?
    - ● (A) 9
    - (B) 10
    - (C) 11
    - (D) 15

20. There were 11 🥛. 5 were broken. How many 🥛 are left?
    - (A) 5
    - ● (B) 6
    - (C) 7
    - (D) 14

Standardized Format • Test Answers

Pretest — Chapter 8, Page 4

21. 7¢ + 4¢
    (A) 2¢ (B) 3¢
    (C) 10¢ (D) 11¢

22. 9¢ − 2¢
    (A) 6¢ (B) 7¢
    (C) 10¢ (D) 11¢

23. 8¢ − 7¢
    (A) 1¢ (B) 3¢
    (C) 4¢ (D) 15¢

24. 6¢ + 6¢
    (A) 0¢ (B) 6¢
    (C) 11¢ (D) 12¢

25. Which number sentence does not belong to the fact family?
    (A) 7 + 4 = 11
    (B) 11 − 7 = 4
    (C) 3 + 4 = 7
    (D) 4 + 7 = 11

26. Which number sentence does not belong to the fact family?
    (A) 9 − 1 = 8
    (B) 1 + 8 = 9
    (C) 9 − 8 = 1
    (D) 7 + 1 = 8

27. Which number sentence belongs to the same fact family as 10 − 6 = 4?
    (A) 6 + 4 = 10
    (B) 10 + 6 = 16
    (C) 6 − 4 = 2
    (D) 10 − 5 = 5

28. Which number sentence belongs to the same fact family as 12 − 7 = 5?
    (A) 2 + 5 = 7
    (B) 7 − 5 = 2
    (C) 6 + 6 = 12
    (D) 5 + 7 = 12

Pretest — Chapter 8, Page 5

Use the graph below to answer questions 29–30.

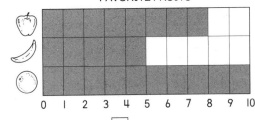

FAVORITE FRUITS

Each ☐ stands for 1 person.

29. How many people like best?
    (A) 5 (B) 6
    (C) 8 (D) 10

30. Which fruit is the favorite of the most people?
    (A) apple (B) banana
    (C) orange

31. There are 8 black puppies. There are 4 white puppies. What do you do to find how many puppies there are in all?
    (A) Add (B) Subtract

32. Roy had 10 pennies. He lost 3 pennies. What do you do to find how many pennies are left?
    (A) Add (B) Subtract

Posttest — Chapter 8, Page 1

Choose the correct answer.

1. Which addition sentence tells how many in all?
   (A) 8 + 3 = 11
   (B) 6 + 3 = 9
   (C) 7 + 3 = 10
   (D) 9 + 3 = 12

2. Which addition sentence tells how many in all?
   (A) 1 + 10 = 11
   (B) 1 + 9 = 10
   (C) 0 + 9 = 9
   (D) 8 + 1 = 9

3. Which addition sentence tells how many in all?
   (A) 4 + 2 = 6
   (B) 6 + 4 = 10
   (C) 6 + 5 = 11
   (D) 4 + 4 = 8

4. Which addition sentence tells how many in all?
   (A) 2 + 5 = 7
   (B) 7 + 1 = 8
   (C) 9 + 1 = 10
   (D) 7 + 2 = 9

5. 8 + 3 = ?
   (A) 5   (B) 10
   (C) 11 (D) 12

6. 9 + 0 = ?
   (A) 0 (B) 6
   (C) 9 (D) 10

Posttest — Chapter 8, Page 2

7. 5 + 5 = ?
   (A) 0 (B) 1
   (C) 5 (D) 10

8. 2 + 8 = ?
   (A) 0   (B) 6
   (C) 11 (D) not here

9. 2 ants are on a hill. 9 more ants come up the hill. How many ants in all?
   (A) 7 ants (B) 10 ants
   (C) 11 ants (D) 12 ants

10. There are 4 red balls. There are 8 green balls. How many balls in all?
    (A) 2 balls (B) 4 balls
    (C) 10 balls (D) 12 balls

11. Which subtraction sentence tells the story?
    (A) 11 − 4 = 7
    (B) 7 − 4 = 3
    (C) 7 − 5 = 2
    (D) 11 − 3 = 8

12. Which subtraction sentence tells the story?
    (A) 5 − 4 = 1
    (B) 8 − 4 = 4
    (C) 10 − 4 = 6
    (D) 9 − 4 = 5

Posttest — Chapter 8, Page 3

13. Which subtraction sentence tells the story?

- (A) 6 − 6 = 0
- (B) 10 − 6 = 4
- **(C) 12 − 6 = 6**
- (D) 12 − 5 = 7

14. Which subtraction sentence tells the story?

- (A) 6 − 5 = 1
- **(B) 11 − 6 = 5**
- (C) 10 − 6 = 4
- (D) 12 − 7 = 5

15. 11 − 3 = ?
- (A) 2
- **(B) 8**
- (C) 9
- (D) not here

16. 10 − 8 = ?
- (A) 12
- (B) 10
- **(C) 2**
- (D) 13

17. 9 − 3 = ?
- (A) 3
- (B) 5
- **(C) 6**
- (D) 12

18. 12 − 7 = ?
- (A) 4
- **(B) 5**
- (C) 6
- (D) not here

19. Mom buys 12 🍎. The children eat 9 🍎. How many 🍎 are left?

- **(A) 3**
- (B) 6
- (C) 9
- (D) 12

20. There were 9 🥛. 8 were broken. How many 🥛 are left?

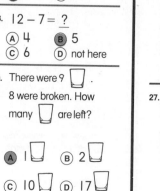

- **(A) 1**
- (B) 2
- (C) 10
- (D) 17

Posttest — Chapter 8, Page 4

21. 5¢ + 6¢
- (A) 1¢
- (B) 2¢
- (C) 10¢
- **(D) 11¢**

22. 12¢ − 4¢
- (A) 2¢
- **(B) 8¢**
- (C) 14¢
- (D) 16¢

23. 11¢ − 9¢
- **(A) 2¢**
- (B) 8¢
- (C) 18¢
- (D) not here

24. 1¢ + 8¢
- (A) 0¢
- (B) 7¢
- (C) 8¢
- **(D) 9¢**

25. Which number sentence does *not* belong to the fact family for 3, 6, and 9?
- (A) 3 + 6 = 9
- (B) 9 − 3 = 6
- **(C) 9 + 3 = 12**
- (D) 9 − 6 = 3

26. Which number sentence does *not* belong to the fact family for 3, 7, and 10?
- (A) 3 + 7 = 10
- **(B) 3 + 4 = 7**
- (C) 10 − 3 = 7
- (D) 10 − 7 = 3

27. Which number sentence belongs to the same fact family as 7 + 4 = 11?
- **(A) 11 − 7 = 4**
- (B) 7 + 0 = 7
- (C) 3 + 4 = 7
- (D) 7 − 4 = 3

28. Which number sentence belongs to the same fact family as 12 − 5 = 7?
- (A) 2 + 5 = 7
- (B) 7 − 5 = 2
- **(C) 7 + 5 = 12**
- (D) 6 + 6 = 12

Posttest — Chapter 8, Page 5

Use the graph below to answer questions 29–30.

FAVORITE LUNCHES

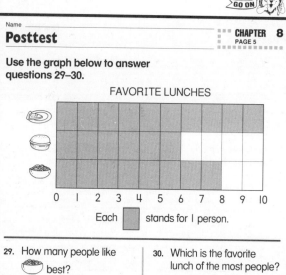

Each ▇ stands for 1 person.

29. How many people like 🍚 best?
- (A) 5
- (B) 6
- **(C) 8**
- (D) 10

30. Which is the favorite lunch of the most people?
- **(A)** 🥪
- (B) 🍔
- (C) 🍚

31. 12 cats were sleeping. 6 cats woke up. What do you do to find how many cats are still sleeping?
- (A) Add
- **(B) Subtract**

32. 8 chickens are sitting on nests. 8 chickens are eating. What do you do to find how many chickens there are in all?
- **(A) Add**
- (B) Subtract

Pretest — Chapter 9, Page 1

Choose the correct answer.

1. Which is the shortest?
- **(A)**
- (B)
- (C)
- (D)

2. Which is the longest?
- (A)
- **(B)**
- (C)
- (D)

3. How long is the fork?
- **(A) 1**
- (B) 2
- (C) 7
- (D) 10

4. How long is the knife?
- (A) 1 centimeter
- (B) 2 centimeters
- **(C) 4 centimeters**
- (D) 5 centimeters

5. How long is the pin?
- **(A) 1 inch**
- (B) 2 inches
- (C) 3 inches
- (D) 4 inches

6. How long is the whistle?
- (A) 1 inch
- (B) 3 inches
- **(C) 2 inches**
- (D) 4 inches

Standardized Format • Test Answers

Pretest — Chapter 9, Page 2

7. How long is the ribbon?
- A) 1 inch
- **B) 2 inches**
- C) 3 inches
- D) 4 inches

8. How long is the comb?
- A) 2 centimeters
- B) 4 centimeters
- C) 6 centimeters
- D) 10 centimeters

9. How long is the paper clip?
- A) 2 centimeters
- B) 3 centimeters
- C) 4 centimeters
- **D) 5 centimeters**

10. How long is the pin?
- A) 1 centimeter
- **B) 3 centimeters**
- C) 5 centimeters
- D) 6 centimeters

11. Which is heavier?
- A) (bird)
- **B) (raccoon)**

12. Which is heavier?
- **A) (toaster)**
- B) (bread)

Pretest — Chapter 9, Page 3

13. Which holds more?
- **A) (kettle)**
- B) (cup)

14. Which is hot?
- A) (vegetables)
- **B) (oven)**

15. Which holds about the same as a pitcher?
- A) (milk carton small)
- B) (juice can)
- **C) (milk carton large)**

16. Which holds about the same as a cup?
- A) (spoon)
- **B) (cup)**
- C) (teapot)

17. Which is about as heavy as a (star)?
- **A) (paper clip)**
- B) (pencil)
- C) (scissors)

18. Which is about as heavy as a (book)?
- A) (envelope)
- **B) (notebook)**
- C) (tricycle)

Pretest — Chapter 9, Page 4

19. Which shows fair shares?
- A)
- B)
- C)
- **D)**

20. Which shows fair shares?
- A)
- **B)**
- C)
- D)

21. Which shows one half?
- A)
- **B)**
- C)
- D)

22. Which shows one third?
- **A)**
- B)
- C)
- D)

23. Which shows one fourth?
- A)
- **B)**
- C)
- D)

24. What fraction does this picture show?
- A) $\frac{1}{5}$
- B) $\frac{1}{4}$
- **C) $\frac{1}{3}$**
- D) $\frac{1}{2}$

Pretest — Chapter 9, Page 5

25. What fraction does this picture show?
- A) $\frac{1}{5}$
- **B) $\frac{1}{4}$**
- C) $\frac{1}{3}$
- D) $\frac{1}{2}$

26. What fraction does this picture show?
- A) $\frac{1}{5}$
- B) $\frac{1}{4}$
- C) $\frac{1}{3}$
- **D) $\frac{1}{2}$**

27. What fraction of the group is small?
- A) $\frac{1}{5}$
- B) $\frac{1}{4}$
- **C) $\frac{1}{3}$**
- D) $\frac{1}{2}$

28. What fraction of the group are girls?
- A) $\frac{1}{5}$
- B) $\frac{1}{4}$
- C) $\frac{1}{3}$
- **D) $\frac{2}{4}$**

Standardized Format • Test Answers

Pretest — Chapter 9, Page 6

Use the table below to answer questions 29–32.

CHILDREN'S MEASUREMENTS (in inches)

Child	Foot	Hand	Arm
Pat	6	5	12
Chris	5	4	10
Bobby	8	6	14

29. How long is Pat's arm?
  - (A) 6 inches
  - (B) 12 inches
  - (C) 14 inches

30. How long is Bobby's hand?
  - (A) 5 inches
  - (B) 6 inches
  - (C) 8 inches

31. Whose arm is shortest?
  - (A) Pat's
  - (B) Chris's
  - (C) Bobby's

32. Whose foot is longest?
  - (A) Pat's
  - (B) Chris's
  - (C) Bobby's

Posttest — Chapter 9, Page 1

Choose the correct answer.

1. Which is the longest?
  - A, B, C, D

2. Which is the shortest?
  - A, B, C, D

3. How long is the worm?
  - (A) 1
  - (B) 2
  - (C) 7
  - (D) 8

4. How long is the frog?
  - (A) 1 centimeter
  - (B) 2 centimeters
  - (C) 4 centimeters
  - (D) 5 centimeters

5. How long is the spoon?
  - (A) 1 inch (B) 2 inches
  - (C) 3 inches (D) 4 inches

6. How long is the paper clip?
  - (A) 1 inch (B) 2 inches
  - (C) 3 inches (D) 4 inches

Posttest — Chapter 9, Page 2

7. How long is the string?
  - (A) 1 inch
  - (B) 2 inches
  - (C) 3 inches
  - (D) 4 inches

8. How long is the bean?
  - (A) 1 centimeter
  - (B) 2 centimeters
  - (C) 3 centimeters
  - (D) 5 centimeters

9. How long is the key?
  - (A) 1 centimeter
  - (B) 3 centimeters
  - (C) 4 centimeters
  - (D) 5 centimeters

10. How long is the eraser?
  - (A) 3 centimeters
  - (B) 5 centimeters
  - (C) 6 centimeters
  - (D) 7 centimeters

11. Which is heavier?
  - A, B

12. Which is heavier?
  - A, B

Posttest — Chapter 9, Page 3

13. Which holds more?
  - A, B

14. Which is hot?
  - A, B

15. Which holds about the same as a ?
  - A, B, C

16. Which holds about the same as a ?
  - A, B, C

17. Which is about as heavy as an ?
  - A, B, C

18. Which is about as heavy as an ?
  - A, B, C

Standardized Format • Test Answers

Posttest — Chapter 9, Page 4

19. Which shows fair shares?

A ⊘ (B) ⊕
C ◯ D ⊕

20. Which shows fair shares?

A △ B △
C △ (D) △

21. Which shows one half?

A ◐ B ⊕
C ● (D) ◐

22. Which shows one fourth?

(A) ⊠ B ▨
C ▮ D ▮

23. Which shows one third?

A ◨
(B) ▯
C ▯
D ▨

24. What fraction does this picture show?

A $\frac{1}{5}$ B $\frac{1}{4}$
C $\frac{1}{3}$ (D) $\frac{1}{2}$

Posttest — Chapter 9, Page 5

25. What fraction does this picture show?

(A) $\frac{1}{5}$ B $\frac{1}{4}$
C $\frac{1}{3}$ D $\frac{1}{2}$

26. What fraction does this picture show?

A $\frac{1}{5}$ (B) $\frac{1}{4}$
C $\frac{1}{3}$ D $\frac{1}{2}$

27. What fraction of the group is large?

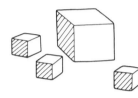

A $\frac{1}{5}$ (B) $\frac{1}{4}$
C $\frac{1}{3}$ D $\frac{1}{2}$

28. What fraction of the group are boys?

A $\frac{1}{5}$ B $\frac{1}{4}$
(C) $\frac{1}{3}$ D $\frac{1}{2}$

Posttest — Chapter 9, Page 6

Use the table below to answer questions 29–32.

LENGTHS (in inches)			
Object	Red	Blue	Yellow
Crayon	2	5	3
Pencil	5	6	4
Chalk	1	6	3

29. Which crayon is shortest?

(A) red
B blue
C yellow

30. Which chalk is longest?

A red
(B) blue
C yellow

31. How long is the yellow pencil?

A 3 inches
(B) 4 inches
C 5 inches

32. How long is the red chalk?

(A) 1 inch
B 2 inches
C 3 inches

Cumulative Test — Chapters 1–9, Page 1

Choose the correct answer.

1. 6 − 1 = ?

A 1   (B) 5
C 6   D 7

2. 5 + 3 = ?

A 2   B 3
C 7   (D) 8

3. Which plane shape matches the face of the solid shape?

(A) ▭   B ◯
C ▭   D △

4. Which plane shape matches the face of the solid shape?

A ▭   B ▭
C ◯   (D) △

5. Which figure is the same size and shape?

A △   B △
(C) △   D △

6. Which shows two parts that match?

(A)
B

Cumulative Test — CHAPTERS 1–9

PAGE 2

7. Which shape comes next in the pattern?

△ □ ○ △ □ ○ △ ?

- (A) △
- (B) □
- (C) ○

8. Which shape comes next in the pattern?

○ ○ □ ○ ○ □ ○ ?

- (A) ○
- (B) ○
- (C) □

9. How many tens and ones are there?

- (A) 2 tens and 7 ones
- (B) 3 tens and 3 ones
- (C) 2 tens and 4 ones
- (D) 10 tens and 7 ones

10. How many tens and ones are there?

- (A) 10 tens and 2 ones
- (B) 3 tens and 0 ones
- (C) 2 tens and 1 one
- (D) 1 ten and 2 ones

11. How many are there in all?

- (A) 8
- (B) 26
- (C) 27
- (D) not here

12. How many are there in all?

- (A) 5
- (B) 6
- (C) 15
- (D) 51

PAGE 3

13. Which number is less?

47   50

- (A) 47
- (B) 50

14. Which number is greater?

18   21

- (A) 18
- (B) 21

15. Which number comes after?

61, ?

- (A) 60
- (B) 62
- (C) 69
- (D) not here

16. Which number comes between?

49, ?, 51

- (A) 40
- (B) 48
- (C) 50
- (D) not here

17. Which one is the 🐄?

first

- (A) second
- (B) third
- (C) first
- (D) fourth

18. Which one is the 👧?

- (A) fourth
- (B) fifth
- (C) second
- (D) third

PAGE 4

19. 4 + 5 = ?
- (A) 9
- (B) 8
- (C) 1
- (D) 10

20. 2 + 4 = ?
- (A) 2
- (B) 4
- (C) 6
- (D) 8

21. 9 + 3 = ?
- (A) 6
- (B) 11
- (C) 12
- (D) 93

22. 6 − 0 = ?
- (A) 0
- (B) 6
- (C) 5
- (D) 9

23. There are 3 red hats. There are 7 blue hats. How many hats are there in all?
- (A) 1 hat
- (B) 4 hats
- (C) 9 hats
- (D) 10 hats

24. Roy has 6 books. Nancy has 5 books. How many books do they have in all?
- (A) 1 book
- (B) 12 books
- (C) 11 books
- (D) 65 books

25. 11 − 7 = ?
- (A) 4
- (B) 6
- (C) 16
- (D) 18

26. There were 12 eggs. 4 were broken. How many are left?
- (A) 2 eggs
- (B) 7 eggs
- (C) 8 eggs
- (D) 16 eggs

27. There were 9 birds. Then 3 flew away. How many birds are left?
- (A) 5 birds
- (B) 6 birds
- (C) 7 birds
- (D) 12 birds

28. 8¢ − 3¢
- (A) 4¢
- (B) 5¢
- (C) 6¢
- (D) 11¢

PAGE 5

29. 6¢ + 3¢
- (A) 1¢
- (B) 8¢
- (C) 9¢
- (D) not here

30. 7¢ + 1¢
- (A) 1¢
- (B) 6¢
- (C) 7¢
- (D) not here

31. Which one is longest?
- (A)
- (B)
- (C)
- (D)

32. How long is the crayon?
- (A) 1 long
- (B) 4 long
- (C) 6 long
- (D) 8 long

33. Use your inch ruler. How long is the worm?
- (A) 1 inch
- (B) 2 inches
- (C) 3 inches
- (D) 4 inches

34. Use your centimeter ruler. How long is the stick?
- (A) 1 centimeter
- (B) 3 centimeters
- (C) 5 centimeters
- (D) 7 centimeters

Standardized Format • Test Answers

Cumulative Test — Chapters 1–9, Page 6

35. Which is heavier, 🚗 or 🚲?
    - (A) car
    - (B) bicycle

36. Which is cold?
    - (A) pitcher
    - (B) kettle

37. Which holds about the same as a mug?
    - (A) milk carton
    - (B) cup
    - (C) spoon

38. Which is about as heavy as an apple?
    - (A) bread
    - (B) chicken
    - (C) orange

39. Which shows fair shares?
    - (A)
    - (B)
    - (C)
    - (D)

40. Which shows one third?
    - (A)
    - (B)
    - (C)
    - (D)

Cumulative Test — Chapters 1–9, Page 7

41. What fraction does this picture show?
    - (A) $\frac{1}{5}$
    - (B) $\frac{1}{4}$
    - (C) $\frac{1}{3}$
    - (D) $\frac{1}{2}$

42. What fraction of the group is small?
    - (A) $\frac{1}{5}$
    - (B) $\frac{1}{4}$
    - (C) $\frac{1}{3}$
    - (D) $\frac{1}{2}$

43. 2 pencils for 1 coin. How many pencils can you buy with 3 coins?
    - (A) 2
    - (B) 3
    - (C) 5
    - (D) 6

44. Which is the number pair? The sum is 7. The difference is 5.
    - (A) 2 and 7
    - (B) 3 and 4
    - (C) 5 and 2
    - (D) 1 and 6

Cumulative Test — Chapters 1–9, Page 8

45. How much did the girl spend?
    - (A) 2¢
    - (B) 5¢
    - (C) 7¢
    - (D) 12¢

46. There are 3 cups. There are 5 spoons. How many more spoons are there than cups?
    - (A) 2 more
    - (B) 3 more
    - (C) 5 more
    - (D) 8 more

Use the table below to answer questions 47–48.

Money to Spend

Children	Dimes	Pennies
Ali	3	2
Barb	1	5
Connie	4	0
Dave	2	3

47. Who has the most dimes?
    - (A) Ali
    - (B) Barb
    - (C) Connie
    - (D) Dave

48. What do you do to find how much money Ali and Barb have in all?
    - (A) Add
    - (B) Subtract

Cumulative Test — Chapters 1–9, Page 9

Use the graph below to answer question 49.

Favorite Colors

Red	🧍🧍🧍
Blue	🧍🧍🧍🧍🧍
Yellow	🧍🧍
Green	🧍🧍🧍🧍

Each 🧍 stands for 1 person.

49. Which color is the favorite of the most people?
    - (A) Blue
    - (B) Yellow
    - (C) Green
    - (D) Red

Use the graph below to answer question 50.

Favorite Toys

Each ▪ stands for 1 person.

50. Which toy is the favorite of the most people?
    - (A) ball
    - (B) bicycle
    - (C) kite
    - (D) drum

Standardized Format • Test Answers

Pretest — Chapter 10, Page 1

Choose the correct answer.

1. Which happened last?
   - A
   - B
   - C
   - D

2. Which happened first?
   - A
   - B
   - C
   - D

3. Which takes more time?
   - A
   - B

4. Which takes less time?
   - A
   - B

Pretest — Chapter 10, Page 2

5. Which one shows 2 o'clock?
   - A
   - B
   - C
   - D

6. Which one shows 7:30?
   - A
   - B
   - C
   - D

7. What time is it?
   - A) 3:00
   - B) 6:00
   - C) 9:00
   - D) 12:00

8. What time is it?
   - A) 8:30
   - B) 9:30
   - C) 7:30
   - D) 6:30

9. What time is it?
   - A) 11:00
   - B) 12:00
   - C) 1:00
   - D) 7:00

10. What time is it?
    - A) 6:00
    - B) 12:30
    - C) 1:30
    - D) not here

Pretest — Chapter 10, Page 3

Use the calendar below to answer questions 11–14.

Sunday	Monday	Tuesday	Wednesday	Thursday	Friday	Saturday
			1	2	3	4
5	6	7	8	9	10	11

11. On what day does this month begin?
    - A) Friday
    - B) Wednesday
    - C) Monday
    - D) Thursday

12. What day of the week is the third day of this month?
    - A) Tuesday
    - B) Friday
    - C) Saturday
    - D) Monday

13. What day comes just after Wednesday?
    - A) Tuesday
    - B) Thursday
    - C) Friday
    - D) Saturday

14. What day comes just before Saturday?
    - A) Friday
    - B) Wednesday
    - C) Monday
    - D) Sunday

Pretest — Chapter 10, Page 4

15. How much money?
    - A) 5¢
    - B) 6¢
    - C) 30¢
    - D) 60¢

16. How much money?
    - A) 3¢
    - B) 5¢
    - C) 15¢
    - D) 30¢

17. How much money?
    - A) 3¢
    - B) 20¢
    - C) 25¢
    - D) 30¢

18. How much money?
    - A) 25¢
    - B) 27¢
    - C) 35¢
    - D) 46¢

Standardized Format • Test Answers

Pretest
CHAPTER 10 PAGE 5

19. How much money?

  Ⓐ 5¢   Ⓑ 14¢
  Ⓒ 24¢ Ⓓ 29¢

20. How much money?

  Ⓐ 17¢   Ⓑ 28¢
  Ⓒ 32¢   Ⓓ 37¢

Use coins to answer questions 21–24.

21. Rita has 1 nickel and 2 pennies. How much money does she have?

  Ⓐ 2¢   Ⓑ 3¢
  Ⓒ 7¢   Ⓓ 12¢

22. Kit has 2 dimes and 2 nickels. How much money does he have?

  Ⓐ 4¢   Ⓑ 10¢
  Ⓒ 20¢ **Ⓓ 30¢**

23. Lois has 5 dimes. How much money does she have?

  Ⓐ 5¢   Ⓑ 10¢
  Ⓒ 25¢ **Ⓓ 50¢**

24. Ned has 1 quarter, 1 dime, and 3 pennies. How much money does he have?

  Ⓐ 5¢   Ⓑ 33¢
  Ⓒ 38¢   Ⓓ 50¢

Pretest
CHAPTER 10 PAGE 6

25. Which group of coins is the same amount as [quarter]?

  Ⓐ (5 pennies)
  Ⓑ (pennies)
  Ⓒ (quarter + pennies)
  Ⓓ (nickels)

26. Which group of coins is the same amount as [quarter]?

  Ⓐ (5 nickels)
  Ⓑ (1 nickel)
  Ⓒ (2 nickels)
  Ⓓ (dimes)

Pretest
CHAPTER 10 PAGE 7

27. Which group of coins is the same amount as [2 dimes]?

  Ⓐ (3 pennies)
  Ⓑ (4 nickels)
  Ⓒ (nickels + pennies)
  Ⓓ (nickels)

28. Which group of coins is the same amount as [2 nickels + pennies]?

  Ⓐ (nickel + penny)
  Ⓑ (nickels)
  Ⓒ (2 dimes)
  Ⓓ (dime)

Pretest
CHAPTER 10 PAGE 8

29. The movie will be over in 1 hour. At what time will the movie be over?

  [clock: 8:00]

  Ⓐ 7:00   Ⓑ 6:00
  Ⓒ 8:30   **Ⓓ 9:00**

30. The plane will take off in 30 minutes. At what time will the plane take off?

  [clock: 3:00]

  Ⓐ 3:00   Ⓑ 2:00
  Ⓒ 3:30   Ⓓ 5:30

31. Karen is going to play for 30 minutes. At what time will she finish?

  [clock: 4:00]

  Ⓐ 5:00   **Ⓑ 4:30**
  Ⓒ 7:30   Ⓓ 7:00

32. Art class will be over in 1 hour. At what time will art class be over?

  [clock: 10:30]

  Ⓐ 11:30   Ⓑ 1:00
  Ⓒ 9:30   Ⓓ 11:00

Standardized Format • Test Answers

Posttest — Chapter 10

Choose the correct answer.

1. Which happened last?
   - A, B, C, D (D selected)

2. Which happened first?
   - A, B, C, D (B selected)

3. Which takes more time?
   - A, B

4. Which takes less time?
   - A, B

5. Which one shows 1 o'clock?
   - A, B, C, D (D selected)

6. Which one shows 5:30?
   - A, B, C, D (C selected)

7. What time is it?
   - Ⓐ 2:00   Ⓑ 4:00
   - Ⓒ 3:00   Ⓓ 12:00

8. What time is it?
   - Ⓐ 4:30 (selected)   Ⓑ 3:30
   - Ⓒ 8:30   Ⓓ 6:30

9. What time is it?
   - Ⓐ 1:00   Ⓑ 12:00
   - Ⓒ 11:00   Ⓓ 2:00

10. What time is it?
    - Ⓐ 5:30   Ⓑ 6:30 (selected)
    - Ⓒ 7:30   Ⓓ not here

Use the calendar below to answer questions 11–14.

	May					
Sunday	Monday	Tuesday	Wednesday	Thursday	Friday	Saturday
					1	2
3	4	5	6	7	8	9

11. On what day does this month begin?
    - Ⓐ Sunday
    - Ⓑ Friday (selected)
    - Ⓒ Saturday
    - Ⓓ Monday

12. What day of the week is the third day of this month?
    - Ⓐ Tuesday
    - Ⓑ Sunday (selected)
    - Ⓒ Saturday
    - Ⓓ Thursday

13. What day comes just after Tuesday?
    - Ⓐ Monday
    - Ⓑ Friday
    - Ⓒ Wednesday (selected)
    - Ⓓ Saturday

14. What day comes just before Friday?
    - Ⓐ Thursday (selected)
    - Ⓑ Saturday
    - Ⓒ Sunday
    - Ⓓ Tuesday

15. How much money?
    - Ⓐ 4¢ (selected)   Ⓑ 5¢
    - Ⓒ 20¢   Ⓓ 40¢

16. How much money?
    - Ⓐ 6¢   Ⓑ 16¢
    - Ⓒ 30¢   Ⓓ 60¢

17. How much money?
    - Ⓐ 5¢   Ⓑ 21¢
    - Ⓒ 30¢   Ⓓ 50¢

18. How much money?
    - Ⓐ 5¢   Ⓑ 17¢
    - Ⓒ 22¢ (selected)   Ⓓ 35¢

Standardized Format • Test Answers

Posttest
CHAPTER 10 PAGE 5

19. How much money?

Ⓐ 4¢   ⦿Ⓑ 28¢
Ⓒ 31¢   Ⓓ 40¢

20. How much money?

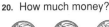

Ⓐ 6¢   Ⓑ 26¢
Ⓒ 41¢   Ⓓ 51¢

Use coins to answer questions 21–24.

21. Rae has 2 nickels and 1 penny. How much money does she have?

Ⓐ 3¢   ⦿Ⓑ 11¢
Ⓒ 15¢   Ⓓ not here

22. Kim has 1 dime and 3 nickels. How much money does she have?

Ⓐ 4¢   Ⓑ 20¢
⦿Ⓒ 25¢   Ⓓ 40¢

23. Lola has 6 dimes. How much money does she have?

Ⓐ 6¢   Ⓑ 10¢
Ⓒ 30¢   ⦿Ⓓ 60¢

24. Nick has 1 quarter, 1 dime, and 1 nickel. How much money does he have?

Ⓐ 3¢   Ⓑ 31¢
Ⓒ 40¢   ⦿Ⓓ 45¢

Posttest
CHAPTER 10 PAGE 6

25. Which group of coins is the same amount as ?

Ⓐ

Ⓑ

Ⓒ

Ⓓ

26. Which group of coins is the same amount as ?

Ⓐ

Ⓑ

Ⓒ

Ⓓ

Posttest
CHAPTER 10 PAGE 7

27. Which group of coins is the same amount as   ?

Ⓐ

Ⓑ

⦿Ⓒ

Ⓓ (coins)

28. Which group of coins is the same amount as 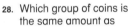 ?

Ⓐ

Ⓑ

Ⓒ

⦿Ⓓ

Posttest
CHAPTER 10 PAGE 8

29. The movie will be over in 1 hour. At what time will the movie be over?

Ⓐ 9:30   Ⓑ 8:00
Ⓒ 10:30   ⦿Ⓓ 10:00

30. The school day will end in 30 minutes. At what time will the school day end?

⦿Ⓐ 2:00   Ⓑ 4:30
Ⓒ 3:30   Ⓓ 3:00

31. Terry is going to play for 30 minutes. At what time will he finish?

Ⓐ 4:30   ⦿Ⓑ 5:30
Ⓒ 8:00   Ⓓ 6:00

32. The show will start in 1 hour. At what time will the show start?

⦿Ⓐ 7:30   Ⓑ 5:30
Ⓒ 7:00   Ⓓ 8:30

Name
Pretest CHAPTER 11 PAGE 1

Choose the letter of the correct answer.

1. $3 + 8 = \underline{?}$
   - (A) 5
   - (B) 10
   - (C) 11
   - (D) 38

2. $6 + 0 = \underline{?}$
   - (A) 0
   - (B) 1
   - (C) 6
   - (D) 60

3. $7 - 5 = \underline{?}$
   - (A) 2
   - (B) 3
   - (C) 12
   - (D) 75

4. $9 - 1 = \underline{?}$
   - (A) 7
   - (B) 8
   - (C) 10
   - (D) 91

5. How many cubes in all?

   - (A) 6 cubes
   - (B) 13 cubes
   - (C) 14 cubes
   - (D) 15 cubes

6. How many cubes in all?

   - (A) 3 cubes
   - (B) 20 cubes
   - (C) 30 cubes
   - (D) 33 cubes

Name
Pretest CHAPTER 11 PAGE 2

7. How many cubes in all?

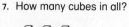

   - (A) 10 cubes
   - (B) 30 cubes
   - (C) 36 cubes
   - (D) 37 cubes

8. How many cubes in all?

   - (A) 46 cubes
   - (B) 60 cubes
   - (C) 64 cubes
   - (D) 66 cubes

9. Which addition sentence tells how many cubes in all?

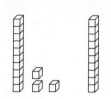

   - (A) $4 + 1 = 5$
   - (B) $10 + 3 = 13$
   - (C) $10 + 10 = 20$
   - (D) $13 + 10 = 23$

10. Which addition sentence tells how many cubes in all?

    - (A) $2 + 6 = 8$
    - (B) $3 + 7 = 10$
    - (C) $20 + 10 = 30$
    - (D) $22 + 16 = 38$

Name
Pretest CHAPTER 11 PAGE 3

11. $\begin{array}{r}10\\+60\\\hline\end{array}$
    - (A) 6
    - (B) 7
    - (C) 70
    - (D) 71

12. $\begin{array}{r}43\\+21\\\hline\end{array}$
    - (A) 22
    - (B) 61
    - (C) 63
    - (D) 64

13. $\begin{array}{r}32\\+54\\\hline\end{array}$
    - (A) 86
    - (B) 87
    - (C) 88
    - (D) not here

14. $\begin{array}{r}33\\+33\\\hline\end{array}$
    - (A) 0
    - (B) 33
    - (C) 66
    - (D) 99

15. Chris has 10 bats. Pat has 12 bats. How many bats in all?
    - (A) 2 bats
    - (B) 10 bats
    - (C) 22 bats
    - (D) 32 bats

16. There are 23 girls and 24 boys playing. How many children are playing?
    - (A) 7 children
    - (B) 27 children
    - (C) 41 children
    - (D) 47 children

17. Tod found 35 nuts. Bob found 14 nuts. How many nuts are there?
    - (A) 9 nuts
    - (B) 19 nuts
    - (C) 37 nuts
    - (D) 49 nuts

18. Sara had 20 berries. She picked 16 more. How many does she have?
    - (A) 4 berries
    - (B) 18 berries
    - (C) 36 berries
    - (D) 81 berries

Name
Pretest CHAPTER 11 PAGE 4

19. Which subtraction sentence tells the story?

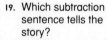

    - (A) $22 - 10 = 12$
    - (B) $12 - 10 = 2$
    - (C) $12 - 2 = 10$
    - (D) $22 - 11 = 11$

20. Which subtraction sentence tells the story?

    - (A) $23 - 13 = 10$
    - (B) $13 - 10 = 3$
    - (C) $23 - 10 = 13$
    - (D) $10 - 3 = 7$

21. $\begin{array}{r}40\\-20\\\hline\end{array}$
    - (A) 0
    - (B) 2
    - (C) 20
    - (D) 60

22. $\begin{array}{r}68\\-41\\\hline\end{array}$
    - (A) 7
    - (B) 27
    - (C) 28
    - (D) 72

23. $\begin{array}{r}54\\-20\\\hline\end{array}$
    - (A) 3
    - (B) 30
    - (C) 32
    - (D) not here

24. $\begin{array}{r}49\\-39\\\hline\end{array}$
    - (A) 10
    - (B) 11
    - (C) 19
    - (D) not here

Standardized Format • Test Answers

Name _____ **CHAPTER 11** PAGE 5
Pretest

25. There were 19 📕.
Then 10 were sold.
How many are left?

- (A) 1 📕
- (B) 9 📕
- (C) 10 📕
- (D) 29 📕

26. Kate had 53 ✏️.
She gave away 22.
How many are left?

- (A) 1 ✏️
- (B) 31 ✏️
- (C) 32 ✏️
- (D) 51 ✏️

27. There were 24 🐥.
Then 11 ran away.
How many are left?

- (A) 11 🐥
- (B) 13 🐥
- (C) 23 🐥
- (D) 35 🐥

28. There were 17 🐞.
Then 12 flew away.
How many are left?

- (A) 5 🐞
- (B) 6 🐞
- (C) 15 🐞
- (D) 29 🐞

Name _____ **CHAPTER 11** PAGE 6
Pretest

29. Mona had 63¢. Then she spent 21¢. <u>About</u> how much money does she have now?

- (A) about 40¢
- (B) about 50¢

30. Sasha spent 48¢. Then she spent 31¢. <u>About</u> how much did she spend in all?

- (A) about 70¢
- (B) about 80¢

31. Ned had 25 ⚾. He gave away 15. How many does he have now?

- (A) 10 ⚾
- (B) 40 ⚾
- (C) 45 ⚾
- (D) 100 ⚾

32. Mike is 4 years old. His sister is 3 years older. How old is Mike's sister?

- (A) 1 year old
- (B) 3 years old
- (C) 7 years old
- (D) 40 years old

Name _____ **CHAPTER 11** PAGE 1
Posttest

Choose the letter of the correct answer.

1. 4 + 7 = ?
- (A) 1 (B) 10
- (C) 11 (D) 47

2. 0 + 7 = ?
- (A) 0 (B) 1
- (C) 7 (D) 8

3. 6 − 6 = ?
- (A) 0 (B) 1
- (C) 6 (D) 12

4. 11 − 8 = ?
- (A) 2 (B) 3
- (C) 4 (D) 19

5. How many cubes in all?

- (A) 7 cubes
- (B) 14 cubes
- (C) 15 cubes
- (D) 16 cubes

6. How many cubes in all?

- (A) 2 cubes
- (B) 10 cubes
- (C) 20 cubes
- (D) 22 cubes

Name _____ **CHAPTER 11** PAGE 2
Posttest

7. How many cubes in all?

- (A) 25 cubes
- (B) 26 cubes
- (C) 52 cubes
- (D) 62 cubes

8. How many cubes in all?

- (A) 38 cubes
- (B) 80 cubes
- (C) 83 cubes
- (D) 88 cubes

9. Which addition sentence tells how many cubes in all?

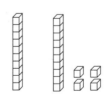

- (A) 1 + 5 = 6
- (B) 10 + 14 = 24
- (C) 9 + 13 = 22
- (D) 10 + 5 = 15

10. Which addition sentence tells how many cubes in all?

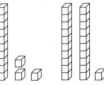

- (A) 10 + 3 = 13
- (B) 13 + 21 = 34
- (C) 4 + 3 = 7
- (D) 13 + 20 = 33

Standardized Format • Test Answers

Posttest — Chapter 11

11. 70 + 10
- (A) 7
- (B) 8
- (C) 80 ●
- (D) 81

12. 54 + 21
- (A) 33
- (B) 71
- (C) 74
- (D) 75

13. 36 + 12
- (A) 48 ●
- (B) 83
- (C) 84
- (D) not here

14. 22 + 22
- (A) 0
- (B) 22
- (C) 44 ●
- (D) not here

15. Carl has 14 tops. Pam has 10 tops. How many tops in all?
- (A) 4 tops
- (B) 10 tops
- (C) 14 tops
- (D) 24 tops

16. There are 42 boys and 32 girls playing. How many children are playing?
- (A) 10 children
- (B) 70 children
- (C) 72 children
- (D) 74 children

17. Toni found 40 rocks. Lee found 13 rocks. How many rocks are there?
- (A) 13 rocks
- (B) 17 rocks
- (C) 43 rocks
- (D) 53 rocks

18. Walt had 36 berries. He picked 23 more. How many does he have?
- (A) 13 berries
- (B) 19 berries
- (C) 59 berries
- (D) 69 berries

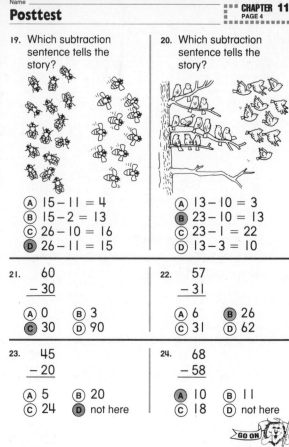

19. Which subtraction sentence tells the story?
- (A) $15 - 11 = 4$
- (B) $15 - 2 = 13$
- (C) $26 - 10 = 16$
- (D) $26 - 11 = 15$ ●

20. Which subtraction sentence tells the story?
- (A) $13 - 10 = 3$
- (B) $23 - 10 = 13$ ●
- (C) $23 - 1 = 22$
- (D) $13 - 3 = 10$

21. 60 − 30
- (A) 0
- (B) 3
- (C) 30 ●
- (D) 90

22. 57 − 31
- (A) 6
- (B) 26 ●
- (C) 31
- (D) 62

23. 45 − 20
- (A) 5
- (B) 20
- (C) 24
- (D) not here

24. 68 − 58
- (A) 10 ●
- (B) 11
- (C) 18
- (D) not here

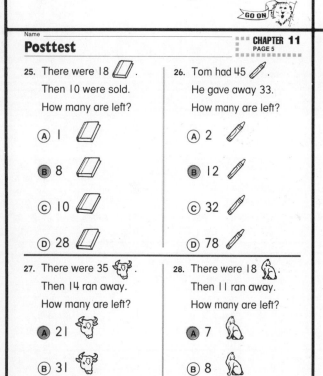

25. There were 18 books. Then 10 were sold. How many are left?
- (A) 1
- (B) 8 ●
- (C) 10
- (D) 28

26. Tom had 45 crayons. He gave away 33. How many are left?
- (A) 2
- (B) 12 ●
- (C) 32
- (D) 78

27. There were 35 cows. Then 14 ran away. How many are left?
- (A) 21 ●
- (B) 31
- (C) 39
- (D) 49

28. There were 18 cats. Then 11 ran away. How many are left?
- (A) 7 ●
- (B) 8
- (C) 17
- (D) 39

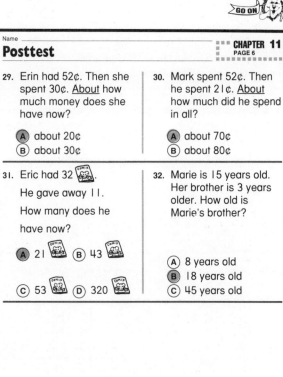

29. Erin had 52¢. Then she spent 30¢. About how much money does she have now?
- (A) about 20¢
- (B) about 30¢

30. Mark spent 52¢. Then he spent 21¢. About how much did he spend in all?
- (A) about 70¢
- (B) about 80¢

31. Eric had 32 comics. He gave away 11. How many does he have now?
- (A) 21 ●
- (B) 43
- (C) 53
- (D) 320

32. Marie is 15 years old. Her brother is 3 years older. How old is Marie's brother?
- (A) 8 years old
- (B) 18 years old ●
- (C) 45 years old

Standardized Format • Test Answers

Pretest — CHAPTER 12, PAGE 1

Choose the letter of the correct answer.

1. Which addition sentence tells how many in all?

   (A) 6 + 8 = 14
   (B) 6 + 2 = 8
   (C) 2 + 4 = 6
   (D) 8 + 2 = 10

2. Which addition sentence tells how many in all?

   (A) 0 + 0 = 0
   (B) 9 + 9 = 18
   (C) 9 + 8 = 17
   (D) 9 + 0 = 9

3. Which addition sentence tells how many in all?

   (A) 1 + 7 = 8
   (B) 8 + 1 = 9
   (C) 8 + 9 = 17
   (D) 7 + 8 = 15

4. Which addition sentence tells how many in all?

   (A) 8 + 9 = 17
   (B) 7 + 1 = 8
   (C) 8 + 1 = 9
   (D) 8 + 8 = 16

Pretest — CHAPTER 12, PAGE 2

5. 5 + 8 = ?
   (A) 12 (B) 13
   (C) 14 (D) 15

6. 8 + 8 = ?
   (A) 0   (B) 8
   (C) 15 (D) 16

7. 9 + 5 = ?
   (A) 4   (B) 13
   (C) 14 (D) 15

8. 6 + 7 = ?
   (A) 1   (B) 11
   (C) 12 (D) not here

9. 9 + 7
   (A) 2   (B) 15
   (C) 16 (D) 17

10. 8 + 6 = ?
    (A) 2   (B) 4
    (C) 13 (D) 14

Use counters to answer questions 11–14.

11. Mary has 9. Karl has 4. How many are there in all?
    (A) 5 (B) 9
    (C) 13 (D) 14

12. There are 7 red on the rack. There are 7 blue on the rack. How many are there in all?
    (A) 0 (B) 2
    (C) 13 (D) 14

Pretest — CHAPTER 12, PAGE 3

13. Mark has 9. Pam has 7. How many in all?
    (A) 2 (B) 9
    (C) 15 (D) 16

14. There are 7 red on a rack. There are 6 blue on the rack. How many in all?
    (A) 1 (B) 12
    (C) 13 (D) 14

15. Which subtraction sentence tells the story?
    (A) 9 − 1 = 8
    (B) 17 − 8 = 9
    (C) 16 − 8 = 8
    (D) 18 − 9 = 9

16. Which subtraction sentence tells the story?
    (A) 7 − 1 = 6
    (B) 14 − 7 = 7
    (C) 6 − 6 = 0
    (D) 13 − 7 = 6

Pretest — CHAPTER 12, PAGE 4

17. Which subtraction sentence tells the story?
    (A) 9 − 7 = 2
    (B) 16 − 7 = 9
    (C) 9 − 8 = 1
    (D) 15 − 9 = 6

18. Which subtraction sentence tells the story?
    (A) 15 − 8 = 7
    (B) 8 − 7 = 1
    (C) 8 − 1 = 7
    (D) 14 − 8 = 6

19. 15 − 6
    (A) 8 (B) 9
    (C) 11 (D) not here

20. 14 − 7
    (A) 3 (B) 6
    (C) 13 (D) not here

21. 13 − 7 = ?
    (A) 4 (B) 6
    (C) 7 (D) 14

22. 16 − 8 = ?
    (A) 2 (B) 7
    (C) 8 (D) 12

Standardized Format • Test Answers

Pretest — CHAPTER 12 PAGE 5

Use counters to answer questions 23 and 24.

23. Sally baked 18 🍪.
 Her friends ate 9 🍪.
 How many are left?
 (A) 0 🍪 (B) 8 🍪
 (C) 9 🍪 (D) 11 🍪

24. There were 15 🎈.
 Then 7 🎈 popped.
 How many are left?
 (A) 2 🎈 (B) 8 🎈
 (C) 9 🎈 (D) not here

25. Which number sentence does not belong to the fact family for 5, 9, and 14?
 (A) $5+4 = 9$
 (B) $9+5 = 14$
 (C) $14-9 = 5$
 (D) $14-5 = 9$

26. Which number sentence does not belong to the fact family for 8, 7, and 15?
 (A) $15-7 = 8$
 (B) $8+7 = 15$
 (C) $15-8 = 7$
 (D) $8-7 = 1$ ●

27. Which number sentence belongs to the same fact family as $13 - 8 = 5$?
 (A) $5 + 3 = 8$
 (B) $8 + 5 = 13$ ●
 (C) $5 - 3 = 2$
 (D) $8 - 5 = 3$

28. Which number sentence belongs to the same fact family as $16 - 7 = 9$?
 (A) $7 + 2 = 9$
 (B) $9 - 7 = 2$
 (C) $14 - 7 = 7$
 (D) $9 + 7 = 16$ ●

Pretest — CHAPTER 12 PAGE 6

Use the table below to answer questions 29 – 30.

FAVORITE SNACKS	Boys	Girls
Popcorn	IIII	IIII III
Fruit	IIII	IIII IIII
Cheese	IIII IIII	IIII I

29. How many boys and girls like cheese best?
 (A) 15 (B) 9
 (C) 12 (D) 3

30. How many boys and girls like fruit best?
 (A) 14 ● (B) 5
 (C) 4 (D) 15

31.
 There are 9 balloons.
 There are 3 children.
 How many balloons does each child get?
 (A) 2 balloons
 (B) 3 balloons
 (C) 4 balloons
 (D) 6 balloons

32.
 There are 16 stars.
 There are 2 children.
 How many stars does each child get?
 (A) 2 stars
 (B) 4 stars
 (C) 8 stars ●
 (D) 14 stars

Posttest — CHAPTER 12 PAGE 1

Choose the letter of the correct answer.

1. Which addition sentence tells how many in all?

 (A) $8 + 7 = 15$ ●
 (B) $8 + 2 = 10$
 (C) $7 + 1 = 8$
 (D) $5 + 2 = 7$

2. Which addition sentence tells how many in all?
 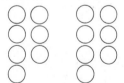
 (A) $0 + 0 = 0$
 (B) $7 + 0 = 7$
 (C) $6 + 7 = 13$
 (D) $7 + 7 = 14$ ●

3. Which addition sentence tells how many in all?

 (A) $4 + 5 = 9$
 (B) $9 + 6 = 15$
 (C) $8 + 5 = 13$
 (D) $9 + 5 = 14$ ●

4. Which addition sentence tells how many in all?

 (A) $5 + 3 = 8$
 (B) $5 + 8 = 13$ ●
 (C) $3 + 8 = 11$
 (D) $5 + 9 = 14$

Posttest — CHAPTER 12 PAGE 2

5. $6 + 7 = \underline{?}$
 (A) 12 (B) 13 ●
 (C) 14 (D) 15

6. $9 + 9 = \underline{?}$
 (A) 0 (B) 9
 (C) 18 ● (D) 19

7. $8 + 6 = \underline{?}$
 (A) 2 (B) 12
 (C) 13 (D) 14 ●

8. $9 + 8 = \underline{?}$
 (A) 1 (B) 16
 (C) 17 ● (D) not here

9. $\begin{array}{r}6\\+9\\\hline\end{array}$
 (A) 3 (B) 15 ●
 (C) 16 (D) 17

10. $8 + 8 = \underline{?}$
 (A) 0 (B) 8
 (C) 14 (D) 16 ●

Use counters to answer questions 11–14.

11. Mark has 7 🌼.
 Pam has 9 🌼.
 How many 🌼 in all?
 (A) 2 🌼 (B) 9 🌼
 (C) 15 🌼 ● (D) 16 🌼

12. There are 6 red 🧤 on a rack. There are 7 blue 🧤 on the rack. How many 🧤 in all?
 (A) 1 🧤 (B) 12 🧤
 (C) 13 🧤 ● (D) 14 🧤

Posttest — Chapter 12

13. Mary has 9 🎈. Karl has 5 🎈. How many 🎈 are there in all?
- (A) 5
- (B) 9
- (C) 13
- (D) 14

14. There are 7 red 🎩 on the rack. There are 7 blue 🎩 on the rack. How many 🎩 are there in all?
- (A) 0
- (B) 2
- (C) 14
- (D) 13

15. Which subtraction sentence tells the story?
- (A) 8 − 8 = 0
- (B) 13 − 5 = 8
- (C) 14 − 5 = 98
- (D) 8 − 5 = 3

16. Which subtraction sentence tells the story?
- (A) 9 − 6 = 3
- (B) 15 − 9 = 6
- (C) 10 − 4 = 6
- (D) 15 − 6 = 9

17. Which subtraction sentence tells the story?
- (A) 8 − 6 = 2
- (B) 14 − 6 = 8
- (C) 15 − 7 = 8
- (D) 13 − 7 = 6

18. Which subtraction sentence tells the story?
- (A) 14 − 5 = 9
- (B) 9 − 5 = 4
- (C) 15 − 9 = 6
- (D) 5 − 5 = 0

19. 16 − 8
- (A) 2
- (B) 8
- (C) 12
- (D) 16

20. 15 − 7
- (A) 2
- (B) 6
- (C) 12
- (D) 8

21. 17 − 8 = ?
- (A) 1
- (B) 9
- (C) 8
- (D) 11

22. 13 − 7 = ?
- (A) 2
- (B) 8
- (C) 6
- (D) not here

Use counters to answer questions 23 and 24.

23. Mac picked 16 🍑. He gave 9 to his friend. How many 🍑 are left?
- (A) 3
- (B) 5
- (C) 7
- (D) 13

24. There were 18 🥚. Then 9 🥚 were broken. How many 🥚 are left?
- (A) 2
- (B) 8
- (C) 9
- (D) 11

25. Which number sentence does not belong to the fact family for 4, 9, and 13?
- (A) 9 − 4 = 5
- (B) 13 − 9 = 4
- (C) 4 + 9 = 13
- (D) 9 + 4 = 13

26. Which number sentence does not belong to the fact family for 7, 8, and 15?
- (A) 8 − 7 = 1
- (B) 15 − 8 = 7
- (C) 7 + 8 = 15
- (D) 8 + 7 = 15

27. Which number sentence belongs to the same fact family as 14 − 6 = 8?
- (A) 8 − 6 = 2
- (B) 14 − 8 = 6
- (C) 4 + 4 = 8
- (D) 2 + 4 = 6

28. Which number sentence belongs to the same fact family as 17 − 8 = 9?
- (A) 9 − 1 = 8
- (B) 8 + 1 = 9
- (C) 8 + 9 = 17
- (D) 18 − 9 = 9

Use the table below to answer questions 29−30.

THINGS CHILDREN LIKE TO DO	Girls	Boys
Jumping	IIII	HHT IIII
Running	HHT	HHT IIII
Swinging	HHT III	HHT I

29. How many girls and boys like running best?
- (A) 14
- (B) 7
- (C) 13
- (D) 0

30. How many girls and boys like jumping best?
- (A) 14
- (B) 13
- (C) 15
- (D) 17

31. There are 8 books. There are 2 children. How many books does each child get?
- (A) 2 books
- (B) 4 books
- (C) 6 books
- (D) 8 books

32. There are 15 boats. There are 5 children. How many boats does each child get?
- (A) 1 boat
- (B) 2 boats
- (C) 3 boats
- (D) 5 boats

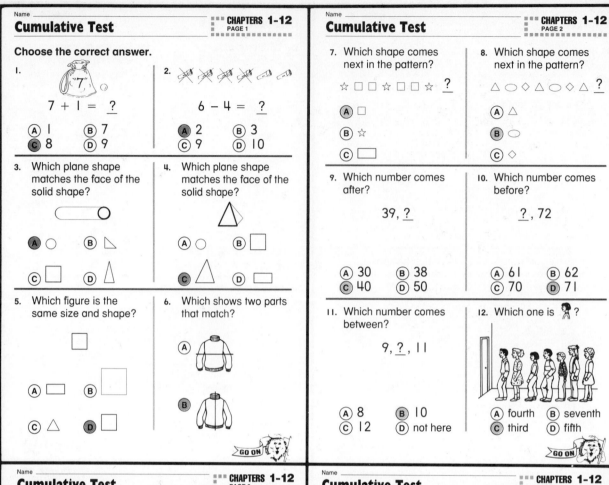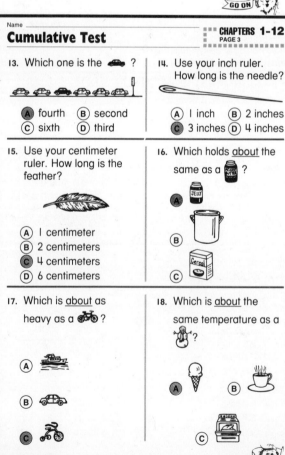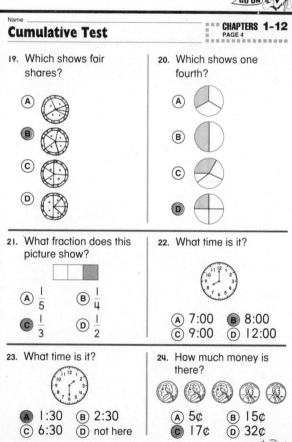

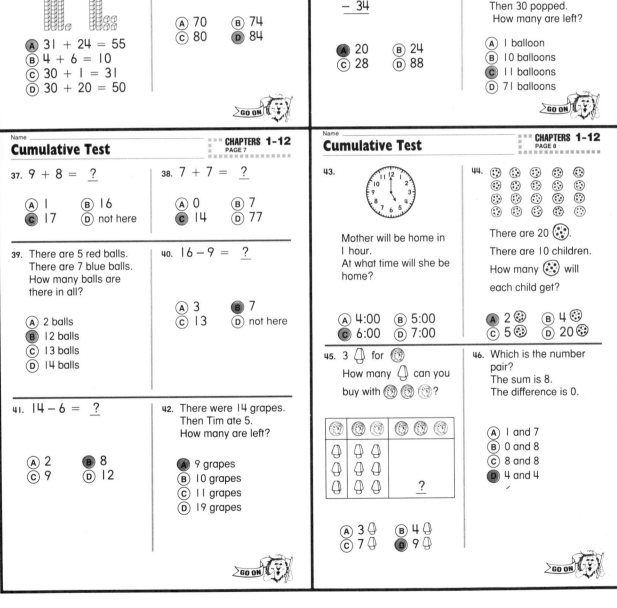

Name _____

Cumulative Test
CHAPTERS 1-12
PAGE 9

Use the graph below to answer question 47.

GLASSES OF MILK IN ONE DAY

Hal	🥛 🥛 🥛
Terri	🥛 🥛 🥛 🥛
Ira	🥛 🥛
Nan	🥛 🥛 🥛 🥛

Each 🥛 stands for 1 glass.

47. How many glasses of milk did Ira drink?
    (A) 2 glasses (B) 3 glasses (C) 4 glasses (D) 5 glasses

Use the graph below to answer question 48.

FAVORITE JUICES

Orange	■ ■ ■
Grape	■ ■ ■ ■ ■
Apple	■ ■
Pineapple	■

Each ■ stands for 1 person.

48. Which juice is the favorite of the most people?
    (A) orange **(B) grape** (C) apple (D) pineapple

Name _____

Cumulative Test
CHAPTERS 1-12
PAGE 10

Use the table below to answer questions 49–50.

	EATING TIMES (in minutes)		
	Breakfast	Lunch	Dinner
Amy	10	15	25
Bob	10	20	30

49. How long did it take Bob to eat lunch?
    (A) 10 minutes
    (B) 15 minutes
    (C) 20 minutes
    (D) 30 minutes

50. What do you do to find how much longer it took Amy to eat dinner than to eat lunch?
    (A) Add   **(B) Subtract**

Name _____

End-of-Book Test
PAGE 1

Choose the correct answer.

1.
   7 − 1 = ?
   (A) 1   **(B) 6**
   (C) 7   (D) 8

2.
   5 + 2 = ?
   (A) 3   **(B) 7**
   (C) 8   (D) 9

3. Which plane shape matches the face of the solid shape?

   (A) □   (B) ○
   (C) □   (D) △

4. Which plane shape matches the face of the solid shape?

   (A) △   (B) □
   (C) □   **(D) ○**

5. Which figure is the same size and shape?
   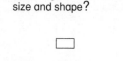
   (A) △   (B) □
   (C) ▭   **(D) ▭**

6. Which shows two parts that match?
    (A)
    (B)

Name _____

End-of-Book Test
PAGE 2

7. Which shape comes next in the pattern?
   ◇ ○ ◇ ○ ?
   (A) ◇   (B) △   (C) ○

8. Which shape comes next in the pattern?
   □ ○ ◇ □ ○ ◇ ?
   (A) ◇   (B) ○   **(C) □**

9. Which number comes just before?
   ?, 31
   (A) 20   (B) 21
   (C) 30   (D) 32

10. Which number comes just after?
    68, ?
    (A) 67   **(B) 69**
    (C) 70   (D) 79

11. Which number comes between?
    44, ?, 46
    (A) 43   **(B) 45**
    (C) 47   (D) not here

12. Which one is 🧍 ?
    (A) second (B) third
    (C) fifth (D) sixth

13. Which one is the 🚗?

    (A) second **(B) third**
    (C) fourth (D) sixth

14. Use your centimeter ruler. How long is the key?
    (A) 1 centimeter
    (B) 2 centimeters
    (C) 3 centimeters
    (D) 4 centimeters

Standardized Format • Test Answers

End-of-Book Test — PAGE 3

15. Use your inch ruler. How long is the leaf?

- (A) 1 inch
- (B) 2 inches
- (C) 3 inches
- (D) 4 inches

16. Which holds about the same as ?

- (A) wheelbarrow
- (B) truck
- (C) bucket

17. Which is about as heavy as a ?

- (A)
- (B)
- (C)

18. Which is about the same temperature as 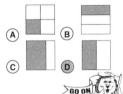?

- (A) coffee
- (B) ice tray
- (C) glass

19. Which shows fair shares?

- (A)
- (B)
- (C)
- (D)

20. Which shows one half?

- (A)
- (B)
- (C)
- (D)

End-of-Book Test — PAGE 4

21. What fraction does this picture show?

- (A) $\frac{1}{5}$
- (B) $\frac{1}{4}$
- (C) $\frac{1}{3}$
- (D) $\frac{1}{2}$

22. What time is it?

- (A) 3:00
- (B) 4:00
- (C) 5:00
- (D) 12:00

23. What time is it?

- (A) 6:30
- (B) 10:30
- (C) 11:30
- (D) not here

24. How much money is there?

- (A) 5¢
- (B) 35¢
- (C) 40¢
- (D) 50¢

25. How much money is there?

- (A) 5¢
- (B) 17¢
- (C) 27¢
- (D) 32¢

26. How much money is there?

- (A) 10¢
- (B) 20¢
- (C) 25¢
- (D) not here

27. How many cubes are there in all?

- (A) 6 cubes
- (B) 7 cubes
- (C) 10 cubes
- (D) 16 cubes

28. How many cubes are there in all?

- (A) 10 cubes
- (B) 17 cubes
- (C) 37 cubes
- (D) 73 cubes

End-of-Book Test — PAGE 5

29. Which addition sentence tells how many cubes there are in all?

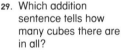

- (A) 23 + 14 = 37
- (B) 3 + 4 = 7
- (C) 5 + 5 = 10
- (D) 20 + 10 = 30

30.
$$45 + 30$$

- (A) 15
- (B) 65
- (C) 70
- (D) 75

31.
$$44 + 44$$

- (A) 0
- (B) 4
- (C) 44
- (D) 88

32. Tina has 25 beans. Bert has 14 beans. How many beans are there in all?

- (A) 11 beans
- (B) 29 beans
- (C) 31 beans
- (D) 39 beans

33. Which subtraction sentence tells the story?

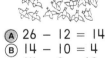

- (A) 26 − 12 = 14
- (B) 14 − 10 = 4
- (C) 14 − 2 = 12
- (D) 12 − 2 = 10

34.
$$56 - 31$$

- (A) 21
- (B) 25
- (C) 27
- (D) 87

End-of-Book Test — PAGE 6

35.
$$63 - 23$$

- (A) 40
- (B) 43
- (C) 46
- (D) 86

36. Fred won 24 stars. Amy won 10 stars. How many more stars did Fred win?

- (A) 4 stars
- (B) 10 stars
- (C) 14 stars
- (D) 34 stars

37. 8 + 5 = ?

- (A) 12
- (B) 13
- (C) 14
- (D) not here

38. 9 + 0 = ?

- (A) 0
- (B) 9
- (C) 18
- (D) 90

39. There are 4 white ✿. There are 7 yellow ✿. How many ✿ in all?

- (A) 3
- (B) 10
- (C) 11
- (D) 12

40. 17 − 8 = ?

- (A) 1
- (B) 9
- (C) 11
- (D) not here

41. 12 − 5 = ?

- (A) 3
- (B) 7
- (C) 8
- (D) not here

42. Use pennies to solve. Pam has 15 pennies. She spends 8 of them. How many pennies does Pam have left?

- (A) 3 pennies
- (B) 7 pennies
- (C) 8 pennies
- (D) 22 pennies

End-of-Book Test — PAGE 7

43. Story time will start in 1 hour. What time will it start?
   - (A) 1:00
   - (B) 2:00
   - **(C) 3:00**
   - (D) 12:00

44. There are 12 eggs. There are 3 children. How many eggs does each child get?
   - (A) 3 eggs
   - **(B) 4 eggs**
   - (C) 6 eggs
   - (D) 12 eggs

45. 4 🧶 for 🪙. How many 🧶 can you buy with 🪙🪙🪙?
   - (A) 4
   - (B) 9
   - (C) 10
   - **(D) 12**

46. Which is the number pair? The sum is 6. The difference is 2.
   - (A) 4 and 6
   - (B) 3 and 3
   - (C) 2 and 2
   - **(D) 2 and 4**

End-of-Book Test — PAGE 8

Use the graph below to answer question 47.

PINE CONES FOUND

Tom	🌰 🌰 🌰 🌰
Sara	🌰 🌰 🌰
Dan	🌰
Alice	🌰 🌰 🌰 🌰 🌰 🌰

Each 🌰 stands for one pine cone.

47. How many 🌰 did Dan find?
   - **(A) 1 pine cone**
   - (B) 3 pine cones
   - (C) 4 pine cones
   - (D) 6 pine cones

Use the graph below to answer question 48.

FLOWERS PICKED

Each ▨ stands for one flower.

48. How many 🌻 were picked?
   - (A) 2
   - (B) 3
   - **(C) 5**
   - (D) 7

End-of-Book Test — PAGE 9

Use the table below to answer questions 49–50.

Number of Children

	School Bus	City Bus
Boys	14	4
Girls	12	5

49. How many girls are on the school bus?
   - (A) 4 girls
   - (B) 5 girls
   - **(C) 12 girls**
   - (D) 17 girls

50. What do you do to find how many children are on the city bus?
   - **(A) Add**
   - (B) Subtract

Standardized Format • Test Answers

Name _____

Inventory Test

PAGE 1

1.

2.

3.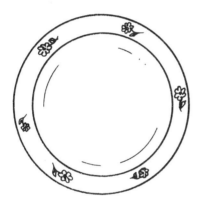

4.

5.

6.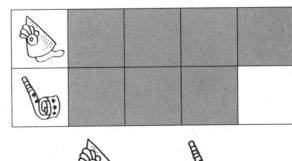

Ring the one that (1) is the right hand; (2) is the same; (3) is the same shape; (4) has the same number; (5) has more; (6) has fewer.

Name _____

Inventory Test

PAGE 2

7.

4   5

8.

9.

10.
 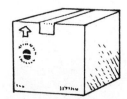

11.
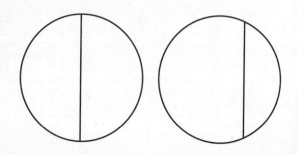

12.
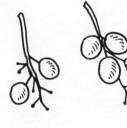 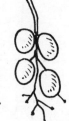 

Ring the number that (7) tells how many. Ring the one that (8) has fewer; (9) is a square; (10) is a cone; (11) shows halves; (12) comes next.

Name _____

Inventory Test

PAGE 3

13.

6   7

14.

8   9

15.

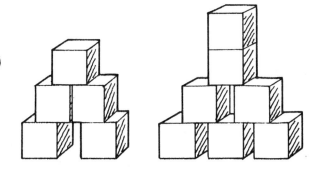

16.

○ ○ △ ○ ○ △ ?

○   △

17.

12   13

18.

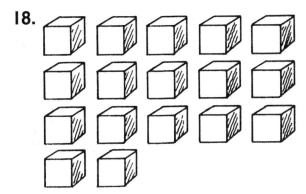

17   18

Ring the number that (13–14) tells how many. Ring the one that (15) has more; (16) comes next. Ring the number that (17–18) tells how many.

Name _____

Inventory Test

PAGE 4

19.

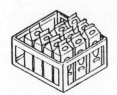 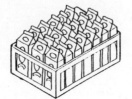

20.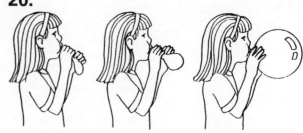

21.

2 o'clock

12 o'clock

22.

1 o'clock

10 o'clock

23.

24.

1¢   10¢

Ring (19) the one that has fewer; (20) what happened last; (21–22) the time that is shown; (23) the one that is a penny; (24) how many cents there are.

Name _____

Inventory Test

PAGE 5

25.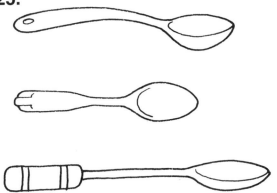

26.

27.

$3 + 1 = \underline{\ ?\ }$

3   4

28.

$5 - 3 = \underline{\ ?\ }$

2   3

29.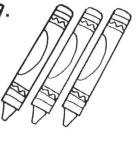

$3 + 2 = 5$

$2 + 1 = 3$

30.

$2 - 1 = 1$

$3 - 1 = 2$

Ring the one that (25) is the longest; (26) holds more. Ring how many (27) in all; (28) are left. Ring the number sentence (29–30) that tells the story.

Pretest

CHAPTER 1
PAGE 1

Read each question. Find the answer.

1. Ring the one that belongs.

2. Ring the one that belongs.

3. Ring the shape that comes next.

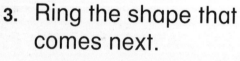 ?

4. Ring the shape that comes next.

 ?

Name _____

Pretest

CHAPTER 1
PAGE 2

5. How many ☐ are there?

_____

6. How many ☐ are there?

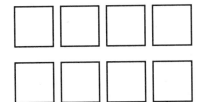

_____

7. How many ☐ are there?

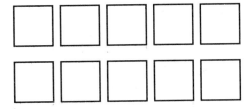

_____

8. How many ☐ are there?

_____

9. How many △ are there?

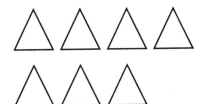

_____

10. How many ○ are there?

_____

GO ON

Name _____

Pretest

CHAPTER 1
PAGE 3

11. How many ☐ are there?

_____

12. How many ☆ are there?

_____

13. Ring the group that has more than this.

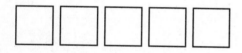

14. Ring the group that has fewer than this.

15. What comes next?

5, 6, 7, _____

16. What comes next?

0, 1, 2, _____

GO ON

Name _____

Pretest

CHAPTER 1
PAGE 4

Use the pictograph below to answer questions 17–20.

Number of Boats

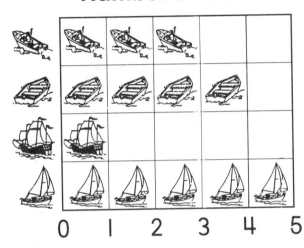

0 1 2 3 4 5

---

17. Ring the one that has more, or .

18. Ring the one that has fewer, or .

---

19. Ring the one that has more than 4.

20. Ring the one that has fewer than 3.

Name _____

Posttest

CHAPTER 1
PAGE 1

Read each question. Find the answer.

1. Ring the one that belongs.

2. Ring the one that belongs.

3. Ring the shape that comes next.

4. Ring the shape that comes next.

Name _____
Posttest

CHAPTER 1
PAGE 2

5. How many ☐ are there?

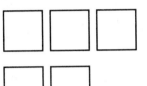

_____

6. How many ☐ are there?

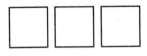

_____

7. How many ☐ are there?

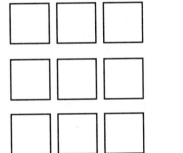

_____

8. How many ☐ are there?

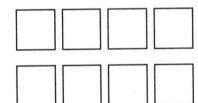

_____

9. How many ○ are there?

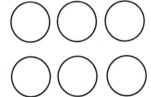

_____

10. How many ☐ are there?

_____

Name _____

Posttest

CHAPTER 1
PAGE 3

11. How many ☆ are there?

_____

12. How many ○ are there?

_____

13. Ring the group that has fewer than this.

14. Ring the group that has more than this.

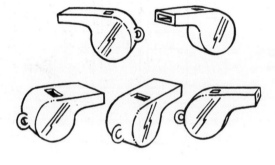

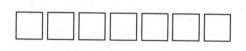

15. What comes next?

2, 3, 4, _____

16. What comes next?

6, 7, 8, _____

Name _____

Posttest

CHAPTER 1
PAGE 4

Use the pictograph below to answer questions 17–20.

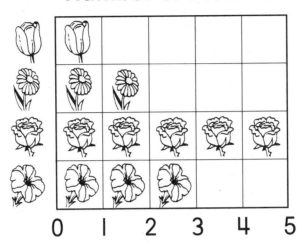

---

17. Ring the one that has fewer, or .

18. Ring the one that has more, or .

---

19. Ring the one that has more than 4.

20. Ring the one that has fewer than 2.

Name _____

Pretest

CHAPTER 2
PAGE 1

Read each question. Find the answer.

1. Ring the addition sentence that tells how many there are.

$4 + 2 = 6$

$4 + 1 = 5$

$1 + 3 = 4$

$2 + 1 = 3$

2. Ring the addition sentence that tells how many there are.

$2 + 1 = 3$

$3 + 3 = 6$

$2 + 3 = 5$

$3 + 1 = 4$

3. Ring the addition sentence that tells how many there are.

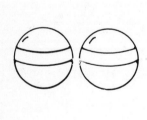 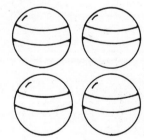

$2 + 4 = 6$

$2 + 2 = 4$

$2 + 3 = 5$

$1 + 1 = 2$

4. Ring the addition sentence that tells how many there are.

$3 + 3 = 6$

$1 + 3 = 4$

$1 + 1 = 2$

$1 + 2 = 3$

GO ON

Name _____

Pretest

CHAPTER 2
PAGE 2

5. Ring the addition sentence that tells how many there are.

$2 + 2 = 4$

$1 + 2 = 3$

$1 + 1 = 2$

$2 + 3 = 5$

6. Ring the addition sentence that tells how many there are.

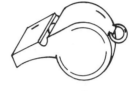

$1 + 3 = 4$

$2 + 1 = 3$

$2 + 3 = 5$

$1 + 1 = 2$

7. $3 + 1 = $ _____

8. $3 + 3 = $ _____

9. $0 + 2 = $ _____

10. $3 + 2 = $ _____

11. 1
   $+5$

12. 2
   $+2$

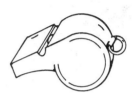

Name _____

Pretest

CHAPTER 2
PAGE 3

13.
$$\begin{array}{r} 5 \\ +0 \\ \hline \end{array}$$

14.
$$\begin{array}{r} 1 \\ +1 \\ \hline \end{array}$$

15. Ring another name for 3.

$3 + 0 \quad 3 + 3$

$1 + 3 \quad 0 + 6$

16. Ring another name for 5.

$1 + 5 \quad 3 + 3$

$2 + 3 \quad 3 + 1$

17. Ring another name for 6.

$1 + 3 \quad 3 + 3$

$3 + 2 \quad 5 + 0$

18. Ring another name for 2.

$1 + 1 \quad 1 + 3$

$2 + 1 \quad 2 + 2$

19. Ring another name for 1.

$1 + 1 \quad 2 + 2$

$1 + 2 \quad 1 + 0$

20. Ring another name for 4.

$3 + 1 \quad 3 + 3$

$4 + 1 \quad 0 + 3$

GO ON

Name _____

Pretest

CHAPTER 2
PAGE 4

21.

4 cats sleep.
2 cats play.

How many cats are there?

_____ cats

22.

1 girl plays.
3 more girls play.

How many girls play?

_____ girls

23.

1 dog sits.
1 dog runs.

How many dogs are there?

_____ dogs

24.

3 boys read.
3 boys write.

How many boys are there?

_____ boys

Name _____

Posttest

CHAPTER 2
PAGE 1

Read each question. Find the answer.

1. Ring the addition sentence that tells how many there are.

4 + 2 = 6
3 + 2 = 5
2 + 2 = 4
2 + 1 = 3

2. Ring the addition sentence that tells how many there are.

1 + 1 = 2
2 + 3 = 5
2 + 2 = 4
4 + 2 = 6

3. Ring the addition sentence that tells how many there are.

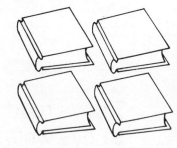

1 + 4 = 5
2 + 4 = 6
2 + 2 = 4
1 + 3 = 4

4. Ring the addition sentence that tells how many there are.

2 + 1 = 3
3 + 1 = 4
3 + 3 = 6
3 + 2 = 5

GO ON

Name _____

Posttest

CHAPTER 2
PAGE 2

5. Ring the addition sentence that tells how many there are.

$3 + 3 = 6$

$1 + 2 = 3$

$2 + 2 = 4$

$3 + 2 = 5$

6. Ring the addition sentence that tells how many there are.

$1 + 3 = 4$

$1 + 2 = 3$

$2 + 3 = 5$

$1 + 1 = 2$

7. $2 + 3 =$ _____

8. $0 + 4 =$ _____

9. $1 + 1 =$ _____

10. $3 + 3 =$ _____

11. $\begin{array}{r} 2 \\ +2 \\ \hline \end{array}$

12. $\begin{array}{r} 5 \\ +1 \\ \hline \end{array}$

GO ON

Name _____

Posttest

CHAPTER 2
PAGE 3

13.
```
   3
 + 0
 ___
```

14.
```
   2
 + 4
 ___
```

15. Ring another name for 4.

3 + 1   3 + 3

2 + 3   0 + 3

16. Ring another name for 1.

0 + 1   1 + 2

1 + 1   2 + 2

17. Ring another name for 5.

1 + 3   3 + 3

3 + 2   4 + 0

18. Ring another name for 6.

1 + 5   3 + 2

5 + 0   2 + 2

19. Ring another name for 3.

1 + 1   2 + 2

0 + 2   3 + 0

20. Ring another name for 2.

3 + 1   1 + 0

0 + 2   2 + 2

GO ON

Name _____

Posttest

CHAPTER 2
PAGE 4

21.

2 frogs jump.
3 frogs sit.

How many frogs are there?

_____ frogs

22.

1 girl plays.
1 girl reads.

How many girls are there?

_____ girls

23.

4 boys smile.
2 more boys smile.

How many boys smile in all?

_____ boys

24.

1 cat sleeps.
5 cats drink.

How many cats are there?

_____ cats

Name _____

Pretest

CHAPTER 3
PAGE 1

Read each question. Find the answer.

1. Ring the subtraction sentence that tells the story.

$6 - 1 = 5$ $6 - 3 = 3$
$6 - 6 = 0$ $6 - 2 = 4$

2. Ring the subtraction sentence that tells the story.

$4 - 1 = 3$ $4 - 2 = 2$
$4 - 0 = 4$ $4 - 1 = 3$

3. Ring the subtraction sentence that tells the story.

$4 - 0 = 4$ $5 - 1 = 4$
$4 - 1 = 3$ $5 - 5 = 0$

4. Ring the subtraction sentence that tells the story.

$3 - 3 = 0$ $6 - 5 = 1$
$3 - 0 = 3$ $6 - 3 = 3$

Name _____

Pretest

CHAPTER 3
PAGE 2

5.

$4 - 2 =$ _____

6.

$5 - 4 =$ _____

7.

$3 - 3 =$ _____

8.

$2 - 1 =$ _____

9. $4 - 3 =$ _____

10. $3 - 0 =$ _____

11. $3 - 1 =$ _____

12. $6 - 5 =$ _____

13. 6
    −4
    ___

14. 1
    −1
    ___

GO ON

Name _____

Pretest

CHAPTER 3
PAGE 3

15.   5
    − 3
    ___

16.   4
    − 2
    ___

17. Ring the subtraction sentence that goes with the addition sentence.

  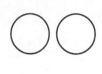

3 + 2 = 5

3 − 3 = 0   5 − 1 = 4

3 − 2 = 1   5 − 2 = 3

18. Ring the subtraction sentence that goes with the addition sentence.

5 + 1 = 6

5 − 1 = 4   6 − 3 = 4

6 − 1 = 5   3 − 3 = 0

19. Ring the number sentence that does <u>not</u> belong.

1 + 3 = 4   4 − 1 = 3

4 − 3 = 1   3 − 1 = 2

20. Ring the number sentence that does <u>not</u> belong.

3 − 2 = 1   5 − 3 = 2

5 − 2 = 3   3 + 2 = 5

GO ON

Name _____

Pretest

CHAPTER 3
PAGE 4

21.

What do you do to find how many in all? Ring the answer.

Add   Subtract

22.

What do you do to find how many are left? Ring the answer.

Add   Subtract

23. Sara sees 4 . Then 1 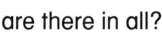 swims away.

Ring the correct question.

How many 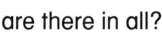 are left?

How many 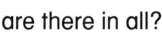 are there in all?

24. Tom sees 1 . Then he sees 1 more .

Ring the correct question.

How many are left?

How many are there in all?

Name _____

Posttest

CHAPTER 3
PAGE 1

Read each question. Find the answer.

1. Ring the subtraction sentence that tells the story.

3 − 2 = 1   4 − 1 = 3
3 − 1 = 2   4 − 2 = 2

2. Ring the subtraction sentence that tells the story.

5 − 1 = 4   3 − 2 = 1
5 − 2 = 3   3 − 3 = 0

3. Ring the subtraction sentence that tells the story.

6 − 3 = 3   3 − 3 = 0
3 − 1 = 2   4 − 1 = 3

4. Ring the subtraction sentence that tells the story.

4 − 1 = 3   5 − 1 = 4
5 − 4 = 1   6 − 1 = 5

GO ON

Name _____

Posttest

CHAPTER 3
PAGE 2

5.

6.

6 – 4 = _____

5 – 3 = _____

7.

8.

4 – 3 = _____

2 – 2 = _____

9. 5 – 4 = _____

10. 6 – 2 = _____

11. 2 – 1 = _____

12. 4 – 0 = _____

13. 4
    – 4
    ____

14. 6
    – 3
    ____

GO ON

Name _____

Posttest

CHAPTER 3
PAGE 3

15.   5
    − 1
    ―――

16.   6
    − 5
    ―――

17. Ring the subtraction sentence that goes with the addition sentence.

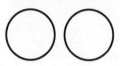

2 + 1 = 3

2 − 1 = 1   2 − 2 = 0

3 − 3 = 0   3 − 1 = 2

18. Ring the subtraction sentence that goes with the addition sentence.

 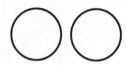

2 + 2 = 4

2 − 0 = 2   3 − 3 = 0

4 − 2 = 2   3 − 2 = 1

19. Ring the number sentence that does <u>not</u> belong.

 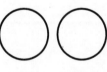

4 + 2 = 6   4 − 2 = 2

6 − 4 = 2   6 − 2 = 4

20. Ring the number sentence that does <u>not</u> belong.

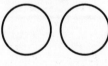 

5 − 1 = 4   5 − 4 = 1

4 + 1 = 5   4 − 1 = 3

GO ON

Name _____

Posttest

CHAPTER 3
PAGE 4

21.

What do you do to find how many in all?

Ring the answer.

Add   Subtract

22.

What do you do to find how many  are left?

Ring the answer.

Add   Subtract

---

23. There are 5 🐝 in this 🐝🏠. Then 2 🐝 fly away.

Ring the correct question.

How many 🐝 are left?

How many 🐝 are there in all?

24. Sara sees 5 🐢 in a pond. Then she sees 1 more 🐢.

Ring the correct question.

How many 🐢 are left?

How many 🐢 are there in all?

Name _____

Cumulative Test

CHAPTERS 1-3
PAGE 1

Read each question. find the answer.

1. How many ◯ are there?

_____

2. How many ▢ are there?

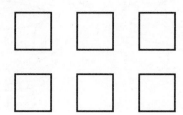

_____

3. How many △ are there?

_____

4. How many ☆ are there?

_____

5. How many ◇ are there?

_____

6. How many ▭ are there?

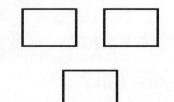

_____

GO ON

Name _____

Cumulative Test

CHAPTERS 1–3
PAGE 2

7. Ring the group that has fewer than this.

8. Ring the group that has fewer than this.

9. Ring the group that has more than this.

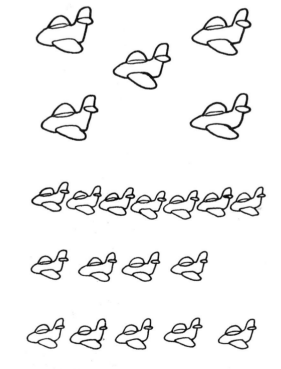

10. Ring the group that has more than this.

Name _____

Cumulative Test

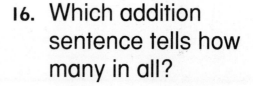

CHAPTERS 1–3
PAGE 3

11. What number comes next?

    3, 4, 5, _____

12. What number comes next?

    7, 8, 9, _____

13. What number comes next?

    0, 1, 2, _____

14. What number comes next?

    4, 5, 6, _____

15. Which addition sentence tells how many in all?

$2 + 1 = 3$

$3 + 2 = 5$

$4 + 2 = 6$

$3 + 1 = 4$

16. Which addition sentence tells how many in all?

$2 + 2 = 4$

$2 + 4 = 6$

$4 + 4 = 8$

$2 + 3 = 5$

GO ON

Name _____

Cumulative Test

CHAPTERS 1–3
PAGE 4

17. Ring the addition sentence that tells how many in all.

2 + 2 = 4
1 + 1 = 2
3 + 0 = 3
4 + 4 = 8

18. Ring the addition sentence that tells how many in all.

1 + 2 = 3
1 + 1 = 2
2 + 3 = 5
2 + 2 = 4

19. Ring the addition sentence that tells how many in all.

2 + 3 = 5
1 + 5 = 6
1 + 0 = 1
2 + 5 = 7

20. Ring the addition sentence that tells how many in all.

2 + 2 = 4
4 + 1 = 5
1 + 1 = 2
3 + 3 = 6

GO ON

Name _____

Cumulative Test

CHAPTERS 1–3
PAGE 5

21.

5 + 1 = ____

22.

2 + 0 = ____

23.

2 + 2 = ____

24.

1 + 4 = ____

25.

```
  0
+ 1
___
```

26.

```
  3
+ 3
___
```

27.

```
  2
+ 3
___
```

28. Ring the subtraction sentence that tells the story.

4 − 2 = 2

6 − 2 = 4

6 − 1 = 5

4 − 3 = 1

GO ON

Name _____

Cumulative Test

CHAPTERS 1–3
PAGE 6

29. Ring the subtraction sentence that tells the story.

5 – 2 = 3
5 – 1 = 4
3 – 2 = 1
3 – 3 = 0

30. Ring the subtraction sentence that tells the story.

3 – 1 = 2
4 – 3 = 1
4 – 2 = 2
3 – 3 = 0

31. Ring the subtraction sentence that tells the story.

2 – 1 = 1
1 – 1 = 0
4 – 1 = 3
3 – 1 = 2

32.

4 – 1 = _____

33.

6 – 3 = _____

34.

5 – 0 = _____

Name _____

Cumulative Test

CHAPTERS 1–3
PAGE 7

35.

4 − 4 = _____

36.

$$\begin{array}{r}5\\-3\\\hline\end{array}$$

37.

6 − 4 = _____

38.

2 − 1 = _____

39.

4 − 0 = _____

40.

6 − 5 = _____

41.

3 − 3 = _____

42.

4 − 1 = _____

43. Jack sees 3 🐟. Then he sees 2 more 🐟. Ring the correct question.

How many 🐟 are there in all?

How many 🐟 are left?

44. Jane sees 5 🐦. Then 4 🐦 fly away. Ring the correct question.

How many 🐦 are there in all?

How many 🐦 are left?

GO ON

Name _____

Cumulative Test

CHAPTERS 1–3
PAGE 8

Use the pictograph below to answer questions 45–50.

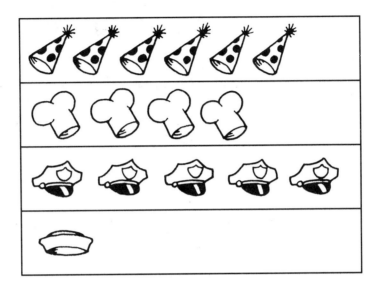

45. Ring the one for the group that has fewer.

46. Ring the one for the group that has more.

47. Ring the one for the group that has more than 5.

48. Ring the one for the group that has fewer than 4.

49. Ring the one that has more than 5.

50. Ring the one that has fewer, or .

Name _____

Pretest

CHAPTER **4**
PAGE 1

Read each question. Find the answer.

1.

   5 + 3 = _____

2.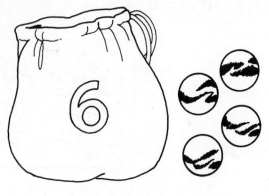

   6 + 4 = _____

3.

   8 + 1 = _____

4.

   5 + 2 = _____

5.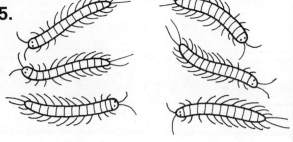

   3 + 3 = _____

6.

   5 + 5 = _____

7. 0 + 2 = _____

8. 5 + 1 = _____

Name _____

Pretest

CHAPTER **4**
PAGE 2

9. 4 + 4 = _____

10. 2 + 7 = _____

11. 4 + 3 = _____

12. 3 + 2 = _____

13. 8 + 2 = _____

14. 2 + 2 = _____

15. Ring another name for 9.

16. Ring another name for 10.

1 + 6   4 + 3
5 + 2   9 + 0

4 + 4   6 + 3
7 + 0   8 + 2

17. Ring the addition sentence that tells how many in all.

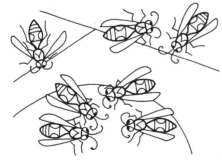

5 + 1 + 1 = 7
4 + 2 + 1 = 7
4 + 1 = 5
2 + 1 = 3

18. Ring the addition sentence that tells how many in all.

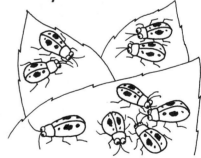

5 + 2 = 7
2 + 2 = 4
5 + 2 + 2 = 9
4 + 2 + 2 = 8

Pretest

CHAPTER 4
PAGE 3

19. $3 + 0 + 3 =$ _____

20. $4 + 1 + 3 =$ _____

21. How much did Kara spend?

$3¢ + 7¢ =$ _____

22. How much did Karl spend?

$4¢ + 2¢ =$ _____

23. Lin sees 5 birds.
Then she sees 3 more birds.
Then she sees 2 cats.
How many birds does she see in all?

Ring the sentence that is <u>not</u> needed.

    Lin sees 5 birds.
    Then she sees 3 more birds.
    Then she sees 2 cats.

24. Tom feeds 4 dogs.
Tom feeds 3 cats.
Then he feeds 2 more cats.
How many cats does he feed in all?

Ring the sentence that is <u>not</u> needed.

    Tom feeds 4 dogs.
    Tom feeds 3 cats.
    Then he feeds 2 more cats.

Name _____

Posttest

CHAPTER 4
PAGE 1

Read each question. Find the answer.

1.

   7 + 1 = _____

2.

   8 + 2 = _____

3.

   6 + 3 = _____

4.

   5 + 2 = _____

5.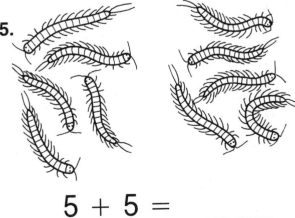

   5 + 5 = _____

6.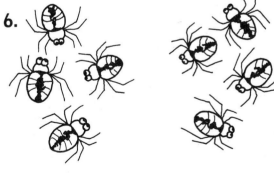

   4 + 4 = _____

7. 0 + 3 = _____

8. 1 + 4 = _____

Name _____

Posttest

CHAPTER 4
PAGE 2

9. 6 + 2 = _____

10. 5 + 3 = _____

11. 6 + 4 = _____

12. 7 + 0 = _____

13. 4 + 5 = _____

14. 3 + 3 = _____

15. Ring another name for 8.

8 + 1   3 + 3
6 + 3   4 + 4

16. Ring another name for 7.

1 + 8   3 + 2
4 + 1   5 + 2

17. Ring the addition sentence that tells how many in all.

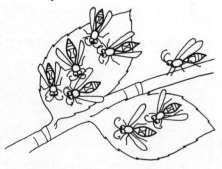

5 + 1 = 6
5 + 1 + 2 = 8
4 + 2 + 1 = 7
1 + 2 = 3

18. Ring the addition sentence that tells how many in all.

4 + 2 + 2 = 8
2 + 2 + 2 = 6
2 + 0 = 2
2 + 2 = 4

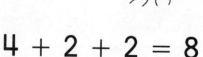

Name _____
Posttest

CHAPTER 4
PAGE 3

19. 4 + 0 + 4 = _____   20. 3 + 2 + 5 = _____

21. How much did Kathy spend?

8¢ + 2¢ = _____

22. How much did Carlos spend?

5¢ + 4¢ = _____

23. Nan sees 4 cows.
Then she sees 3 more cows.
Then she sees 2 dogs.
How many cows does she see in all?

Ring the sentence that is <u>not</u> needed.

  Nan sees 4 cows.
  Then she sees 3 more cows.
  Then she sees 2 dogs.

24. Sam feeds 6 fish.
Sam feeds 4 birds.
Then he feeds 2 more birds.
How many birds does he feed in all?

Ring the sentence that is <u>not</u> needed.

  Sam feeds 6 fish.
  Sam feeds 4 birds.
  Then he feeds 2 more birds.

Pretest

CHAPTER 5
PAGE 1

Read each question. Find the answer.

1. Subtract.
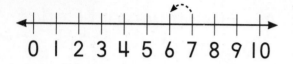
$7 - 1 =$ _____

2. Subtract.

$8 - 3 =$ _____

3. Subtract.
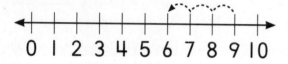
$9 - 3 =$ _____

4. Subtract.
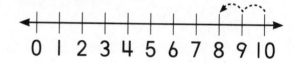
$10 - 2 =$ _____

5. Subtract.
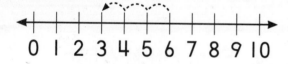
$6 - 3 =$ _____

6. Subtract.
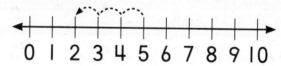
$5 - 3 =$ _____

7. 7
   −2

8. 10
   −7

9. 8
   −4

10. 6
    −6

11. $9 - 2 =$ _____

12. $8 - 6 =$ _____

13. $7 - 0 =$ _____

14. $9 - 8 =$ _____

GO ON

Name _____

Pretest

CHAPTER 5
PAGE 2

15. Use counters to solve.
Dale has 6 .
Then 4 run away.
Ring how many are left.

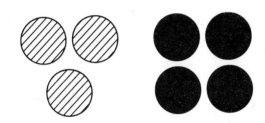

16. Use counters to solve.
There are 8 cars.
Then 1 leaves.
Ring how many are left.

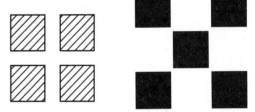

17.

$3 + 4 = 7$

$7 - 4 = $ _____

18.

$4 + 5 = 9$

$9 - 5 = $ _____

19. Ring the subtraction fact that goes with this addition fact.

$2 + 7 = 9$

$7 - 5 = 2$ $9 - 5 = 4$

$7 - 2 = 5$ $9 - 7 = 2$

20. Ring the subtraction fact that goes with this addition fact.

$6 + 4 = 10$

$6 - 4 = 2$ $10 - 6 = 4$

$4 - 2 = 2$ $10 - 2 = 8$

GO ON

Name _____

Pretest

CHAPTER 5
PAGE 3

21. Ring the correct number pair.

    The sum is 6.
    The difference is 4.

    4 and 2   5 and 1

    3 and 3   6 and 2

22. Ring the correct number pair.

    The sum is 7.
    The difference is 7.

    4 and 3   7 and 7

    7 and 0   6 and 1

23.

There are 5 .

There is 1 .

Ring how many more than there are.

3 more   5 more

4 more   6 more

24.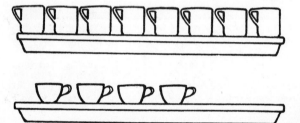

There are 8 🍵 .

There are 4 ☕ .

Ring how many more 🍵 than ☕ there are.

4 more   8 more

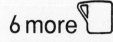
6 more   10 more

Name _____

Posttest

CHAPTER 5
PAGE 1

Read each question. Find the answer.

1. Subtract.

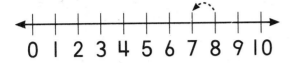

8 − 1 = _____

2. Subtract.

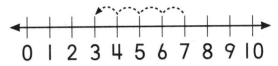

7 − 4 = _____

3. Subtract.

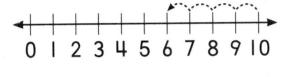

10 − 4 = _____

4. Subtract.

6 − 2 = _____

5. Subtract.

9 − 1 = _____

6. Subtract.

8 − 3 = _____

7. 9
   −2
   ___

8. 10
   − 8
   ___

9. 7
   −5
   ___

10. 5
    −5
    ___

11. 10 − 3 = _____

12. 8 − 5 = _____

13. 6 − 0 = _____

14. 9 − 6 = _____

GO ON

Name _____

Posttest

CHAPTER 5
PAGE 2

15. Use counters to solve.
There are 5 🚗.
Then 1 leaves.
Ring how many are left.

   1 🚗   5 🚗

   4 🚗   6 🚗

16. Use counters to solve.
There are 9 🐦.
Then 4 fly away.
Ring how many are left.

   4 🐦   5 🐦

   6 🐦   7 🐦

17.

$4 + 3 = 7$

$7 - 3 = $ _____

18.

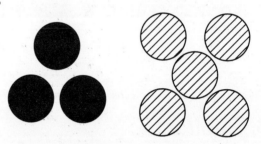

$3 + 5 = 8$

$8 - 5 = $ _____

19. Ring the subtraction fact that goes with this addition fact.

$6 + 1 = 7$

$6 - 1 = 5$ $8 - 1 = 7$

$1 - 1 = 0$ $7 - 1 = 6$

20. Ring the subtraction fact that goes with this addition fact.

$5 + 5 = 10$

$5 - 0 = 5$ $10 - 5 = 5$

$5 - 5 = 0$ $10 - 0 = 10$

GO ON

Name _____

Posttest

CHAPTER 5
PAGE 3

21. Ring the correct number pair.

   The sum is 5.
   The difference is 3.

   3 and 3   3 and 2

   5 and 0   4 and 1

22. Ring the correct number pair.

   The sum is 9.
   The difference is 9.

   8 and 1   3 and 3

   6 and 3   9 and 0

23.

There are 6 🚚.

There are 3 🚗.

Ring how many more 🚚 than 🚗 there are.

3 more 🚚   5 more 🚚

4 more 🚚   8 more 🚚

24.

There are 4 🐝.

There are 2 🌹.

Ring how many more 🐝 than 🌹 there are.

1 more   3 more

2 more 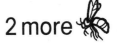   6 more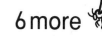

Name _____

Pretest

CHAPTER 6
PAGE 1

Read each question. Find the answer.

1. Ring the one that is the same shape.

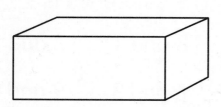

2. Ring the one that is the same shape.

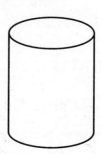

3. Ring the plane shape that matches the face of the solid shape.

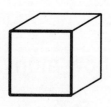

4. Ring the plane shape that matches the face of the solid shape.

Name _____

Pretest

CHAPTER 6
PAGE 2

5. Ring the picture that shows the star <u>inside</u> the square.

6. Ring the picture that shows the circle <u>on</u> the triangle.

7. Ring the open figure.

 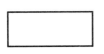

8. Ring the closed figure.

9. How many corners are there?

_____

10. How many corners are there?

_____

Name _____

Pretest

CHAPTER 6
PAGE 3

11. How many sides are there?

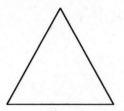

_____

12. How many sides are there?

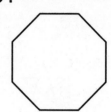

_____

13. Ring the figure that is the same size and shape as this one.

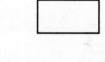

14. Ring the figure that is the same size and shape as this one.

15. Ring the picture that shows two parts that match.

16. Ring the picture that shows two parts that match.

GO ON

Name _____

Pretest

CHAPTER 6
PAGE 4

17. Ring the pattern.

18. Ring the pattern.

19. Ring the shape that comes next in the pattern.

20. Ring the shape that comes next in the pattern.

Name _____

Pretest

CHAPTER 6
PAGE 5

21. Ring the missing shape.

□ ○ □ ○ ? ○

□   ○

22. Ring the missing shape.

△ △ ○ △ ? ○

○   △

Use the graph to answer questions 23–24.

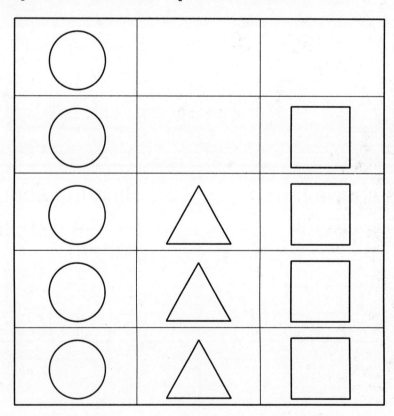

23. How many ○ are there?

_____

24. How many □ are there?

_____

Name _____

Posttest

CHAPTER 6
PAGE 1

Read each question. Find the answer.

1. Ring the one that is the same shape.

2. Ring the one that is the same shape.

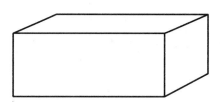

3. Ring the plane shape that matches the face of the solid shape.

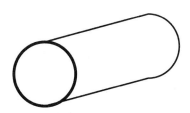

4. Ring the plane shape that matches the face of the solid shape.

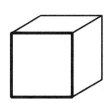

Posttest

CHAPTER 6 — PAGE 2

5. Ring the picture that shows the star <u>inside</u> the circle.

6. Ring the picture that shows the square <u>outside</u> the circle.

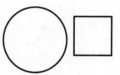

7. Ring the open figure.

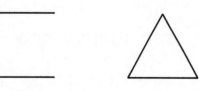

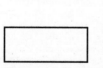

8. Ring the closed figure.

9. How many corners are there?

_____

10. How many corners are there?

_____

Name _____

Posttest

CHAPTER 6
PAGE 3

11. How many sides are there?

_____

12. How many sides are there?

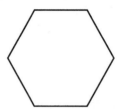

_____

13. Ring the figure that is the same size and shape as this one.

14. Ring the figure that is the same size and shape as this one.

15. Ring the picture that shows two parts that match.

16. Ring the picture that shows two parts that match.

Name _____

Posttest

CHAPTER 6
PAGE 4

17. Ring the pattern.

18. Ring the pattern.

19. Ring the shape that comes next in the pattern.

20. Ring the shape that comes next in the pattern.

GO ON

Name _____

Posttest

CHAPTER 6
PAGE 5

21. Ring the missing shape.

22. Ring the missing shape.

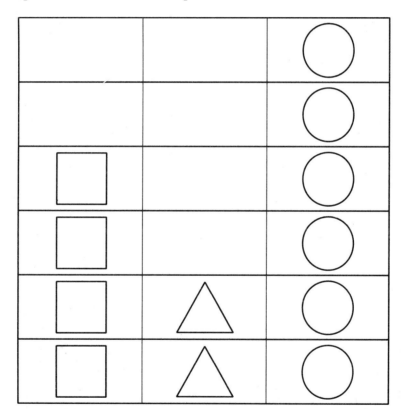

Use the graph to answer questions 23–24.

23. How many △ are there?

_____

24. How many ☐ are there?

_____

Name _____

Cumulative Test

CHAPTERS 1-6
PAGE 1

Read each question. Find the answer.

1. How many  are there?

   ☐ ☐ ☐
   ☐ ☐ ☐

   _____

2. How many ☆ are there?

   ☆ ☆ ☆ ☆
   ☆ ☆ ☆

   _____

3. How many are there?

   ○ ○
   ○ ○

   _____

4. How many △ are there?

   △ △ △ △ △
   △ △ △ △

   _____

5. Ring the group that has fewer than this.

6. Ring the group that has more than this.

GO ON

Name _____

Cumulative Test

 **CHAPTERS 1-6**
PAGE 2

7. Write the number that comes next.

2, 3, 4, _____

8. Write the number that comes next.

6, 7, 8, _____

9. Ring the addition sentence that tells how many in all.

2 + 2 = 4
2 + 0 = 2
4 + 4 = 8
4 + 2 = 6

10. Ring the subtraction sentence that tells the story.

3 − 1 = 2
2 − 1 = 1
3 − 3 = 0
4 − 1 = 3

11.

7 + 2 = _____

12.

6 − 4 = _____

13.

1 + 4 = _____

14.

0 + 3 = _____

Name _____

Cumulative Test

CHAPTERS 1–6
PAGE 3

15.

$6 + 2 =$ _____

16.

$5 + 5 =$ _____

17.

8
$+1$
___

18.

3
$+4$
___

19. Ring another name for 10.

$3 + 4 \qquad 9 + 1$

$6 + 1 \qquad 9 + 0$

20. Ring another name for 8.

$3 + 3 \qquad 6 + 1$

$4 + 5 \qquad 7 + 1$

21.

4
-2
___

22.

10
-6
___

23.

$7 - 1 =$ _____

24.

$2 - 0 =$ _____

GO ON

Name _____

Cumulative Test

CHAPTERS 1–6
PAGE 4

25.

8 – 3 = _____

26.

6 – 5 = _____

27. 5 dogs are playing.
I dog runs away.
How many are left?

_____ dogs

28. There are 9 buses.
Then 2 buses leave.
How many are left?

_____ buses

29. Ring the one that is the same shape.

30. Ring the one that is the same shape.

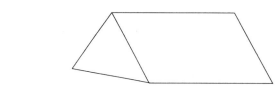

Name _____

Cumulative Test

CHAPTERS 1–6
PAGE 5

31. Ring the plane shape that matches the face of the solid shape.

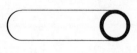

32. Ring the plane shape that matches the face of the solid shape.

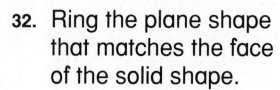

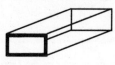

33. Ring the plane shape that matches the face of the solid shape.

34. Ring the one that is the same size and shape.

35. Ring the one that is the same size and shape.

36. Ring the picture that shows two parts that match.

GO ON

Name _____

Cumulative Test

CHAPTERS 1–6
PAGE 6

37. Ring the picture that shows two parts that match.

 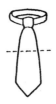

38. Ring the pattern.

△ △ ○ △ △ ○

△ △

△ ○

△ ○

○ △ ○

39. Ring the pattern.

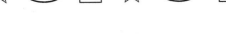

40. Ring the shape that comes next in the pattern.

 ?

△

☆

41. Ring the shape that comes next in the pattern.

 ?

42. Ring the shape that comes next in the pattern.

 ?

Name _____

Cumulative Test

CHAPTERS 1–6
PAGE 7

43. Ring the missing shape.

44. Ring the number pair.
The sum is 5.
The difference is 3.

2 and 3   3 and 6
1 and 4   3 and 5

45. Ring what you do to find how many 🐟 are left.

Add   Subtract

46. Ring what you do to find how many 🐕 in all.

Add   Subtract

47. How much did Rusty spend?

_____

48.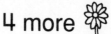

There are 4 🌷.

There are 5 🌼.

Ring how many more 🌼 than 🌷 there are.

1 more 🌼

4 more 🌼

5 more 🌼

9 more 🌼

Name _____

Cumulative Test

CHAPTERS 1-6
PAGE 8

Use the pictograph below to answer questions 49–50.

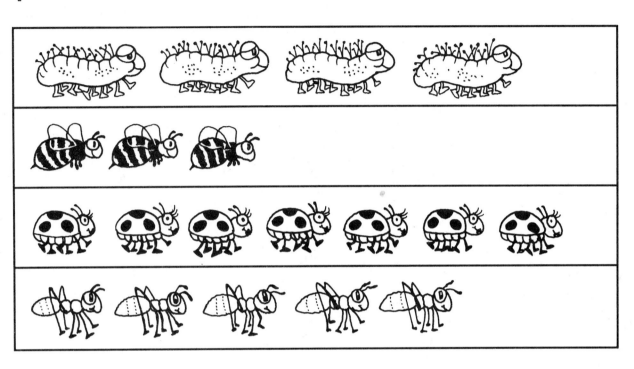

49. Ring the one for the group that has fewer.

50. Ring the one for the group that has more than 6.

Name _____

Pretest

CHAPTER **7**
PAGE 1

Read each question. Find the answer.

1. Ring how many tens and ones there are.

   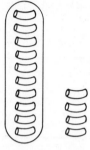

   1 ten 4 ones

   4 tens 1 one

   4 tens 10 ones

   10 tens 1 one

2. Ring how many tens and ones there are.

   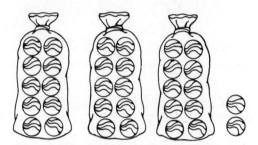

   30 tens 2 ones

   1 ten 3 ones

   3 tens 2 ones

   2 tens 3 ones

3. Ring how many tens and ones there are.

   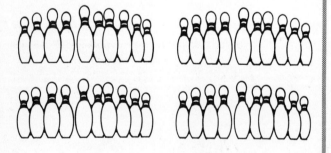

   4 tens 0 ones

   0 tens 4 ones

   1 ten 4 ones

   10 tens 1 one

4. Ring how many tens and ones there are.

   3 tens 0 ones

   20 tens 1 one

   1 ten 2 ones

   2 tens 1 one

GO ON

Name _____

Pretest

CHAPTER 7
PAGE 2

5. Ring how many there are in all.

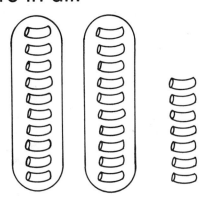

9   17   27   72

6. Ring how many there are in all.

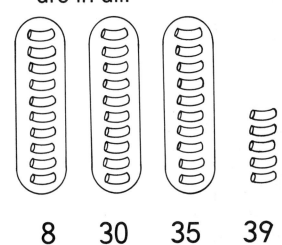

8   30   35   39

7. Ring the greater number.

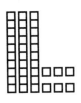   

36   55

8. Ring the number that is less.

29   30

9. Write the number that comes before.

_____ , 47

10. Write the number that comes after.

70, _____

11. Write the number that comes between.

86, _____ , 88

12. Write the number that comes between.

69, _____ , 71

GO ON

Pretest

CHAPTER 7
PAGE 3

13. Ring the one that is third.

first

14. Ring the one that is fifth.

first

15. Ring the one that is second.

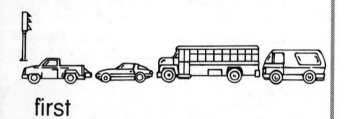
first

16. Ring the one that is sixth.

17. Count by twos. Write the number that is missing.

| 2 | 4 | 6 | ___ |

18. Count by tens. Write the number that is missing.

| 10 | 20 | 30 | ___ |

19. Count by fives. Write the number that is missing.

| 20 | 25 | 30 | ___ |

20. Count by twos. Write the number that is missing.

| 44 | 46 | 48 | ___ |

Name _____

Pretest

CHAPTER 7
PAGE 4

21. Ring the best estimate.

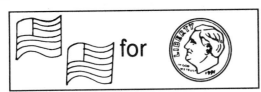

40   50   60   90

22. Ring the best estimate.

30   40   50   60

23. Ring how many 🏴 you can buy with 🪙🪙🪙.

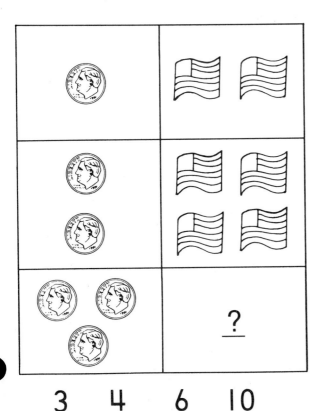

3   4   6   10

24. Ring how many ☺ you can buy with 🪙🪙🪙.

11   12   15   20

Name _____

Posttest

CHAPTER 7
PAGE 1

Read each question. Find the answer.

1. Ring how many tens and ones there are.

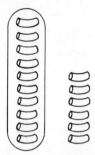

1 ten 6 ones

6 tens 1 one

6 tens 10 ones

10 tens 1 one

2. Ring how many tens and ones there are.

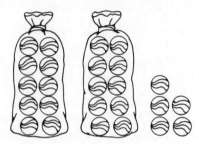

20 tens 5 ones

50 tens 2 ones

5 tens 2 ones

2 tens 5 ones

3. Ring how many tens and ones there are.

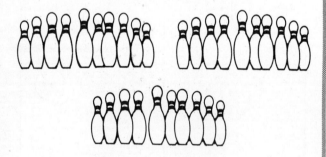

3 tens 0 ones

0 tens 3 ones

1 ten 3 ones

30 tens 0 ones

4. Ring how many tens and ones there are.

3 tens 0 ones

10 tens 1 one

1 ten 4 ones

3 tens 4 ones

GO ON

Name _____

Posttest

CHAPTER 7
PAGE 2

5. Ring how many there are in all.

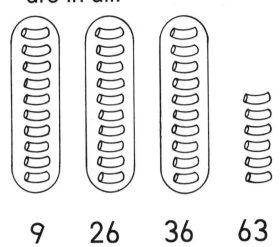

9   26   36   63

6. Ring how many there are in all.

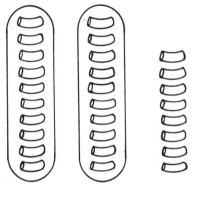

10   20   18   28

7. Ring the greater number.

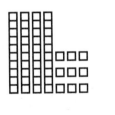

49   50

8. Ring the number that is less.

   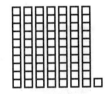

38   71

9. Write the number that comes before.

_____ , 86

10. Write the number that comes after.

50, _____

11. Write the number that comes between.

44, _____ , 46

12. Write the number that comes between.

79, _____ , 81

Name _____

Posttest

CHAPTER 7
PAGE 3

13. Ring the one that is fourth.

first

14. Ring the one that is sixth.

first

15. Ring the one that is second.

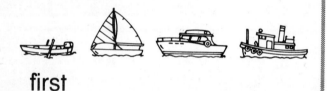

first

16. Ring the one that is fifth.

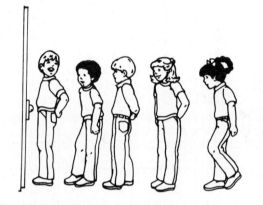

17. Count by fives. Write the number that is missing.

| 5 | 10 | 15 | ___ |

18. Count by tens. Write the number that is missing.

| 10 | 20 | 30 | ___ |

19. Count by twos. Write the number that is missing.

| 54 | 56 | 58 | ___ |

20. Count by fives. Write the number that is missing.

| 15 | 20 | 25 | ___ |

Name _____

Posttest

CHAPTER 7
PAGE 4

21. Ring the best estimate.

3   10   30   40

22. Ring the best estimate.

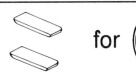

10   50   60   70

23. for

Ring how many  you can buy with .

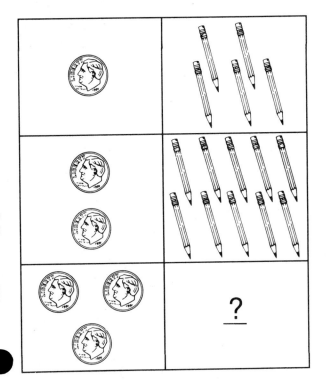

6   7   15   20

24. for

Ring how many you can buy with .

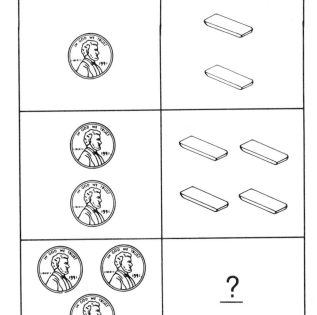

5   6   8   10

STOP

Name _____

Pretest

CHAPTER 8
PAGE 1

Read each question. Find the answer.

1. Ring the addition sentence that tells how many in all.

$3 + 5 = 8$

$7 + 2 = 9$

$1 + 3 = 4$

$8 + 3 = 11$

2. Ring the addition sentence that tells how many in all.

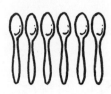 

$3 + 3 = 6$

$7 + 7 = 14$

$6 + 6 = 12$

$5 + 5 = 10$

3. Ring the addition sentence that tells how many in all.

$4 + 1 = 5$

$4 + 5 = 9$

$5 + 5 = 10$

$3 + 4 = 7$

4. Ring the addition sentence that tells how many in all.

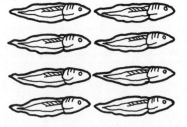

$8 + 1 = 9$

$7 + 1 = 8$

$9 + 1 = 10$

$9 + 9 = 18$

Name _____

Pretest

CHAPTER 8
PAGE 2

5. $2 + 9 =$ _____

6. $0 + 6 =$ _____

7. $7 + 3 =$ _____

8. $8 + 4 =$ _____

9. Sam catches 5 fish. Dad catches 7 fish. How many fish do they catch in all?

   _____ fish

10. Meg has 5 pet birds. She has 5 pet fish. How many pets does she have in all?

    _____ pets

11. Ring the subtraction sentence that tells the story.

$11 - 3 = 8$ $5 - 3 = 2$
$8 - 3 = 5$ $7 - 3 = 4$

12. Ring the subtraction sentence that tells the story.

$9 - 3 = 6$ $12 - 9 = 3$
$6 - 3 = 3$ $11 - 8 = 3$

Name _____

Pretest

CHAPTER 8
PAGE 3

13. Ring the subtraction sentence that tells the story.

$8 - 2 = 6$

$6 - 2 = 4$

$12 - 2 = 10$

$10 - 2 = 8$

14. Ring the subtraction sentence that tells the story.

$6 - 3 = 3$

$6 - 6 = 0$

$9 - 3 = 6$

$12 - 6 = 6$

15. $11 - 6 =$ _____

16. $9 - 0 =$ _____

17. $10 - 1 =$ _____

18. $12 - 8 =$ _____

19. Mom buys 12 🍎. Pam eats 3 🍎. How many 🍎 are left?

20. There were 11 🥤. 5 were broken. How many 🥤 are left?

GO ON

Name _____

Pretest

CHAPTER **8**
PAGE 4

21. 7¢
   + 4¢

22. 9¢
   − 2¢

23. 8¢
   − 7¢

24. 6¢
   + 6¢

25. Ring the number sentence that does <u>not</u> belong to the fact family.

$$7 + 4 = 11$$
$$11 - 7 = 4$$
$$3 + 4 = 7$$
$$4 + 7 = 11$$

26. Ring the number sentence that does <u>not</u> belong to the fact family.

$$9 - 1 = 8$$
$$1 + 8 = 9$$
$$9 - 8 = 1$$
$$7 + 1 = 8$$

27. Ring the number sentence that belongs to the same fact family as $10 - 6 = 4$.

$$6 + 4 = 10$$
$$10 + 6 = 16$$
$$6 - 4 = 2$$
$$10 - 5 = 5$$

28. Ring the number sentence that belongs to the same fact family as $12 - 7 = 5$.

$$2 + 5 = 7$$
$$7 - 5 = 2$$
$$6 + 6 = 12$$
$$5 + 7 = 12$$

GO ON

Name _____

Pretest

CHAPTER 8
PAGE 5

Use the graph below to answer questions 29–30.

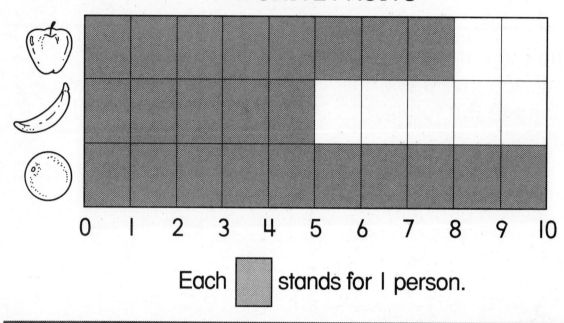

29. How many people like  best?

    _____ people

30. Ring the fruit that is the favorite of the most people.

31. There are 8 black puppies. There are 4 white puppies. What do you do to find how many puppies there are in all?

    Add   Subtract

32. Roy had 10 pennies. He lost 3 pennies. What do you do to find how many pennies are left?

    Add   Subtract

Name _____

Posttest

CHAPTER 8
PAGE 1

Read each question. Find the answer.

1. Ring the addition sentence that tells how many in all.

   $8 + 3 = 11$
   $6 + 3 = 9$
   $7 + 3 = 10$
   $9 + 3 = 12$

2. Ring the addition sentence that tells how many in all.

   $1 + 10 = 11$
   $1 + 9 = 10$
   $0 + 9 = 9$
   $8 + 1 = 9$

3. Ring the addition sentence that tells how many in all.

   $4 + 2 = 6$
   $6 + 4 = 10$
   $6 + 5 = 11$
   $4 + 4 = 8$

4. Ring the addition sentence that tells how many in all.

   $2 + 5 = 7$
   $7 + 1 = 8$
   $9 + 1 = 10$
   $7 + 2 = 9$

GO ON

Name _____

Posttest

CHAPTER 8
PAGE 2

5. $8 + 3 =$ _____

6. $9 + 0 =$ _____

7. $5 + 5 =$ _____

8. $2 + 8 =$ _____

9. 2 ants are on a hill. 9 more ants come up the hill. How many ants in all?

   _____ ants

10. There are 4 red balls. There are 8 green balls. How many balls in all?

    _____ balls

11. Ring the subtraction sentence that tells the story.

   $11 - 4 = 7$ $7 - 5 = 2$
   $7 - 4 = 3$ $11 - 3 = 8$

12. Ring the subtraction sentence that tells the story.

   $5 - 4 = 1$ $10 - 4 = 6$
   $8 - 4 = 4$ $9 - 4 = 5$

GO ON

Posttest

CHAPTER 8
PAGE 3

13. Ring the subtraction sentence that tells the story.

$6 - 6 = 0$
$10 - 6 = 4$
$12 - 6 = 6$
$12 - 5 = 7$

14. Ring the subtraction sentence that tells the story.

$6 - 5 = 1$
$11 - 6 = 5$
$10 - 6 = 4$
$12 - 7 = 5$

15. $11 - 3 =$ _____

16. $10 - 8 =$ _____

17. $9 - 3 =$ _____

18. $12 - 7 =$ _____

19. Mom buys 12 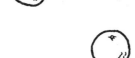. The children eat 9 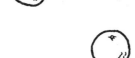. How many 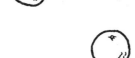 are left?

_____

20. There were 9 . 8 were broken. How many are left?

_____

GO ON

Posttest

CHAPTER 8 — PAGE 4

21. 5¢
   + 6¢

22. 12¢
    − 4¢

23. 11¢
    − 9¢

24.   1¢
   + 8¢

25. Ring the number sentence that does **not** belong to the fact family.

   3 + 6 = 9
   9 − 3 = 6
   9 + 3 = 12
   9 − 6 = 3

26. Ring the number sentence that does **not** belong to the fact family.

   3 + 7 = 10
   3 + 4 = 7
   10 − 3 = 7
   10 − 7 = 3

27. Ring the number sentence that belongs to the same fact family as 7 + 4 = 11.

   11 − 7 = 4
   7 + 0 = 7
   3 + 4 = 7
   7 − 4 = 3

28. Ring the number sentence that belongs to the same fact family as 12 − 5 = 7.

   2 + 5 = 7
   7 − 5 = 2
   7 + 5 = 12
   6 + 6 = 12

GO ON

Name _____

Posttest

CHAPTER **8**
PAGE 5

Use the graph below to answer questions 29–30.

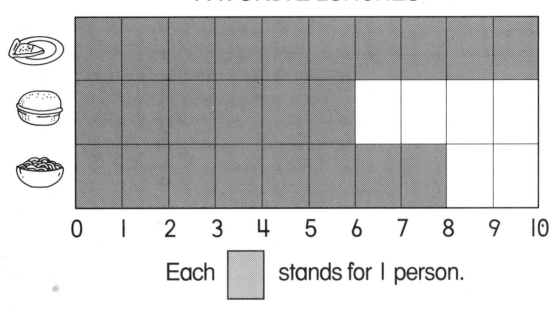

---

29. How many people like  best?

_____ people

30. Ring the favorite lunch of the most people.

---

31. 12 cats were sleeping. 6 cats woke up. What do you do to find how many cats are still sleeping?

Add   Subtract

32. 8 chickens are sitting on nests. 8 chickens are eating. What do you do to find how many chickens there are in all?

Add   Subtract

Name _____

Pretest

CHAPTER **9**
PAGE 1

Read each question. Find the answer.

1. Ring the one that is shortest.

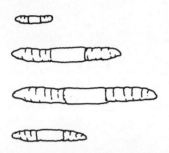

2. Ring the one that is longest.

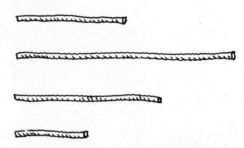

3. How long is the fork?

_____

4. How long is the knife?

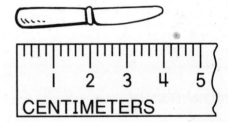

_____ centimeters

5. How long is the pin?

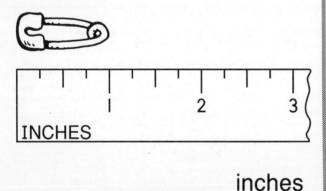

_____ inches

6. How long is the whistle?

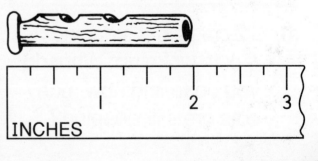

_____ inches

Name _____

Pretest

CHAPTER 9
PAGE 2

7. How long is the ribbon?

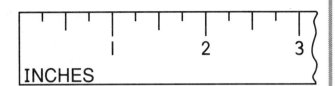

_____ inches

8. How long is the comb?

_____ centimeters

9. How long is the paper clip?

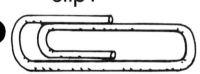

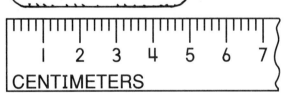

_____ centimeters

10. How long is the pin?

_____ centimeters

11. Ring the one that is heavier.

12. Ring the one that is heavier.

Name _____

Pretest

CHAPTER 9
PAGE 3

13. Ring the one that holds more.

14. Ring the one that is hot.

15. Ring the one that holds <u>about</u> the same as a .

16. Ring the one that holds <u>about</u> the same as a .

17. Ring the one that is <u>about</u> as heavy as a .

18. Ring the one that is <u>about</u> as heavy as a .

19. Ring the one that shows fair shares.

20. Ring the one that shows fair shares.

Pretest

CHAPTER 9
PAGE 4

21. Ring the one that shows one half.

22. Ring the one that shows one third.

23. Ring the one that shows one fourth.

24. Ring the fraction that matches this picture.

$\frac{1}{5}$ $\frac{1}{4}$ $\frac{1}{3}$ $\frac{1}{2}$

25. Ring the fraction that shows the shaded part.

$\frac{1}{5}$ $\frac{1}{4}$ $\frac{1}{3}$ $\frac{1}{2}$

26. Ring the fraction that shows the shaded part.

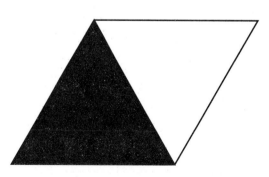

$\frac{1}{5}$ $\frac{1}{4}$ $\frac{1}{3}$ $\frac{1}{2}$

GO ON

Name _____

Pretest

CHAPTER 9
PAGE 5

27. What fraction of the group is small? Ring the fraction.

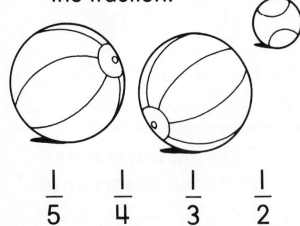

$\frac{1}{5}$ $\frac{1}{4}$ $\frac{1}{3}$ $\frac{1}{2}$

28. What fraction of the group are girls? Ring the fraction.

$\frac{1}{5}$ $\frac{1}{4}$ $\frac{1}{3}$ $\frac{2}{4}$

Use the table below to answer questions 29–32.

| CHILDREN'S MEASUREMENTS (in inches) ||||
Child	Foot	Hand	Arm
Pat	6	5	12
Chris	5	4	10
Bobby	8	6	14

29. How long is Pat's arm?

_____ inches

30. How long is Bobby's hand?

_____ inches

31. Whose arm is shortest?

_____

32. Whose foot is longest?

_____

Name _____

Posttest

CHAPTER **9**
PAGE 1

Read each question. Find the answer.

1. Ring the one that is longest.

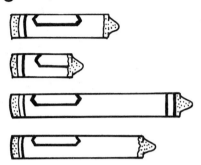

2. Ring the one that is shortest.

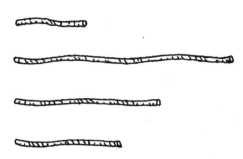

3. How long is the worm?

_____

4. How long is the frog?

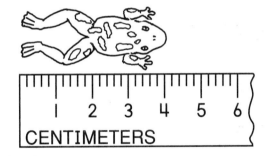

_____ centimeters

5. How long is the spoon?

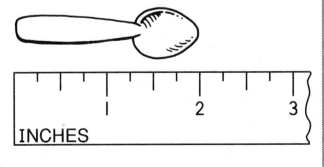

_____ inches

6. How long is the paper clip?

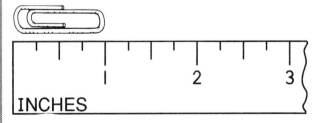

_____ inches

Name _____

Posttest

CHAPTER 9
PAGE 2

7. How long is the string?

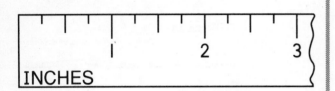

_____ inches

8. How long is the bean?

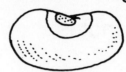

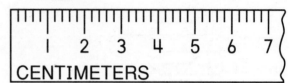

_____ centimeters

9. How long is the key?

_____ centimeters

10. How long is the eraser?

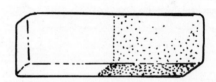

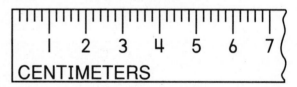

_____ centimeters

11. Ring the one that is heavier.

12. Ring the one that is heavier.

Name _____

Posttest

CHAPTER 9
PAGE 3

13. Ring the one that holds more.

14. Ring the one that is hot.

15. Ring the one that holds about the same as a .

16. Ring the one that holds about the same as a ☕.

17. Ring the one that is about as heavy as an 🍎.

18. Ring the one that is about as heavy as an ▭.

19. Ring the one that shows fair shares.

20. Ring the one that shows fair shares.

Name _____

Posttest

CHAPTER 9
PAGE 4

21. Ring the one that shows one half.

22. Ring the one that shows one fourth.

23. Ring the one that shows one third.

24. Ring the fraction that matches this picture.

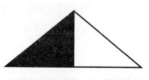

$\frac{1}{5}$ $\frac{1}{4}$ $\frac{1}{3}$ $\frac{1}{2}$

25. Ring the fraction that matches this picture.

$\frac{1}{5}$ $\frac{1}{4}$ $\frac{1}{3}$ $\frac{1}{2}$

26. Ring the fraction that matches this picture.

$\frac{1}{5}$ $\frac{1}{4}$ $\frac{1}{3}$ $\frac{1}{2}$

GO ON

Name _____

Posttest

CHAPTER 9
PAGE 5

27. What fraction of the group is large? Ring the fraction.

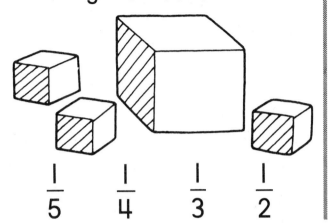

$\frac{1}{5}$ $\frac{1}{4}$ $\frac{1}{3}$ $\frac{1}{2}$

28. What fraction of the group are boys? Ring the fraction.

$\frac{1}{5}$ $\frac{1}{4}$ $\frac{1}{3}$ $\frac{1}{2}$

Use the table below to answer questions 29–32.

LENGTHS (in inches)			
Object	Red	Blue	Yellow
Crayon	2	5	3
Pencil	5	6	4
Chalk	1	6	3

29. Which crayon is shortest?

_____

30. Which chalk is longest?

_____

31. How long is the yellow pencil?

_____ inches

32. How long is the red chalk?

_____ inches

Name _____

Cumulative Test

CHAPTERS 1–9
PAGE 1

Read each question. Find the answer.

1.

    6 − 1 = _____

2.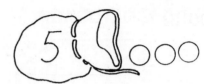

    5 + 3 = _____

3. Ring the plane shape that matches the face of the solid shape.

4. Ring the plane shape that matches the face of the solid shape.

5. Ring the figure that is the same size and shape.

6. Ring the picture that shows two parts that match.

Name _____

Cumulative Test

CHAPTERS 1–9
PAGE 2

7. Ring the shape that comes next in the pattern.

△ □ ○ △ □ ○ △ ?

△   □   ○

8. Ring the shape that comes next in the pattern.

○ ○ □ ○ ○ □ ○ ?

○   ○   □

9. Ring how many tens and ones there are.

2 tens and 7 ones

3 tens and 3 ones

2 tens and 4 ones

10 tens and 7 ones

10. Ring how many tens and ones there are.

10 tens and 2 ones

3 tens and 0 ones

2 tens and 1 one

1 ten and 2 ones

11. Ring how many there are in all.

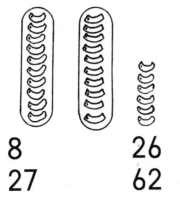

8   26
27   62

12. Ring how many there are in all.

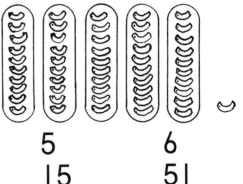

5   6
15   51

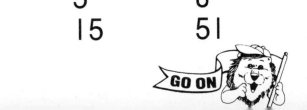
GO ON

Name _____

Cumulative Test

CHAPTERS 1-9
PAGE 3

13. Ring the number that is less.

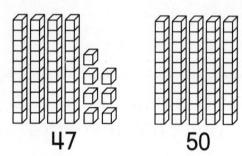

47   50

47   50

14. Ring the number that is greater.

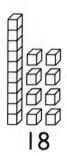 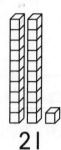

18   21

18   21

15. Write the number that comes after.

61, _____

16. Write the number that comes between.

49, _____, 51

17. Ring the one that is third.

first

18. Ring the one who is second.

GO ON

Name _____

Cumulative Test

CHAPTERS 1–9
PAGE 4

19.

4 + 5 = _____

20.

2 + 4 = _____

21.

9 + 3 = _____

22.

6 − 0 = _____

23. There are 3 red hats. There are 7 blue hats. How many hats are there in all?

_____ hats

24. Roy has 6 books. Nancy has 5 books. How many books do they have in all?

_____ books

25.

11 − 7 = _____

26. There were 12 eggs. 4 were broken. How many are left?

_____ eggs

27. There were 9 birds. Then 3 flew away. How many birds are left?

_____ birds

28.

8¢
− 3¢
———

GO ON

Name _____

Cumulative Test

CHAPTERS 1–9
PAGE 5

29.

    6¢
+ 3¢
―――

30.

    7¢
+ 1¢
―――

31. Ring the one that is longest.

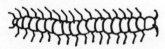

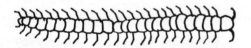

32. How long is the  ?

_____  long

33. Use your inch ruler. How long is the worm?

_____ inches

34. Use your centimeter ruler. How long is the stick?

_____ centimeters

Name _____

Cumulative Test

CHAPTERS 1–9
PAGE 6

35. Ring the one that is heavier, or .

36. Ring the one that is cold.

37. Ring the one that holds about the same as a .

38. Ring the one that is about as heavy as an .

39. Ring the one that shows fair shares.

40. Ring the one that shows one third.

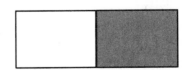

Name _____

Cumulative Test

CHAPTERS 1–9
PAGE 7

41. Ring the fraction that matches this picture.

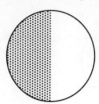

$\frac{1}{5}$ $\frac{1}{4}$

$\frac{1}{3}$ $\frac{1}{2}$

42. What fraction of the group is small? Ring the answer.

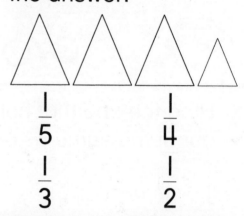

$\frac{1}{5}$ $\frac{1}{4}$

$\frac{1}{3}$ $\frac{1}{2}$

43. 2 for

Ring how many you can buy with .

2 ✏️

3 ✏️

5 ✏️

6 ✏️

44. Ring the number pair. The sum is 7. The difference is 5.

2 and 7
3 and 4
5 and 2
1 and 6

GO ON

Name _____

Cumulative Test

CHAPTERS 1–9
PAGE 8

45. Ring how much the girl spent.

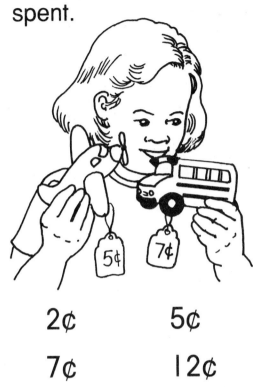

2¢   5¢

7¢   12¢

46. There are 3 🥫.
There are 5 ✂.
How many more ✂ are there than 🥫?

_____ more ✂

Use the table below to answer questions 47–48.

Money to Spend

Children	Dimes	Pennies
Ali	3	2
Barb	1	5
Connie	4	0
Dave	2	3

47. Who has the most dimes? Ring the name.

Ali   Barb

Connie   Dave

48. Ring what you do to find how much money Ali and Barb have in all.

Add   Subtract

Cumulative Test

CHAPTERS 1-9
PAGE 9

Use the graph below to answer question 49.

Favorite Colors

Red	👧	👧	👧	👧			
Blue	👧	👧	👧	👧	👧	👧	
Yellow	👧	👧	👧				
Green	👧	👧	👧	👧	👧		

Each 👧 stands for 1 person.

49. Ring the color that is the favorite of the most people.

    Blue   Yellow   Green   Red

Use the graph below to answer question 50.

Favorite Toys

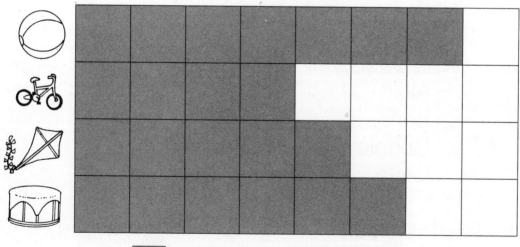

Each ■ stands for 1 person.

50. Ring the toy that is the favorite of the most people.

Name _____

Pretest

CHAPTER **10**
PAGE 1

Read each question. Find the answer.

1. Ring which happened last.

2. Ring which happened first.

3. Ring which takes more time.

4. Ring which takes less time.

Name _____

Pretest

CHAPTER 10
PAGE 2

5. Ring the one that shows 2 o'clock.

6. Ring the one that shows 7:30.

7. Ring the time it is.

3:00   6:00

9:00   12:00

8. Ring the time it is.

8:30   9:30

7:30   6:30

9. Ring the time it is.

11:00   12:00

1:00   7:00

10. Ring the time it is.

6:00   12:30

1:30   not here

Name _____

Pretest

CHAPTER 10
PAGE 3

Use the calendar below to answer questions 11–14.

April						
Sunday	Monday	Tuesday	Wednesday	Thursday	Friday	Saturday
			1	2	3	4
5	6	7	8	9	10	11

11. Ring the day on which this month begins.

   Friday   Monday

   Wednesday Thursday

12. Ring the day of the week that is the third day of this month.

   Tuesday   Saturday

   Friday   Monday

13. Ring the day that comes just after Wednesday.

   Tuesday   Friday

   Thursday   Saturday

14. Ring the day that comes just before Saturday.

   Friday   Monday

   Wednesday Sunday

Name _____
Pretest

CHAPTER 10
PAGE 4

15. Ring how much money there is.

5¢   6¢
30¢   60¢

16. Ring how much money there is.

3¢   5¢
15¢   30¢

17. Ring how much money there is.

3¢   20¢
25¢   30¢

18. Ring how much money there is.

25¢   27¢
35¢   46¢

Name _____

Pretest

CHAPTER 10
PAGE 5

19. Ring how much money there is.

5¢   14¢
24¢   29¢

20. Ring how much money there is.

17¢   28¢
32¢   37¢

Use coins to answer questions 21–24.

21. Rita has 1 nickel and 2 pennies. How much money does she have?

_____ ¢

22. Kit has 2 dimes and 2 nickels. How much money does he have?

_____ ¢

23. Lois has 5 dimes. How much money does she have?

_____ ¢

24. Ned has 1 quarter, 1 dime, and 3 pennies. How much money does he have?

_____ ¢

Name _____

Pretest

CHAPTER 10
PAGE 6

25. Ring the group of coins that is the same amount as .

26. Ring the group of coins that is the same amount as .

Name _____

Pretest

CHAPTER 10
PAGE 7

27. Ring the group of coins that is the same amount as .

28. Ring the group of coins that is the same amount as

Name _____

Pretest

CHAPTER 10
PAGE 8

29. The movie will be over in 1 hour. Ring the time the movie will be over.

7:00   6:00
8:30   9:00

30. The plane will take off in 30 minutes. Ring the time the plane will take off.

3:00   2:00
3:30   5:30

31. Karen is going to play for 30 minutes. Ring the time she will finish.

5:00   4:30
7:30   7:00

32. Art class will be over in 1 hour. Ring the time art class will be over.

11:30   1:00
9:30   11:00

Name _____

Posttest

CHAPTER 10
PAGE 1

Read each question. Find the answer.

1. Ring which happened last.

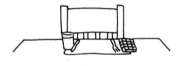

2. Ring which happened first.

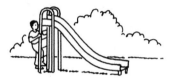

3. Ring which takes more time.

4. Ring which takes less time.

GO ON

Name _____

Posttest

CHAPTER 10
PAGE 2

5. Ring the one that shows 1 o'clock.

6. Ring the one that shows 5:30.

7. Ring the time it is.

2:00   4:00

3:00   12:00

8. Ring the time it is.

4:30   3:30

8:30   6:30

9. Ring the time it is.

1:00   12:00

11:00   2:00

10. Ring the time it is.

5:30   6:30

7:30   not here

GO ON

Name _____

Posttest

CHAPTER 10
PAGE 3

Use the calendar below to answer questions 11–14.

May						
Sunday	Monday	Tuesday	Wednesday	Thursday	Friday	Saturday
					1	2
3	4	5	6	7	8	9

11. Ring the day on which this month begins.

    Sunday   Friday

    Saturday   Monday

12. Ring the day of the week that is the third day of this month.

    Tuesday   Sunday

    Saturday   Thursday

13. Ring the day that comes just after Tuesday.

    Monday   Friday

    Wednesday   Saturday

14. Ring the day that comes just before Friday.

    Thursday   Saturday

    Sunday   Tuesday

GO ON

Posttest

CHAPTER 10
PAGE 4

15. Ring how much money there is.

4¢   5¢

20¢   40¢

16. Ring how much money there is.

6¢   16¢

30¢   60¢

17. Ring how much money there is.

5¢   21¢

30¢   50¢

18. Ring how much money there is.

5¢   17¢

22¢   35¢

Name _____

Posttest

CHAPTER 10
PAGE 5

19. Ring how much money there is.

4¢   28¢
31¢   40¢

20. Ring how much money there is.

6¢   26¢
41¢   51¢

Use coins to answer questions 21–24.

21. Rae has 2 nickels and 1 penny. How much money does she have?

_____ ¢

22. Kim has 1 dime and 3 nickels. How much money does she have?

_____ ¢

---

23. Lola has 6 dimes. How much money does she have?

_____ ¢

24. Nick has 1 quarter, 1 dime, and 1 nickel. How much money does he have?

_____ ¢

GO ON

Name _____

Posttest

CHAPTER 10
PAGE 6

25. Ring the group of coins that is the same amount as .

26. Ring the group of coins that is the same amount as  .

Name _____

Posttest

CHAPTER 10
PAGE 7

27. Ring the group of coins that is the same amount as

28. Ring the group of coins that is the same amount as

Name _____

Posttest

CHAPTER 10
PAGE 8

29. The movie will be over in 1 hour. Ring the time the movie will be over.

9:30   8:00

10:30   10:00

30. The school day will end in 30 minutes. Ring the time the school day will end.

2:00   4:30

3:30   3:00

31. Terry is going to play for 30 minutes. Ring the time he will finish.

4:30   5:30

8:00   6:00

32. The show will start in 1 hour. Ring the time the show will start.

7:30   5:30

7:00   8:30

Name _____

Pretest

CHAPTER 11
PAGE 1

Read each question. Find the answer.

1. 3 + 8 = _____

2. 6 + 0 = _____

3. 7 − 5 = _____

4. 9 − 1 = _____

5. Ring the number of cubes in all.

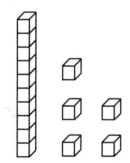

6 cubes   14 cubes

13 cubes   15 cubes

6. Ring the number of cubes in all.

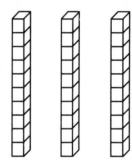

3 cubes   30 cubes

20 cubes   33 cubes

7. Ring the number of cubes in all.

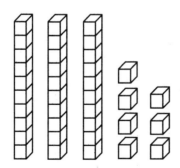

10 cubes   36 cubes

30 cubes   37 cubes

8. Ring the number of cubes in all.

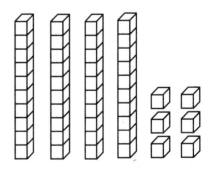

46 cubes   64 cubes

60 cubes   66 cubes

GO ON

Name _____

Pretest

CHAPTER 11
PAGE 2

9. Ring the addition sentence that tells how many cubes in all.

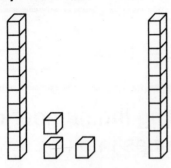

4 + 1 = 5
10 + 3 = 13
10 + 10 = 20
13 + 10 = 23

10. Ring the addition sentence that tells how many cubes in all.

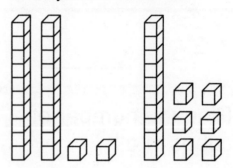

2 + 6 = 8
3 + 7 = 10
20 + 10 = 30
22 + 16 = 38

11.
```
   10
+ 60
```

12.
```
   43
+ 21
```

13.
```
   32
+ 54
```

14.
```
   33
+ 33
```

15. Chris has 10 bats. Pat has 12 bats. How many bats in all?

_____ bats

16. There are 23 girls and 24 boys playing. How many children are playing?

_____ children

Name _____

Pretest

CHAPTER 11
PAGE 3

17. Tod found 35 nuts. Bob found 14 nuts. How many nuts are there?

_____ nuts

18. Sara had 20 berries. She picked 16 more. How many does she have?

_____ berries

19. Ring the subtraction sentence that tells the story.

22 − 10 = 12

12 − 10 = 2

12 − 2 = 10

22 − 11 = 11

20. Ring the subtraction sentence that tells the story.

23 − 13 = 10

13 − 10 = 3

23 − 10 = 13

10 − 3 = 7

21. 40
   − 20
   ─────

22. 68
   − 41
   ─────

Name _____

Pretest

CHAPTER 11
PAGE 4

23. 54
    −20

24. 49
    −39

25. There were 19 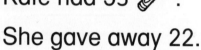.
Then 10 were sold.
How many are left?

_____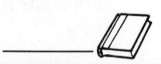

26. Kate had 53 .
She gave away 22.
How many are left?

_____

27. There were 24 .
Then 11 ran away.
How many are left?

_____

28. There were 17 .
Then 12 flew away.
How many are left?

_____

29. Mona had 63¢. Then she spent 21¢. Ring about how much money she has now.

about 40¢   about 50¢

30. Sasha spent 48¢. Then she spent 31¢. Ring about how much she spent in all.

about 70¢   about 80¢

31. Ned had 25 .
He gave away 15.
How many does he have now?

_____

32. Mike is 4 years old. His sister is 3 years older. How old is Mike's sister?

_____ years old

Name _____

Posttest

CHAPTER 11
PAGE 1

Read each question. Find the answer.

1. 4 + 7 = _____

2. 0 + 7 = _____

3. 6 − 6 = _____

4. 11 − 8 = _____

5. Ring the number of cubes in all.

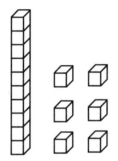

   7 cubes   15 cubes

   14 cubes   16 cubes

6. Ring the number of cubes in all.

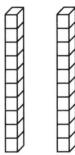

   2 cubes   20 cubes

   10 cubes   22 cubes

7. Ring the number of cubes in all.

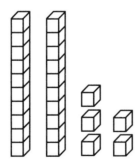

   25 cubes   52 cubes

   26 cubes   62 cubes

8. Ring the number of cubes in all.

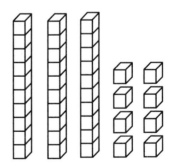

   38 cubes   83 cubes

   80 cubes   88 cubes

Name _____

Posttest

CHAPTER 11
PAGE 2

9. Ring the addition sentence that tells how many cubes in all.

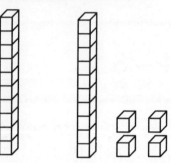

1 + 5 = 6
10 + 14 = 24
9 + 13 = 22
10 + 5 = 15

10. Ring the addition sentence that tells how many cubes in all.

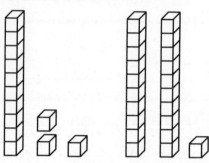

10 + 3 = 13
13 + 21 = 34
4 + 3 = 7
13 + 20 = 33

11. 70
    +10

12. 54
    +21

13. 36
    +12

14. 22
    +22

15. Carl has 14 tops. Pam has 10 tops. How many tops in all?

_____ tops

16. There are 42 boys and 32 girls playing. How many children are playing?

_____ children

Name _____

Posttest

CHAPTER 11
PAGE 3

17. Toni found 40 rocks. Lee found 13 rocks. How many rocks are there?

_____ rocks

18. Walt had 36 berries. He picked 23 more. How many does he have?

_____ berries

19. Ring the subtraction sentence that tells the story.

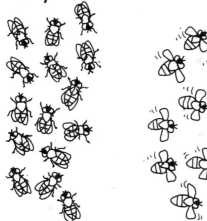

$15 - 11 = 4$
$15 - 2 = 13$
$26 - 10 = 16$
$26 - 11 = 15$

20. Ring the subtraction sentence that tells the story.

$13 - 10 = 3$
$23 - 10 = 13$
$23 - 1 = 22$
$13 - 3 = 10$

21.
  60
− 30
———

22.
  57
− 31
———

GO ON

Name _____

Posttest

CHAPTER 11
PAGE 4

23. 45
    −20

24. 68
    −58

25. There were 18 .
Then 10 were sold.
How many are left?

_____

26. Tom had 45 .
He gave away 33.
How many are left?

_____

27. There were 35 .
Then 14 ran away.
How many are left?

_____

28. There were 18 .
Then 11 ran away.
How many are left?

_____

29. Erin had 52¢. Then she spent 30¢. Ring <u>about</u> how much money she has now.

about 20¢   about 30¢

30. Mark spent 52¢. Then he spent 21¢. Ring <u>about</u> how much he spent in all.

about 70¢   about 80¢

31. Eric had 32 .
He gave away 11.
How many does he have now?

_____

32. Marie is 15 years old. Her brother is 3 years older. How old is Marie's brother?

_____ years old

Name _____

Pretest

CHAPTER 12
PAGE 1

Read each question. Find the answer.

1. Ring the addition sentence that tells how many in all.

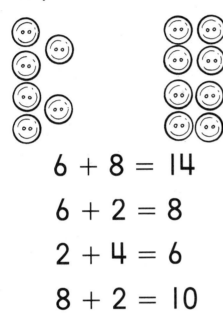

6 + 8 = 14
6 + 2 = 8
2 + 4 = 6
8 + 2 = 10

2. Ring the addition sentence that tells how many in all.

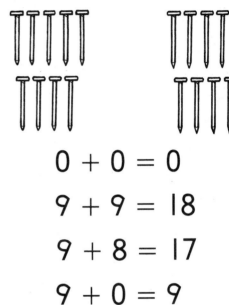

0 + 0 = 0
9 + 9 = 18
9 + 8 = 17
9 + 0 = 9

3. Ring the addition sentence that tells how many in all.

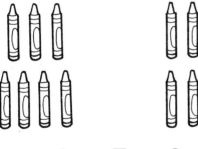

1 + 7 = 8
8 + 1 = 9
8 + 9 = 17
7 + 8 = 15

4. Ring the addition sentence that tells how many in all.

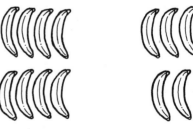

8 + 9 = 17
7 + 1 = 8
8 + 1 = 9
8 + 8 = 16

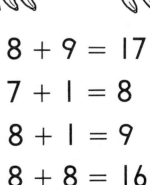

Name _____

Pretest

CHAPTER 12
PAGE 2

5.

$5 + 8 =$ _____

6.

$8 + 8 =$ _____

7.

$9 + 5 =$ _____

8.

$6 + 7 =$ _____

9.

$$\begin{array}{r} 9 \\ +7 \\ \hline \end{array}$$

10. $8 + 6 =$ _____

Use counters to answer questions 11–14.

11. Mary has 9 🎈.
Karl has 4 🎈.
Write how many 🎈
there are in all.

_____ 🎈

12. There are 7 red 🎩 on the rack. There are 7 blue 🎩 on the rack. Write how many 🎩 there are in all.

_____ 🎩

GO ON

Pretest

CHAPTER 12
PAGE 3

13. Mark has 9 .
    Pam has 7 .
    Write how many
    there are in all.

    _____

14. There are 7 red  on a rack. There are 6 blue  on the rack. Write how many there are in all.

    _____

15. Ring the subtraction sentence that tells the story.

    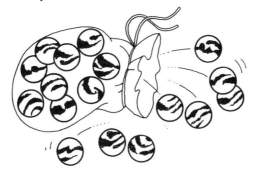

    $9 - 1 = 8$
    $17 - 8 = 9$
    $16 - 8 = 8$
    $18 - 9 = 9$

16. Ring the subtraction sentence that tells the story.

    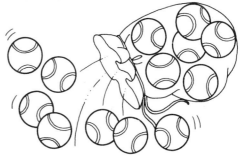

    $7 - 1 = 6$
    $14 - 7 = 7$
    $6 - 6 = 0$
    $13 - 7 = 6$

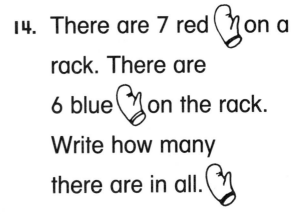

Pretest

CHAPTER 12
PAGE 4

17. Ring the subtraction sentence that tells the story.

$9 - 7 = 2$

$16 - 7 = 9$

$9 - 8 = 1$

$15 - 9 = 6$

18. Ring the subtraction sentence that tells the story.

$15 - 8 = 7$

$8 - 7 = 1$

$8 - 1 = 7$

$14 - 8 = 6$

19.
$$\begin{array}{r} 15 \\ -6 \\ \hline \end{array}$$

20.
$$\begin{array}{r} 14 \\ -7 \\ \hline \end{array}$$

21. $13 - 7 = $ _____

22. $16 - 8 = $ _____

GO ON

Name _____

Pretest

CHAPTER 12
PAGE 5

Use counters to answer questions 23 and 24.

23. Sally baked 18 .
Her friends ate 9 .
How many are left?

_____

24. There were 15 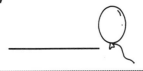.
Then 7 🎈 popped.
How many are left?

_____

25. Ring the number sentence that does <u>not</u> belong to the fact family for 5, 9, and 14.

$5 + 4 = 9$

$9 + 5 = 14$

$14 - 9 = 5$

$14 - 5 = 9$

26. Ring the number sentence that does <u>not</u> belong to the fact family for 8, 7, and 15.

$15 - 7 = 8$

$8 + 7 = 15$

$15 - 8 = 7$

$8 - 7 = 1$

27. Ring the number sentence that belongs to the same fact family as $13 - 8 = 5$.

$5 + 3 = 8$

$8 + 5 = 13$

$5 - 3 = 2$

$8 - 5 = 3$

28. Ring the number sentence that belongs to the same fact family as $16 - 7 = 9$.

$7 + 2 = 9$

$9 - 7 = 2$

$14 - 7 = 7$

$9 + 7 = 16$

GO ON

Pretest

CHAPTER 12
PAGE 6

Use the table below to answer questions 29–30.

FAVORITE SNACKS	Boys	Girls
Popcorn	IIII	IIII III
Fruit	IIII	IIII IIII
Cheese	IIII IIII	IIII I

29. How many boys and girls like cheese best?

_____

30. How many boys and girls like fruit best?

_____

31.

There are 9 balloons.
There are 3 children.
How many balloons does each child get?

_____ balloons

32.

There are 16 stars.
There are 2 children.
How many stars does each child get?

_____ stars

Name _____

Posttest

CHAPTER 12
PAGE 1

Read each question. Find the answer.

1. Ring the addition sentence that tells how many in all.

   8 + 7 = 15
   7 + 1 = 8
   8 + 2 = 10
   5 + 2 = 7

2. Ring the addition sentence that tells how many in all.

   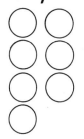 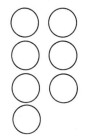

   0 + 0 = 0
   7 + 0 = 7
   6 + 7 = 13
   7 + 7 = 14

3. Ring the addition sentence that tells how many in all.

   4 + 5 = 9
   9 + 6 = 15
   8 + 5 = 13
   9 + 5 = 14

4. Ring the addition sentence that tells how many in all.

   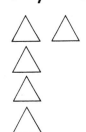 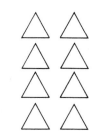

   5 + 3 = 8
   5 + 8 = 13
   3 + 8 = 11
   5 + 9 = 14

GO ON

Name _____

Posttest

CHAPTER 12
PAGE 2

5.

6 + 7 = _____

6.

9 + 9 = _____

7.

8 + 6 = _____

8.

9 + 8 = _____

9.

  6
+ 9
___

10.

8 + 8 = _____

Use counters to answer questions 11–14.

11. Mark has 7 .
Pam has 9 🌼.
Write how many 🌼
there are in all.

12. There are 6 red 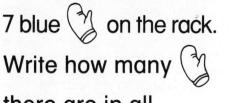 on a rack. There are 7 blue 🧤 on the rack. Write how many 🧤 there are in all.

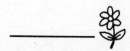

Posttest

CHAPTER 12
PAGE 3

13. Mary has 9 🎈.
Karl has 5 🎈.
Write how many 🎈
there are in all.

_____ 🎈

14. There are 7 red 🎩 on the rack. There are 7 blue 🎩 on the rack. Write how many 🎩 there are in all.

_____ 🎩

15. Ring the subtraction sentence that tells the story.

$8 - 8 = 0$
$13 - 5 = 8$
$14 - 5 = 9$
$8 - 5 = 3$

16. Ring the subtraction sentence that tells the story.

$9 - 6 = 3$
$15 - 6 = 9$
$10 - 4 = 6$
$15 - 9 = 6$

GO ON

Name _____

Posttest

CHAPTER 12
PAGE 4

17. Ring the subtraction sentence that tells the story.

$14 - 6 = 8$

$8 - 6 = 2$

$15 - 7 = 8$

$13 - 7 = 6$

18. Ring the subtraction sentence that tells the story.

$14 - 5 = 9$

$9 - 5 = 4$

$15 - 9 = 6$

$5 - 5 = 0$

19.
$$\begin{array}{r} 16 \\ -8 \\ \hline \end{array}$$

20.
$$\begin{array}{r} 15 \\ -7 \\ \hline \end{array}$$

21. $17 - 8 = $ _____

22. $13 - 7 = $ _____

GO ON

Name _____

Posttest

CHAPTER 12
PAGE 5

Use counters to answer questions 23 and 24.

23. Mac picked 16 🍑. He gave 9 to his friend. How many 🍑 are left?

    _____ 🍑

24. There were 18 🥚. Then 9 🥚 were broken. How many 🥚 are left?

    _____ 🥚

---

25. Ring the number sentence that does <u>not</u> belong to the fact family for 4, 9, and 13.

    $9 - 4 = 5$

    $13 - 9 = 4$

    $4 + 9 = 13$

    $9 + 4 = 13$

26. Ring the number sentence that does <u>not</u> belong to the fact family for 7, 8, and 15.

    $8 - 7 = 1$

    $15 - 8 = 7$

    $7 + 8 = 15$

    $8 + 7 = 15$

---

27. Ring the number sentence that belongs to the same fact family as $14 - 6 = 8$.

    $8 - 6 = 2$

    $14 - 8 = 6$

    $4 + 4 = 8$

    $2 + 4 = 6$

28. Ring the number sentence that belongs to the same fact family as $17 - 8 = 9$.

    $9 - 1 = 8$

    $8 + 1 = 9$

    $8 + 9 = 17$

    $18 - 9 = 9$

GO ON

Name _____

Posttest

CHAPTER 12
PAGE 6

Use the table below to answer questions 29–30.

THINGS CHILDREN LIKE TO DO	Girls	Boys
Jumping	IIII	HHT IIII
Running	HHT	HHT IIII
Swinging	HHT III	HHT I

29. How many girls and boys like running best?

_____

30. How many girls and boys like jumping best?

_____

31.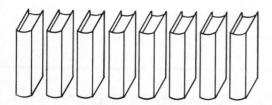

There are 8 books.
There are 2 children.
How many books does each child get?

_____ books

32.

There are 15 boats.
There are 5 children.
How many boats does each child get?

_____ boats

Name _____

Cumulative Test

CHAPTERS 1–12
PAGE 1

Read each question. Find the answer.

1.

   7 + 1 = _____

2.

   6 – 4 = _____

3. Ring the plane shape that matches the face of the solid shape.

4. Ring the plane shape that matches the face of the solid shape.

5. Ring the figure that is the same size and shape.

6. Ring the picture that shows two parts that match.

Name _____

Cumulative Test

CHAPTERS 1–12
PAGE 2

7. Ring the shape that comes next in the pattern.

 ?

8. Ring the shape that comes next in the pattern.

9. Write the number that comes after.

39, _____

10. Write the number that comes before.

_____ , 72

11. Write the number that comes between.

9, _____ , 11

12. Ring the one who is third.

Name _____

Cumulative Test

CHAPTERS 1–12
PAGE 3

13. Ring the one that is fourth.

14. Use your inch ruler. How long is the needle?

_____ inches

15. Use your centimeter ruler. How long is the feather?

_____ centimeters

16. Ring the one that holds about the same as a .

17. Ring the one that is about as heavy as a .

18. Ring the one that is about the same temperature as a .

Name _____

Cumulative Test

CHAPTERS **1–12**
PAGE 4

19. Ring the one that shows fair shares.

20. Ring the one that shows one fourth.

---

21. Ring the fraction that matches this picture.

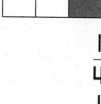

$\frac{1}{5}$ $\frac{1}{4}$

$\frac{1}{3}$ $\frac{1}{2}$

22. Ring the correct time.

7:00   8:00
9:00   12:00

---

23. Ring the correct time.

1:30   2:30
6:30   not here

24. How much money is there?

_____

Name _____

Cumulative Test

CHAPTERS 1-12
PAGE 5

25. How much money is there?

_____

26. How much money is there?

_____

27. Ring how many cubes there are in all.

3 cubes
4 cubes
10 cubes
13 cubes

28. Ring how many cubes there are in all.

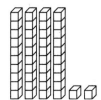

4 cubes
6 cubes
24 cubes
42 cubes

29. Ring the addition sentence that tells how many cubes in all.

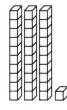 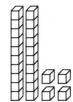

31 + 24 = 55
4 + 6 = 10
30 + 1 = 31
30 + 20 = 50

30. 70
      + 14
      ____

GO ON

Cumulative Test

CHAPTERS 1–12
PAGE 6

31.

$$\begin{array}{r} 22 \\ +22 \\ \hline \end{array}$$

32. Ella has 46 beads. Gus has 21 beads. How many beads are there in all?

_____ beads

33. Ring the subtraction sentence that tells the story.

23 − 10 = 13
13 − 10 = 3
13 − 3 = 10
23 − 0 = 23

34.

$$\begin{array}{r} 73 \\ -21 \\ \hline \end{array}$$

35.

$$\begin{array}{r} 54 \\ -34 \\ \hline \end{array}$$

36. There were 41 balloons. Then 30 popped. How many are left?

_____ balloons

Name _____
Cumulative Test

CHAPTERS 1-12
PAGE 7

37.

9 + 8 = _____

38.

7 + 7 = _____

39. There are 5 red balls.
There are 7 blue balls.
How many balls are
there in all?

_____ balls

40.

16 − 9 = _____

41.

14 − 6 = _____

42. There were 14 grapes.
Then Tim ate 5.
How many are left?

_____ grapes

Name _____

Cumulative Test

CHAPTERS 1–12
PAGE 8

43.

Mother will be home in 1 hour.
Ring what time she will be home.

4:00   5:00
6:00   7:00

44.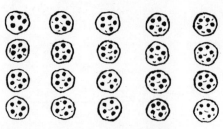

There are 20 🍪.
There are 10 children.
How many 🍪 will each child get?

_____ 🍪

45. 3 🔔 for 🪙
Ring how many 🔔 you can buy with 🪙🪙🪙.

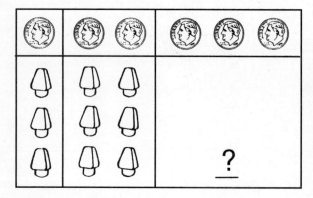

3 🔔   4 🔔
7 🔔   9 🔔

46. Ring the number pair.
The sum is 8.
The difference is 0.

1 and 7

0 and 8

8 and 8

4 and 4

GO ON

Name _____

Cumulative Test

CHAPTERS 1-12
PAGE 9

Use the graph below to answer question 47.

GLASSES OF MILK IN ONE DAY

Hal	🥛 🥛 🥛
Terri	🥛 🥛 🥛 🥛 🥛
Ira	🥛 🥛
Nan	🥛 🥛 🥛

Each 🥛 stands for 1 glass.

47. Ring how many glasses of milk Ira drank.

    2 glasses   3 glasses   4 glasses   5 glasses

Use the graph below to answer question 48.

FAVORITE JUICES

Orange	■	■	■			
Grape	■	■	■	■	■	■
Apple	■	■	■	■	■	
Pineapple	■	■				

Each ■ stands for 1 person.

48. Ring the juice that is the favorite of the most people.

    Orange   Grape   Apple   Pineapple

Cumulative Test

CHAPTERS 1-12
PAGE 10

Use the table below to answer questions 49–50.

EATING TIMES (in minutes)			
	Breakfast	Lunch	Dinner
Amy	10	15	25
Bob	10	20	30

49. Ring how long it took Bob to eat lunch.

    10 minutes
    15 minutes
    20 minutes
    30 minutes

50. Ring what you do to find how much longer it took Amy to eat dinner than to eat lunch.

    Add   Subtract

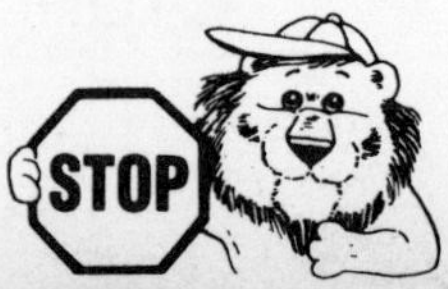

Name _____

End-of-Book Test

PAGE 1

Read each question. Find the answer.

1.

    7 − 1 = _____

2.

    5 + 2 = _____

3. Ring the plane shape that matches the face of the solid shape.

    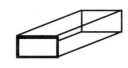

4. Ring the plane shape that matches the face of the solid shape.

5. Ring the figure that is the same size and shape.

6. Ring the picture that shows two parts that match.

Name _____

End-of-Book Test

PAGE 2

7. Ring the shape that comes next in the pattern.

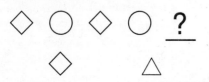

8. Ring the shape that comes next in the pattern.

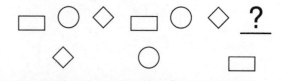

9. Write the number that comes just before.

_____, 31

10. Write the number that comes just after.

68, _____

11. Write the number that comes between.

44, _____, 46

12. Ring the one who is fifth.

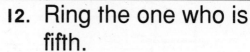

13. Ring the one that is third.

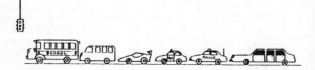

14. Use your centimeter ruler. How long is the key?

_____ centimeters

Name _____

End-of-Book Test

PAGE 3

15. Use your inch ruler. How long is the leaf?

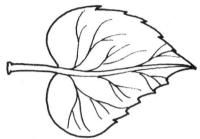

_____ inches

16. Ring the one that holds about the same as .

17. Ring the one that is about as heavy as a .

18. Ring the one that is about the same temperature as .

19. Ring the one that shows fair shares.

20. Ring the one that shows one half.

GO ON

Name _____

End-of-Book Test

PAGE 4

21. Ring the fraction that matches this picture.

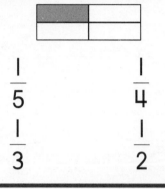

$\frac{1}{5}$ $\frac{1}{4}$

$\frac{1}{3}$ $\frac{1}{2}$

22. Ring the correct time.

3:00   4:00

5:00   12:00

23. Ring the correct time.

6:30   10:30

11:30   9:30

24. Ring how much money there is.

5¢   35¢

40¢   50¢

25. Ring how much money there is.

5¢   17¢

27¢   32¢

26. Ring how much money there is.

10¢   20¢

25¢   35¢

27. Ring how many cubes there are in all.

6 cubes   7 cubes

10 cubes   16 cubes

28. Ring how many cubes there are in all.

10 cubes   17 cubes

37 cubes   73 cubes

Name _____

End-of-Book Test

PAGE 5

29. Ring the addition sentence that tells how many cubes there are in all.

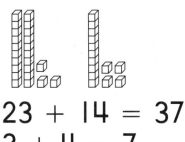

23 + 14 = 37
3 + 4 = 7
5 + 5 = 10
20 + 10 = 30

30. 45
      + 30
      ────

31. 44
      + 44
      ────

32. Tina has 25 beans. Bert has 14 beans. How many beans are there in all?

_____ beans

33. Ring the subtraction sentence that tells the story.

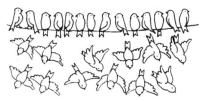

26 − 12 = 14
14 − 10 = 4
14 − 2 = 12
12 − 2 = 10

34. 56
      − 31
      ────

GO ON

Name _____

End-of-Book Test

PAGE 6

35.

$$63 - 23$$

36. Fred won 24 stars. Amy won 10 stars. How many more stars did Fred win?

_____ more stars

37.

8 + 5 = _____

38.

9 + 0 = _____

39. There are 4 white 🌼.
There are 7 yellow 🌼.
How many 🌼 in all?

_____

40.

17 − 8 = _____

41.

12 − 5 = _____

42. Use pennies to solve. Pam has 15 pennies. She spends 8 of them. How many pennies does Pam have left?

_____ pennies

GO ON

Name _____

End-of-Book Test

PAGE 7

43. Story time will start in 1 hour. Ring the time it will start.

1:00   2:00
3:00   12:00

44. There are 12 eggs. There are 3 children. How many eggs does each child get?

_____ eggs

45. 4  for

Ring how many  you can buy with

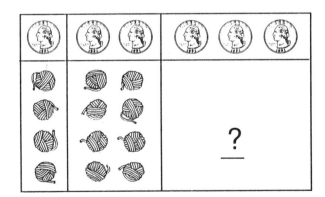

4   9
10   12

46. Ring the number pair.
The sum is 6.
The difference is 2.

4 and 6

3 and 3

2 and 2

2 and 4

Name _____
End-of-Book Test

PAGE 8

Use the graph below to answer question 47.

PINE CONES FOUND

Tom	🌰 🌰 🌰 🌰
Sara	🌰 🌰 🌰
Dan	🌰
Alice	🌰 🌰 🌰 🌰 🌰 🌰

Each 🌰 stands for one pine cone.

47. Ring how many 🌰 Dan found.

    1 pine cone   3 pine cones
    4 pine cones   6 pine cones

Use the graph below to answer question 48.

FLOWERS PICKED

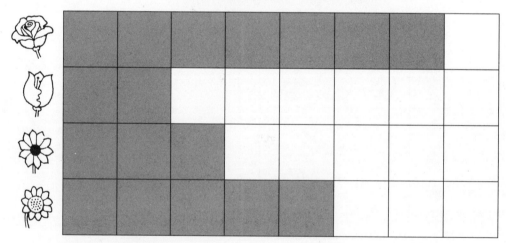

Each stands for one flower.

48. Ring how many were picked.

   2   3   5   7

End-of-Book Test

Use the table below to answer questions 49–50.

Number of Children

	School Bus	City Bus
Boys	14	4
Girls	12	5

49. How many girls are on the school bus?

    _____ girls

50. Ring what you do to find how many children are on the city bus.

    Add   Subtract

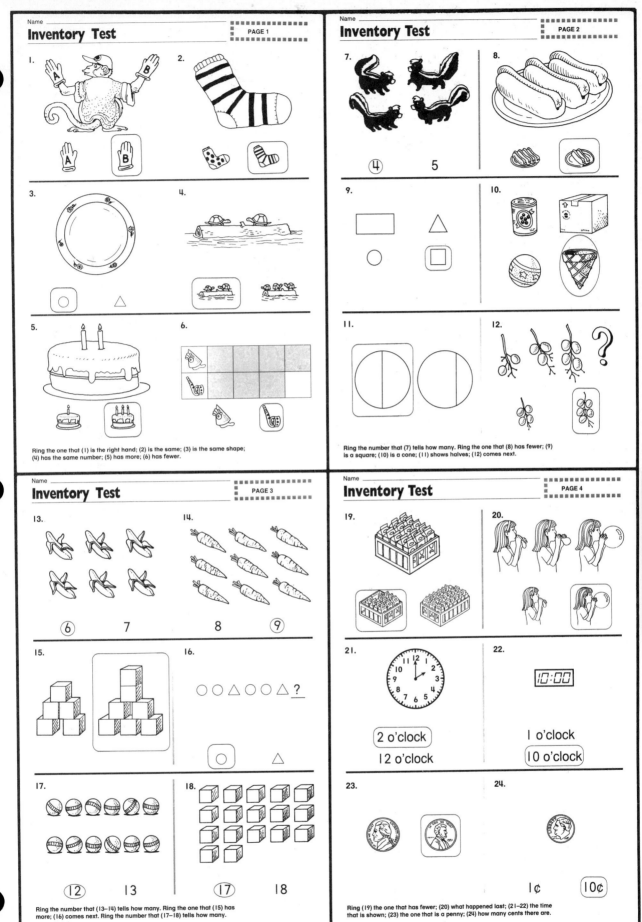

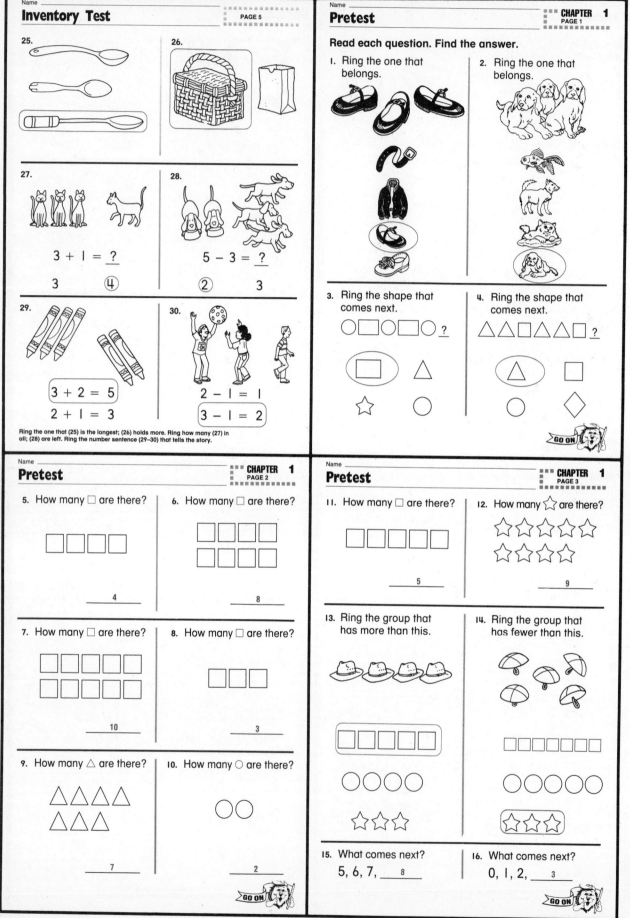

Pretest

CHAPTER 1 — PAGE 4

Use the pictograph below to answer questions 17–20.

Number of Boats

0 1 2 3 4 5

17. Ring the one that has more.

18. Ring the one that has fewer.

19. Ring the one that has more than 4.

20. Ring the one that has fewer than 3.

Posttest

CHAPTER 1 — PAGE 1

Read each question. Find the answer.

1. Ring the one that belongs.

2. Ring the one that belongs.

3. Ring the shape that comes next.
   △ ○ △ ○ △ ?

4. Ring the shape that comes next.
   □ ○ ○ □ ○ ○ ?

Posttest

CHAPTER 1 — PAGE 2

5. How many □ are there? __5__

6. How many □ are there? __3__

7. How many □ are there? __9__

8. How many □ are there? __8__

9. How many ○ are there? __6__

10. How many □ are there? __1__

Posttest

CHAPTER 1 — PAGE 3

11. How many ☆ are there? __8__

12. How many ○ are there? __7__

13. Ring the group that has fewer than this.

14. Ring the group that has more than this.

15. What comes next?
    2, 3, 4, __5__

16. What comes next?
    6, 7, 8, __9__

B. TEST ANSWERS — FREE RESPONSE FORMAT

Free Response Format • Test Answers

Name _____

Posttest
CHAPTER 1 · PAGE 4

Use the pictograph below to answer questions 17–20.

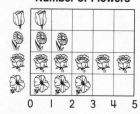

Number of Flowers

17. Ring the one that has fewer, or .

18. Ring the one that has more, or .

19. Ring the one that has more than 4.

20. Ring the one that has fewer than 2.

Name _____

Pretest
CHAPTER 2 · PAGE 1

Read each question. Find the answer.

1. Ring the addition sentence that tells how many there are.

4 + 2 = 6
(4 + 1 = 5)
1 + 3 = 4
2 + 1 = 3

2. Ring the addition sentence that tells how many there are.

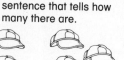

2 + 1 = 3
3 + 3 = 6
(2 + 3 = 5)
3 + 1 = 4

3. Ring the addition sentence that tells how many there are.

(2 + 4 = 6)
2 + 2 = 4
2 + 3 = 5
1 + 1 = 2

4. Ring the addition sentence that tells how many there are.

3 + 3 = 6
1 + 3 = 4
1 + 1 = 2
(1 + 2 = 3)

Name _____

Pretest
CHAPTER 2 · PAGE 2

5. Ring the addition sentence that tells how many there are.

(2 + 2 = 4)
1 + 2 = 3
1 + 1 = 2
2 + 3 = 5

6. Ring the addition sentence that tells how many there are.

1 + 3 = 4
(2 + 1 = 3)
2 + 3 = 5
1 + 1 = 2

7. 3 + 1 = __4__

8. 3 + 3 = __6__

9. 0 + 2 = __2__

10. 3 + 2 = __5__

11. 1
    +5
    ――
     6

12. 2
    +2
    ――
     4

Name _____

Pretest
CHAPTER 2 · PAGE 3

13. 5
    +0
    ――
     5

14. 1
    +1
    ――
     2

15. Ring another name for 3.

(3 + 0)   3 + 3
1 + 3   0 + 6

16. Ring another name for 5.

1 + 5   3 + 3
(2 + 3)   3 + 1

17. Ring another name for 6.

1 + 3   (3 + 3)
3 + 2   5 + 0

18. Ring another name for 2.

(1 + 1)   1 + 3
2 + 1   2 + 2

19. Ring another name for 1.

1 + 1   2 + 2
1 + 2   (1 + 0)

20. Ring another name for 4.

(3 + 1)   3 + 3
4 + 1   0 + 3

Free Response Format • Test Answers

Pretest — Chapter 2, Page 4

21.
4 cats sleep.
2 cats play.
How many cats are there?

_____6_____ cats

22.
1 girl plays.
3 more girls play.
How many girls play?

_____4_____ girls

23.
1 dog sits.
1 dog runs.
How many dogs are there?

_____2_____ dogs

24.
3 boys read.
3 boys write.
How many boys are there?

_____6_____ boys

Posttest — Chapter 2, Page 1

Read each question. Find the answer.

1. Ring the addition sentence that tells how many there are.

$4 + 2 = 6$
$(3 + 2 = 5)$
$2 + 2 = 4$
$2 + 1 = 3$

2. Ring the addition sentence that tells how many there are.

$1 + 1 = 2$
$2 + 3 = 5$
$(2 + 2 = 4)$
$4 + 2 = 6$

3. Ring the addition sentence that tells how many there are.

$(1 + 4 = 5)$
$2 + 4 = 6$
$2 + 2 = 4$
$1 + 3 = 4$

4. Ring the addition sentence that tells how many there are.

$2 + 1 = 3$
$(3 + 1 = 4)$
$3 + 3 = 6$
$3 + 2 = 5$

Posttest — Chapter 2, Page 2

5. Ring the addition sentence that tells how many there are.

$(3 + 3 = 6)$
$1 + 2 = 3$
$2 + 2 = 4$
$3 + 2 = 5$

6. Ring the addition sentence that tells how many there are.

$1 + 3 = 4$
$(1 + 2 = 3)$
$2 + 3 = 5$
$1 + 1 = 2$

7. $2 + 3 =$ __5__

8. $0 + 4 =$ __4__

9. $1 + 1 =$ __2__

10. $3 + 3 =$ __6__

11. $\begin{array}{r} 2 \\ +2 \\ \hline 4 \end{array}$

12. $\begin{array}{r} 5 \\ +1 \\ \hline 6 \end{array}$

Posttest — Chapter 2, Page 3

13. $\begin{array}{r} 3 \\ +0 \\ \hline 3 \end{array}$

14. $\begin{array}{r} 2 \\ +4 \\ \hline 6 \end{array}$

15. Ring another name for 4.

$(3 + 1)$ $3 + 3$
$2 + 3$ $0 + 3$

16. Ring another name for 1.

$(0 + 1)$ $1 + 2$
$1 + 1$ $2 + 2$

17. Ring another name for 5.

$1 + 3$ $3 + 3$
$(3 + 2)$ $4 + 0$

18. Ring another name for 6.

$(1 + 5)$ $3 + 2$
$5 + 0$ $2 + 2$

19. Ring another name for 3.

$1 + 1$ $2 + 2$
$0 + 2$ $(3 + 0)$

20. Ring another name for 2.

$3 + 1$ $1 + 0$
$(0 + 2)$ $2 + 2$

Free Response Format • Test Answers

Posttest — Chapter 2, Page 4

21.
2 frogs jump.
3 frogs sit.
How many frogs are there?
___5___ frogs

22.
1 girl plays.
1 girl reads.
How many girls are there?
___2___ girls

23.
4 boys smile.
2 more boys smile.
How many boys smile in all?
___6___ boys

24.
1 cat sleeps.
5 cats drink.
How many cats are there?
___6___ cats

Pretest — Chapter 3, Page 1

Read each question. Find the answer.

1. Ring the subtraction sentence that tells the story.

6 − 1 = 5   6 − 3 = 3
6 − 6 = 0   (6 − 2 = 4)

2. Ring the subtraction sentence that tells the story.

4 − 1 = 3   4 − 2 = 2
(4 − 0 = 4)   4 − 1 = 3

3. Ring the subtraction sentence that tells the story.

4 − 0 = 4   (5 − 1 = 4)
4 − 1 = 3   5 − 5 = 0

4. Ring the subtraction sentence that tells the story.

3 − 3 = 0   6 − 5 = 1
3 − 0 = 3   (6 − 3 = 3)

Pretest — Chapter 3, Page 2

5.
4 − 2 = ___2___

6.
5 − 4 = ___1___

7.
3 − 3 = ___0___

8.
2 − 1 = ___1___

9. 4 − 3 = ___1___

10. 3 − 0 = ___3___

11. 3 − 1 = ___2___

12. 6 − 5 = ___1___

13.
  6
− 4
―――
  2

14.
  1
− 1
―――
  0

Pretest — Chapter 3, Page 3

15.
  5
− 3
―――
  2

16.
  4
− 2
―――
  2

17. Ring the subtraction sentence that goes with the addition sentence.

3 + 2 = 5
3 − 3 = 0   5 − 1 = 4
3 − 2 = 1   (5 − 2 = 3)

18. Ring the subtraction sentence that goes with the addition sentence.

5 + 1 = 6
5 − 1 = 4   6 − 3 = 4
(6 − 1 = 5)   3 − 3 = 0

19. Ring the number sentence that does not belong.

1 + 3 = 4   4 − 1 = 3
4 − 3 = 1   (3 − 1 = 2)

20. Ring the number sentence that does not belong.

(3 − 2 = 1)   5 − 3 = 2
5 − 2 = 3   3 + 2 = 5

Free Response Format • Test Answers

Pretest

CHAPTER 3 — PAGE 4

21.

What do you do to find how many in all?

Ring the answer.

(Add)   Subtract

22.

What do you do to find how many  are left? Ring the answer.

Add   (Subtract)

23. Sara sees 4 . Then 1 swims away.

Ring the correct question.

[How many are left?]

How many are there in all?

24. Tom sees 1 . Then he sees 1 more .

Ring the correct question.

How many are left?

[How many are there in all?]

Posttest

CHAPTER 3 — PAGE 1

Read each question. Find the answer.

1. Ring the subtraction sentence that tells the story.

   3 − 2 = 1   (4 − 1 = 3)
   3 − 1 = 2   4 − 2 = 2

2. Ring the subtraction sentence that tells the story.

   5 − 1 = 4   3 − 2 = 1
   (5 − 2 = 3)   3 − 3 = 0

3. Ring the subtraction sentence that tells the story.

   6 − 3 = 3   (3 − 3 = 0)
   3 − 1 = 2   4 − 1 = 3

4. Ring the subtraction sentence that tells the story.

   4 − 1 = 3   5 − 1 = 4
   5 − 4 = 1   (6 − 1 = 5)

Posttest

CHAPTER 3 — PAGE 2

5.    6 − 4 = __2__

6.    5 − 3 = __2__

7.    4 − 3 = __1__

8.    2 − 2 = __0__

9. 5 − 4 = __1__

10. 6 − 2 = __4__

11. 2 − 1 = __1__

12. 4 − 0 = __4__

13. 4
    −4
    ―
     0

14. 6
    −3
    ―
     3

Posttest

CHAPTER 3 — PAGE 3

15. 5
    −1
    ―
     4

16. 6
    −5
    ―
     1

17. Ring the subtraction sentence that goes with the addition sentence.

   ○ ○   ○
   2 + 1 = 3

   2 − 1 = 1   2 − 2 = 0
   3 − 3 = 0   (3 − 1 = 2)

18. Ring the subtraction sentence that goes with the addition sentence.

   ○ ○   ○ ○
   2 + 2 = 4

   2 − 0 = 2   3 − 3 = 0
   (4 − 2 = 2)   3 − 2 = 1

19. Ring the number sentence that does <u>not</u> belong.

   ○ ○
   ○ ○

   4 + 2 = 6   (4 − 2 = 2)
   6 − 4 = 2   6 − 2 = 4

20. Ring the number sentence that does <u>not</u> belong.

   ○ ○ ○
   ○ ○

   5 − 1 = 4   5 − 4 = 1
   4 + 1 = 5   (4 − 1 = 3)

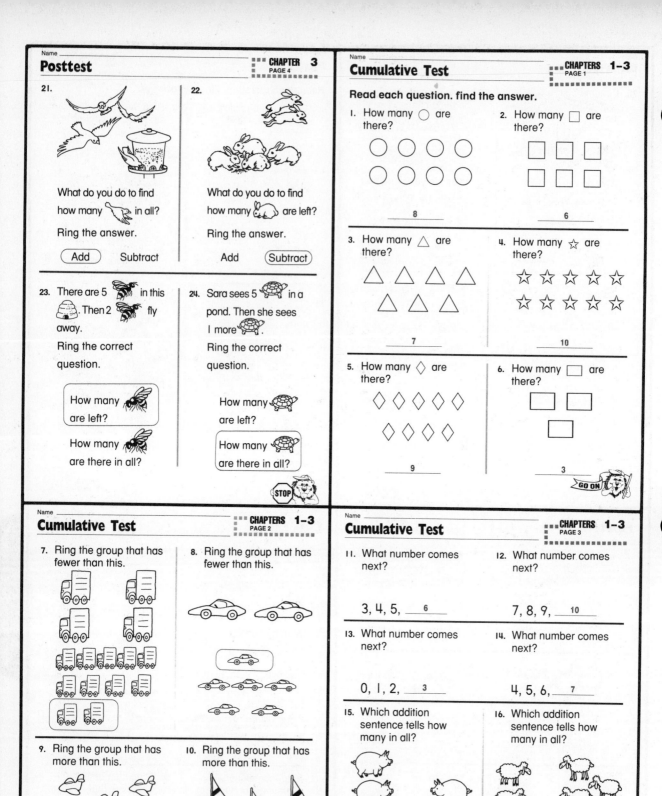

Cumulative Test — CHAPTERS 1–3 — PAGE 4

17. Ring the addition sentence that tells how many in all.

(2 + 2 = 4)
1 + 1 = 2
3 + 0 = 3
4 + 4 = 8

18. Ring the addition sentence that tells how many in all.

1 + 2 = 3
1 + 1 = 2
(2 + 3 = 5)
2 + 2 = 4

19. Ring the addition sentence that tells how many in all.

2 + 3 = 5
(1 + 5 = 6)
1 + 0 = 1
2 + 5 = 7

20. Ring the addition sentence that tells how many in all.

2 + 2 = 4
4 + 1 = 5
1 + 1 = 2
(3 + 3 = 6)

Cumulative Test — CHAPTERS 1–3 — PAGE 5

21. 5 + 1 = __6__

22. 2 + 0 = __2__

23. 2 + 2 = __4__

24. 1 + 4 = __5__

25. 0
    +1

     1

26. 3
    +3

     6

27. 2
    +3

     5

28. Ring the subtraction sentence that tells the story.

4 − 2 = 2
(6 − 2 = 4)
6 − 1 = 5
4 − 3 = 1

Cumulative Test — CHAPTERS 1–3 — PAGE 6

29. Ring the subtraction sentence that tells the story.

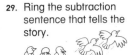

(5 − 2 = 3)
5 − 1 = 4
3 − 2 = 1
3 − 3 = 0

30. Ring the subtraction sentence that tells the story.

3 − 1 = 2
(4 − 3 = 1)
4 − 2 = 2
3 − 3 = 0

31. Ring the subtraction sentence that tells the story.

2 − 1 = 1
1 − 1 = 0
4 − 1 = 3
(3 − 1 = 2)

32.

4 − 1 = __3__

33.

6 − 3 = __3__

34.

5 − 0 = __5__

Cumulative Test — CHAPTERS 1–3 — PAGE 7

35. 4 − 4 = __0__

36. 5
    −3

     2

37. 6 − 4 = __2__

38. 2 − 1 = __1__

39. 4 − 0 = __4__

40. 6 − 5 = __1__

41. 3 − 3 = __0__

42. 4 − 1 = __3__

43. Jack sees 3 🐟. Then he sees 2 more 🐟. Ring the correct question.

  (How many 🐟 are there in all?)
  How many 🐟 are left?

44. Jane sees 5 🐦. Then 4 🐦 fly away. Ring the correct question.

  How many 🐦 are there in all?
  (How many 🐦 are left?)

Name _____
Cumulative Test
CHAPTERS 1–3
PAGE 8

Use the pictograph below to answer questions 45–50.

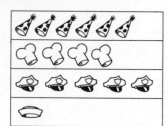

45. Ring the one for the group that has fewer.

46. Ring the one for the group that has more.

47. Ring the one for the group that has more than 5.

48. Ring the one for the group that has fewer than 4.

49. Ring the one that has more than 5.

50. Ring the one that has fewer, or .

STOP

Name _____
Pretest
CHAPTER 4
PAGE 1

Read each question. Find the answer.

1.
   5 + 3 = __8__

2.
   6 + 4 = __10__

3.
   8 + 1 = __9__

4.
   5 + 2 = __7__

5.
   3 + 3 = __6__

6.
   5 + 5 = __10__

7. 0 + 2 = __2__

8. 5 + 1 = __6__

GO ON

Name _____
Pretest
CHAPTER 4
PAGE 2

9. 4 + 4 = __8__
10. 2 + 7 = __9__
11. 4 + 3 = __7__
12. 3 + 2 = __5__
13. 8 + 2 = __10__
14. 2 + 2 = __4__

15. Ring another name for 9.
   1 + 6   4 + 3
   5 + 2   (9 + 0)

16. Ring another name for 10.
   4 + 4   6 + 3
   7 + 0   (8 + 2)

17. Ring the addition sentence that tells how many in all.

   5 + 1 + 1 = 7
   (4 + 2 + 1 = 7)
   4 + 1 = 5
   2 + 1 = 3

18. Ring the addition sentence that tells how many in all.

   5 + 2 = 7
   2 + 2 = 4
   (5 + 2 + 2 = 9)
   4 + 2 + 2 = 8

GO ON

Name _____
Pretest
CHAPTER 4
PAGE 3

19. 3 + 0 + 3 = __6__
20. 4 + 1 + 3 = __8__

21. How much did Kara spend?

   3¢ + 7¢ = __10¢__

22. How much did Karl spend?
   4¢ + 2¢ = __6¢__

23. Lin sees 5 birds. Then she sees 3 more birds. Then she sees 2 cats. How many birds does she see in all?

Ring the sentence that is <u>not</u> needed.

   Lin sees 5 birds.
   Then she sees 3 more birds.
   (Then she sees 2 cats)

24. Tom feeds 4 dogs. Then he feeds 3 cats. Then he feeds 2 more cats. How many cats does he feed in all?

Ring the sentence that is <u>not</u> needed.

   (Tom feeds 4 dogs)
   Tom feeds 3 cats.
   Then he feeds 2 more cats.

STOP

Free Response Format • Test Answers

Posttest — Chapter 4, Page 1

Read each question. Find the answer.

1.  7 + 1 = __8__
2. 8 + 2 = __10__
3. 6 + 3 = __9__
4. 5 + 2 = __7__
5. 5 + 5 = __10__
6. 4 + 4 = __8__
7. 0 + 3 = __3__
8. 1 + 4 = __5__

Posttest — Chapter 4, Page 2

9. 6 + 2 = __8__
10. 5 + 3 = __8__
11. 6 + 4 = __10__
12. 7 + 0 = __7__
13. 4 + 5 = __9__
14. 3 + 3 = __6__

15. Ring another name for 8.

   8 + 1   3 + 3
   6 + 3   (4 + 4)

16. Ring another name for 7.

   1 + 8   3 + 2
   4 + 1   (5 + 2)

17. Ring the addition sentence that tells how many in all.

   5 + 1 = 6
   (5 + 1 + 2 = 8)
   4 + 2 + 1 = 7
   1 + 2 = 3

18. Ring the addition sentence that tells how many in all.

   (4 + 2 + 2 = 8)
   2 + 2 + 2 = 6
   2 + 0 = 2
   2 + 2 = 4

Posttest — Chapter 4, Page 3

19. 4 + 0 + 4 = __8__
20. 3 + 2 + 5 = __10__

21. How much did Kathy spend?
   8¢ + 2¢ = __10¢__

22. How much did Carlos spend?
   5¢ + 4¢ = __9¢__

23. Nan sees 4 cows. Then she sees 3 more cows. Then she sees 2 dogs. How many cows does she see in all?

Ring the sentence that is <u>not</u> needed.

   Nan sees 4 cows.
   Then she sees 3 more cows.
   (Then she sees 2 dogs.)

24. Sam feeds 6 fish. Sam feeds 4 birds. Then he feeds 2 more birds. How many birds does he feed in all?

Ring the sentence that is <u>not</u> needed.

   (Sam feeds 6 fish.)
   Sam feeds 4 birds.
   Then he feeds 2 more birds.

Pretest — Chapter 5, Page 1

Read each question. Find the answer.

1. Subtract.
   7 − 1 = __6__

2. Subtract.
   8 − 3 = __5__

3. Subtract.
   9 − 3 = __6__

4. Subtract.
   10 − 2 = __8__

5. Subtract.
   6 − 3 = __3__

6. Subtract.
   5 − 3 = __2__

7. 7 − 2 = 5
8. 10 − 7 = 3
9. 8 − 4 = 4
10. 6 − 6 = 0
11. 9 − 2 = __7__
12. 8 − 6 = __2__
13. 7 − 0 = __7__
14. 9 − 8 = __1__

Free Response Format • Test Answers

Pretest — Chapter 5, Page 2

15. Use counters to solve.
Dale has 6 🐱.
Then 4 run away.
Ring how many are left.
(2) 4 6 10

16. Use counters to solve.
There are 8 🚗.
Then 1 leaves.
Ring how many are left.
6 (7) 8 9

17.
3 + 4 = 7
7 − 4 = __3__

18.
4 + 5 = 9
9 − 5 = __4__

19. Ring the subtraction fact that goes with this addition fact.
2 + 7 = 9
7 − 5 = 2   9 − 5 = 4
7 − 2 = 5   (9 − 7 = 2)

20. Ring the subtraction fact that goes with this addition fact.
6 + 4 = 10
6 − 4 = 2   (10 − 6 = 4)
4 − 2 = 2   10 − 2 = 8

Pretest — Chapter 5, Page 3

21. Ring the correct number pair.
The sum is 6.
The difference is 4.
4 and 2   (5 and 1)
3 and 3   6 and 2

22. Ring the correct number pair.
The sum is 7.
The difference is 7.
4 and 3   7 and 7
(7 and 0)   6 and 1

23. There are 5 👧.
There is 1 👦.
Ring how many more 👧 than 👦 there are.
3 more   5 more
(4 more)   6 more

24. There are 8 🥛.
There are 4 ☕.
Ring how many more 🥛 than ☕ there are.
(4 more)   8 more
6 more   10 more

Posttest — Chapter 5, Page 1

Read each question. Find the answer.

1. Subtract.
0 1 2 3 4 5 6 7 8 9 10
8 − 1 = __7__

2. Subtract.
0 1 2 3 4 5 6 7 8 9 10
7 − 4 = __3__

3. Subtract.
0 1 2 3 4 5 6 7 8 9 10
10 − 4 = __6__

4. Subtract.
0 1 2 3 4 5 6 7 8 9 10
6 − 2 = __4__

5. Subtract.
0 1 2 3 4 5 6 7 8 9 10
9 − 1 = __8__

6. Subtract.
0 1 2 3 4 5 6 7 8 9 10
8 − 3 = __5__

7. 9 − 2 = 7

8. 10 − 8 = 2

9. 7 − 5 = 2

10. 5 − 5 = 0

11. 10 − 3 = __7__

12. 8 − 5 = __3__

13. 6 − 0 = __6__

14. 9 − 6 = __3__

Posttest — Chapter 5, Page 2

15. Use counters to solve.
There are 5 🚗.
Then 1 leaves.
Ring how many are left.
1   5
(4) 6

16. Use counters to solve.
There are 9 🐦.
Then 4 fly away.
Ring how many are left.
4   (5)
6   7

17.
4 + 3 = 7
7 − 3 = __4__

18.
3 + 5 = 8
8 − 5 = __3__

19. Ring the subtraction fact that goes with this addition fact.
6 + 1 = 7
6 − 1 = 5   8 − 1 = 7
1 − 1 = 0   (7 − 1 = 6)

20. Ring the subtraction fact that goes with this addition fact.
5 + 5 = 10
5 − 0 = 5   (10 − 5 = 5)
5 − 5 = 0   10 − 0 = 10

Free Response Format • Test Answers

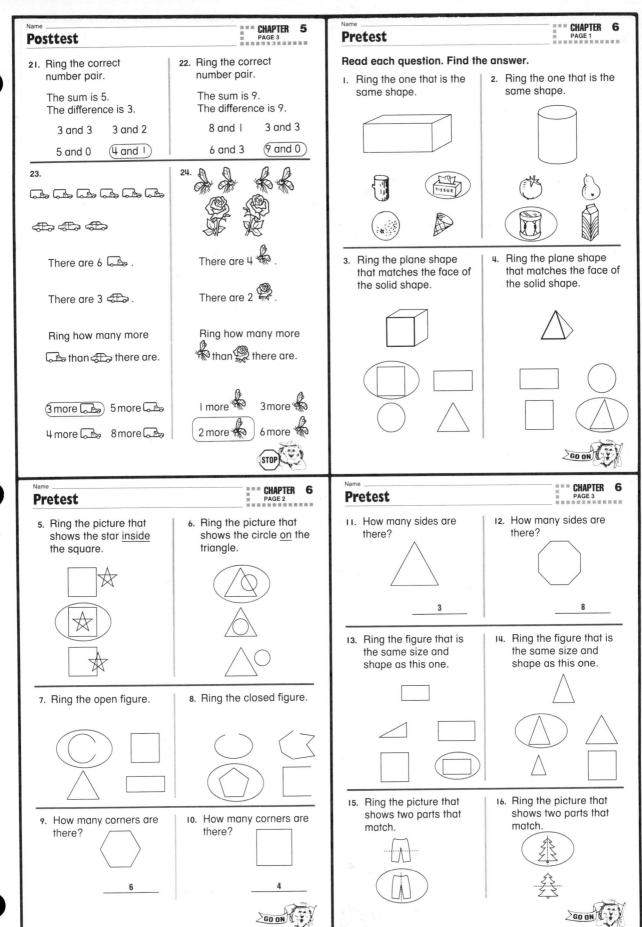

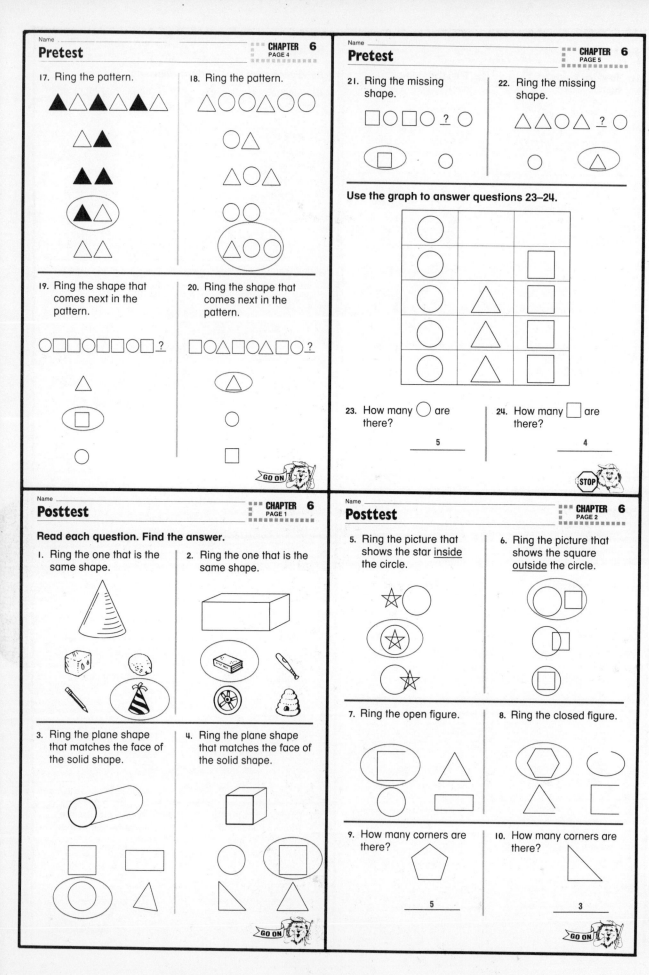

Posttest — Chapter 6, Page 3

11. How many sides are there?

    4

12. How many sides are there?

    6

13. Ring the figure that is the same size and shape as this one.

14. Ring the figure that is the same size and shape as this one.

15. Ring the picture that shows two parts that match.

16. Ring the picture that shows two parts that match.

Posttest — Chapter 6, Page 4

17. Ring the pattern.

18. Ring the pattern.

19. Ring the shape that comes next in the pattern.

20. Ring the shape that comes next in the pattern.

Posttest — Chapter 6, Page 5

21. Ring the missing shape.

22. Ring the missing shape.

Use the graph to answer questions 23–24.

23. How many △ are there?

    2

24. How many ▢ are there?

    4

Cumulative Test — Chapters 1–6, Page 1

Read each question. Find the answer.

1. How many ▢ are there?

    6

2. How many ☆ are there?

    7

3. How many ○ are there?

    4

4. How many △ are there?

    9

5. Ring the group that has fewer than this.

6. Ring the group that has more than this.

Free Response Format • Test Answers

Cumulative Test — Chapters 1–6, Page 2

7. Write the number that comes next.

2, 3, 4, __5__

8. Write the number that comes next.

6, 7, 8, __9__

9. Ring the addition sentence that tells how many in all.

(2 + 2 = 4)
2 + 0 = 2
4 + 4 = 8
4 + 2 = 6

10. Ring the subtraction sentence that tells the story.

3 − 1 = 2
2 − 1 = 1
3 − 3 = 0
(4 − 1 = 3)

11. 7 + 2 = __9__

12. 6 − 4 = __2__

13. 1 + 4 = __5__

14. 0 + 3 = __3__

Cumulative Test — Chapters 1–6, Page 3

15. 6 + 2 = __8__

16. 5 + 5 = __10__

17.
  8
 +1
 ―
  9

18.
  3
 +4
 ―
  7

19. Ring another name for 10.

3 + 4   (9 + 1)
6 + 1   9 + 0

20. Ring another name for 8.

3 + 3   6 + 1
4 + 5   (7 + 1)

21.
  4
 −2
 ―
  2

22.
 10
 −6
 ―
  4

23. 7 − 1 = __6__

24. 2 − 0 = __2__

Cumulative Test — Chapters 1–6, Page 4

25. 8 − 3 = __5__

26. 6 − 5 = __1__

27. 5 dogs are playing. 1 dog runs away. How many are left?

__4__ dogs

28. There are 9 buses. Then 2 buses leave. How many are left?

__7__ buses

29. Ring the one that is the same shape.

30. Ring the one that is the same shape.

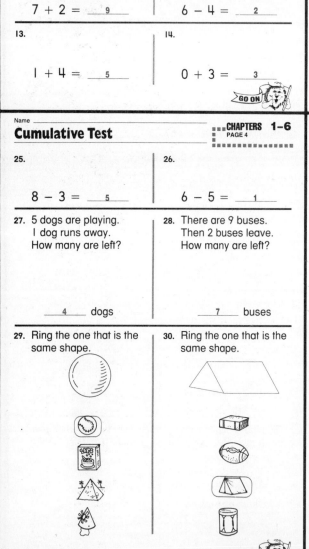

Cumulative Test — Chapters 1–6, Page 5

31. Ring the plane shape that matches the face of the solid shape.

32. Ring the plane shape that matches the face of the solid shape.

33. Ring the plane shape that matches the face of the solid shape.

34. Ring the one that is the same size and shape.

35. Ring the one that is the same size and shape.

36. Ring the picture that shows two parts that match.

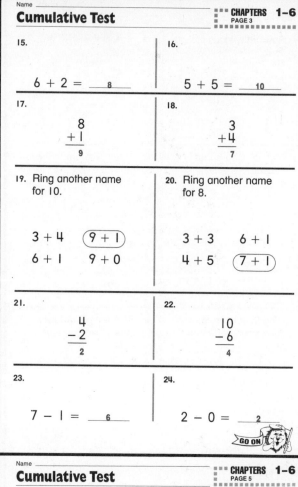
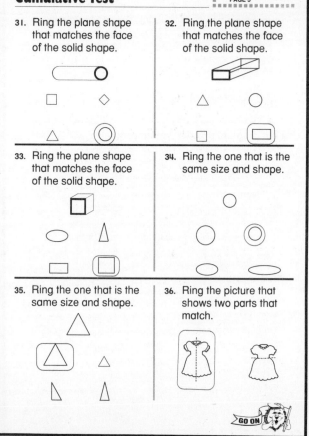

Free Response Format • Test Answers

Cumulative Test — CHAPTERS 1–6 PAGE 6

37. Ring the picture that shows two parts that match.

38. Ring the pattern.

39. Ring the pattern.

40. Ring the shape that comes next in the pattern.

41. Ring the shape that comes next in the pattern.

42. Ring the shape that comes next in the pattern.

Cumulative Test — CHAPTERS 1–6 PAGE 7

43. Ring the missing shape.

44. Ring the number pair. The sum is 5. The difference is 3.

2 and 3   3 and 6
(1 and 4)   3 and 5

45. Ring what you do to find how many 🐟 are left.

Add   (Subtract)

46. Ring what you do to find how many 🐕 in all.

(Add)   Subtract

47. How much did Rusty spend?

10¢

48. There are 4 🌷. There are 5 🌼. Ring how many more 🌼 than 🌷 there are.

(1 more 🌼)
4 more 🌼
5 more 🌼
9 more 🌼

Cumulative Test — CHAPTERS 1–6 PAGE 8

Use the pictograph below to answer questions 49–50.

49. Ring the one for the group that has fewer.

50. Ring the one for the group that has more than 6.

Pretest — CHAPTER 7 PAGE 1

Read each question. Find the answer.

1. Ring how many tens and ones there are.

(1 ten 4 ones)
4 tens 1 one
4 tens 10 ones
10 tens 1 one

2. Ring how many tens and ones there are.

30 tens 2 ones
1 ten 3 ones
(3 tens 2 ones)
2 tens 3 ones

3. Ring how many tens and ones there are.

(4 tens 0 ones)
0 tens 4 ones
1 ten 4 ones
10 tens 1 one

4. Ring how many tens and ones there are.

3 tens 0 ones
20 tens 1 one
1 ten 2 ones
(2 tens 1 one)

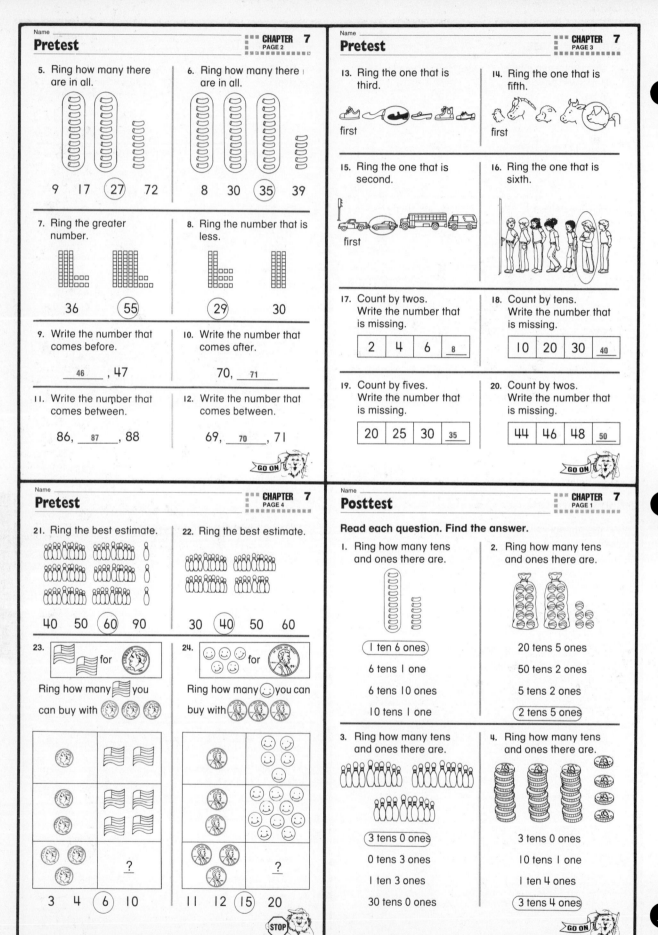

Posttest — Chapter 7, Page 2

5. Ring how many there are in all.

9   26   (36)   63

6. Ring how many there are in all.

10   20   18   (28)

7. Ring the greater number.
49   (50)

8. Ring the number that is less.
(38)   71

9. Write the number that comes before.
___85___ , 86

10. Write the number that comes after.
50, ___51___

11. Write the number that comes between.
44, ___45___ , 46

12. Write the number that comes between.
79, ___80___ , 81

Posttest — Chapter 7, Page 3

13. Ring the one that is fourth.

first

14. Ring the one that is sixth.

first

15. Ring the one that is second.

first

16. Ring the one that is fifth.
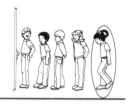

17. Count by fives. Write the number that is missing.
| 5 | 10 | 15 | _20_ |

18. Count by tens. Write the number that is missing.
| 10 | 20 | 30 | _40_ |

19. Count by twos. Write the number that is missing.
| 54 | 56 | 58 | _60_ |

20. Count by fives. Write the number that is missing.
| 15 | 20 | 25 | _30_ |

Posttest — Chapter 7, Page 4

21. Ring the best estimate.

3   10   (30)   40

22. Ring the best estimate.

10   50   60   (70)

23. 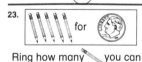 for
Ring how many you can buy with .
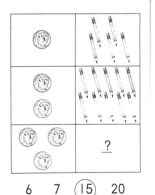
6   7   (15)   20

24. for
Ring how many  you can buy with .
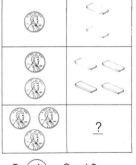
5   (6)   8   10

Pretest — Chapter 8, Page 1

Read each question. Find the answer.

1. Ring the addition sentence that tells how many in all.
3 + 5 = 8
7 + 2 = 9
1 + 3 = 4
(8 + 3 = 11)

2. Ring the addition sentence that tells how many in all.
3 + 3 = 6
7 + 7 = 14
(6 + 6 = 12)
5 + 5 = 10

3. Ring the addition sentence that tells how many in all.
4 + 1 = 5
(4 + 5 = 9)
5 + 5 = 10
3 + 4 = 7

4. Ring the addition sentence that tells how many in all.
(8 + 1 = 9)
7 + 1 = 8
9 + 1 = 10
9 + 9 = 18

Free Response Format • Test Answers

Pretest — Chapter 8, Page 2

5. $2 + 9 = \underline{11}$
6. $0 + 6 = \underline{6}$
7. $7 + 3 = \underline{10}$
8. $8 + 4 = \underline{12}$

9. Sam catches 5 fish. Dad catches 7 fish. How many fish do they catch in all?

    $\underline{12}$ fish

10. Meg has 5 pet birds. She has 5 pet fish. How many pets does she have in all?

    $\underline{10}$ pets

11. Ring the subtraction sentence that tells the story.

    $(11 - 3 = 8)$ $5 - 3 = 2$
    $8 - 3 = 5$ $7 - 3 = 4$

12. Ring the subtraction sentence that tells the story.

    $9 - 3 = 6$ $(12 - 9 = 3)$
    $6 - 3 = 3$ $11 - 8 = 3$

Pretest — Chapter 8, Page 3

13. Ring the subtraction sentence that tells the story.

    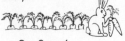

    $8 - 2 = 6$
    $6 - 2 = 4$
    $12 - 2 = 10$
    $(10 - 2 = 8)$

14. Ring the subtraction sentence that tells the story.

    $6 - 3 = 3$
    $6 - 6 = 0$
    $(9 - 3 = 6)$
    $12 - 6 = 6$

15. $11 - 6 = \underline{5}$
16. $9 - 0 = \underline{9}$
17. $10 - 1 = \underline{9}$
18. $12 - 8 = \underline{4}$

19. Mom buys 12 apples. Pam eats 3 apples. How many apples are left?

    $\underline{9}$

20. There were 11 cups. 5 were broken. How many cups are left?

    $\underline{6}$

Pretest — Chapter 8, Page 4

21. $7¢ + 4¢ = 11¢$
22. $9¢ - 2¢ = 7¢$
23. $8¢ - 7¢ = 1¢$
24. $6¢ + 6¢ = 12¢$

25. Ring the number sentence that does **not** belong to the fact family.

    $7 + 4 = 11$
    $11 - 7 = 4$
    $(3 + 4 = 7)$
    $4 + 7 = 11$

26. Ring the number sentence that does **not** belong to the fact family.

    $9 - 1 = 8$
    $1 + 8 = 9$
    $9 - 8 = 1$
    $(7 + 1 = 8)$

27. Ring the number sentence that belongs to the same fact family as $10 - 6 = 4$.

    $(6 + 4 = 10)$
    $10 + 6 = 16$
    $6 - 4 = 2$
    $10 - 5 = 5$

28. Ring the number sentence that belongs to the same fact family as $12 - 7 = 5$.

    $2 + 5 = 7$
    $7 - 5 = 2$
    $6 + 6 = 12$
    $(5 + 7 = 12)$

Pretest — Chapter 8, Page 5

Use the graph below to answer questions 29–30.

FAVORITE FRUITS

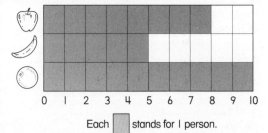

Each ▪ stands for 1 person.

29. How many people like apple best?

    $\underline{8}$ people

30. Ring the fruit that is the favorite of the most people.

    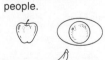 (orange ringed)

31. There are 8 black puppies. There are 4 white puppies. What do you do to find how many puppies there are in all?

    (Add) Subtract

32. Roy had 10 pennies. He lost 3 pennies. What do you do to find how many pennies are left?

    Add (Subtract)

Free Response Format • Test Answers

Name _____

Posttest CHAPTER 8 PAGE 1

Read each question. Find the answer.

1. Ring the addition sentence that tells how many in all.

   8 + 3 = 11
   6 + 3 = 9
   7 + 3 = 10
   (9 + 3 = 12)

2. Ring the addition sentence that tells how many in all.

   1 + 10 = 11
   (1 + 9 = 10)
   0 + 9 = 9
   8 + 1 = 9

3. Ring the addition sentence that tells how many in all.

   4 + 2 = 6
   (6 + 4 = 10)
   6 + 5 = 11
   4 + 4 = 8

4. Ring the addition sentence that tells how many in all.

   2 + 5 = 7
   7 + 1 = 8
   9 + 1 = 10
   (7 + 2 = 9)

Name _____

Posttest CHAPTER 8 PAGE 2

5. 8 + 3 = __11__   6. 9 + 0 = __9__

7. 5 + 5 = __10__   8. 2 + 8 = __10__

9. 2 ants are on a hill. 9 more ants come up the hill. How many ants in all?

   __11__ ants

10. There are 4 red balls. There are 8 green balls. How many balls in all?

    __12__ balls

11. Ring the subtraction sentence that tells the story.

    (11 − 4 = 7)   7 − 5 = 2
    7 − 4 = 3   11 − 3 = 8

12. Ring the subtraction sentence that tells the story.

    5 − 4 = 1   10 − 4 = 6
    8 − 4 = 4   (9 − 4 = 5)

Name _____

Posttest CHAPTER 8 PAGE 3

13. Ring the subtraction sentence that tells the story.

    6 − 6 = 0
    10 − 6 = 4
    (12 − 6 = 6)
    12 − 5 = 7

14. Ring the subtraction sentence that tells the story.

    6 − 5 = 1
    (11 − 6 = 5)
    10 − 6 = 4
    12 − 7 = 5

15. 11 − 3 = __8__

16. 10 − 8 = __2__

17. 9 − 3 = __6__

18. 12 − 7 = __5__

19. Mom buys 12. The children eat 9. How many are left?

    __3__

20. There were 9. 8 were broken. How many are left?

    __1__

Name _____

Posttest CHAPTER 8 PAGE 4

21. 5¢
   + 6¢

    11¢

22. 12¢
   − 4¢

    8¢

23. 11¢
   − 9¢

    2¢

24. 1¢
   + 8¢

    9¢

25. Ring the number sentence that does not belong to the fact family.

    3 + 6 = 9
    9 − 3 = 6
    (9 + 3 = 12)
    9 − 6 = 3

26. Ring the number sentence that does not belong to the fact family.

    3 + 7 = 10
    (3 + 4 = 7)
    10 − 3 = 7
    10 − 7 = 3

27. Ring the number sentence that belongs to the same fact family as 7 + 4 = 11.

    (11 − 7 = 4)
    7 + 0 = 7
    3 + 4 = 7
    7 − 4 = 3

28. Ring the number sentence that belongs to the same fact family as 12 − 5 = 7.

    2 + 5 = 7
    7 − 5 = 2
    (7 + 5 = 12)
    6 + 6 = 12

Free Response Format • Test Answers

Posttest — Chapter 8, Page 5

Use the graph below to answer questions 29–30.

FAVORITE LUNCHES

0 1 2 3 4 5 6 7 8 9 10

Each ▩ stands for 1 person.

29. How many people like 🥣 best?

___8___ people

30. Ring the favorite lunch of the most people.

31. 12 cats were sleeping. 6 cats woke up. What do you do to find how many cats are still sleeping?

Add (Subtract)

32. 8 chickens are sitting on nests. 8 chickens are eating. What do you do to find how many chickens there are in all?

(Add) Subtract

Pretest — Chapter 9, Page 1

Read each question. Find the answer.

1. Ring the one that is shortest.

2. Ring the one that is longest.

3. How long is the fork?

___7___

4. How long is the knife?

___4___ centimeters

5. How long is the pin?

___1___ inches

6. How long is the whistle?

___2___ inches

Pretest — Chapter 9, Page 2

7. How long is the ribbon?

___2___ inches

8. How long is the comb?

___6___ centimeters

9. How long is the paper clip?

___5___ centimeters

10. How long is the pin?

___3___ centimeters

11. Ring the one that is heavier.

12. Ring the one that is heavier.

Pretest — Chapter 9, Page 3

13. Ring the one that holds more.

14. Ring the one that is hot.

15. Ring the one that holds about the same as a 🫙.

16. Ring the one that holds about the same as a 🥤.

17. Ring the one that is about as heavy as a 🎲.

18. Ring the one that is about as heavy as a ▭.

19. Ring the one that shows fair shares.

20. Ring the one that shows fair shares.

Free Response Format • Test Answers

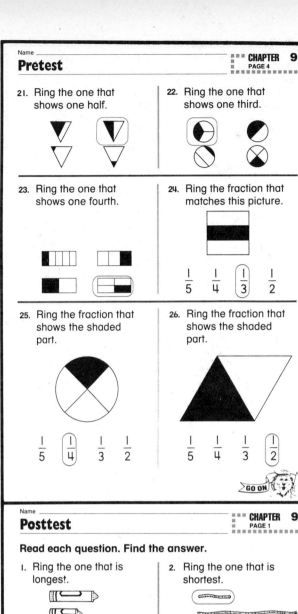
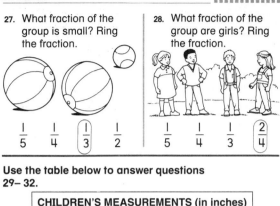
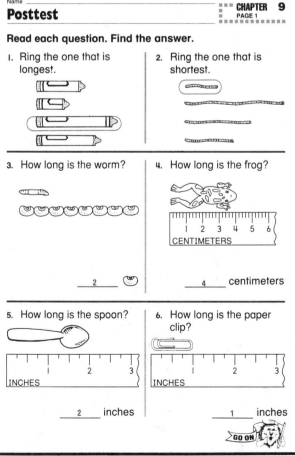
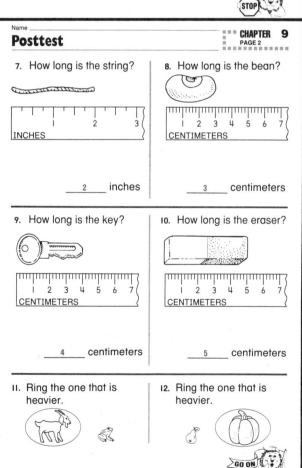

Posttest — Chapter 9, Page 3

13. Ring the one that holds more.
14. Ring the one that is hot.
15. Ring the one that holds about the same as a bottle.
16. Ring the one that holds about the same as a mug.
17. Ring the one that is about as heavy as an apple.
18. Ring the one that is about as heavy as an eraser.
19. Ring the one that shows fair shares.
20. Ring the one that shows fair shares.

Posttest — Chapter 9, Page 4

21. Ring the one that shows one half.
22. Ring the one that shows one fourth.
23. Ring the one that shows one third.
24. Ring the fraction that matches this picture.
    $\frac{1}{5}$ $\frac{1}{4}$ $\frac{1}{3}$ $\boxed{\frac{1}{2}}$
25. Ring the fraction that matches this picture.
    $\frac{1}{5}$ $\frac{1}{4}$ $\boxed{\frac{1}{3}}$ $\frac{1}{2}$
26. Ring the fraction that matches this picture.
    $\frac{1}{5}$ $\boxed{\frac{1}{4}}$ $\frac{1}{3}$ $\frac{1}{2}$

Posttest — Chapter 9, Page 5

27. What fraction of the group is large? Ring the fraction.
    $\frac{1}{5}$ $\boxed{\frac{1}{4}}$ $\frac{1}{3}$ $\frac{1}{2}$
28. What fraction of the group are boys? Ring the fraction.
    $\frac{1}{5}$ $\frac{1}{4}$ $\boxed{\frac{1}{3}}$ $\frac{1}{2}$

Use the table below to answer questions 29–32.

LENGTHS (in inches)			
Object	Red	Blue	Yellow
Crayon	2	5	3
Pencil	5	6	4
Chalk	1	6	3

29. Which crayon is shortest? __red__
30. Which chalk is longest? __blue__
31. How long is the yellow pencil? __4__ inches
32. How long is the red chalk? __1__ inches

Cumulative Test — Chapters 1–9, Page 1

Read each question. Find the answer.

1. $6 - 1 = \underline{5}$
2. $5 + 3 = \underline{8}$
3. Ring the plane shape that matches the face of the solid shape.
4. Ring the plane shape that matches the face of the solid shape.
5. Ring the figure that is the same size and shape.
6. Ring the picture that shows two parts that match.

Free Response Format • Test Answers

Cumulative Test — Chapters 1-9, Page 2

7. Ring the shape that comes next in the pattern.
△ ▢ ○ △ ▢ ○ △ ?
△ ⬛(ringed) ○

8. Ring the shape that comes next in the pattern.
○ ○ ▢ ○ ○ ○ ?
⬭(ringed) ○ ▢

9. Ring how many tens and ones there are.
2 tens and 7 ones
3 tens and 3 ones
(2 tens and 4 ones) ← ringed
10 tens and 7 ones

10. Ring how many tens and ones there are.
10 tens and 2 ones
3 tens and 0 ones
2 tens and 1 one
(1 ten and 2 ones) ← ringed

11. Ring how many there are in all.
8
(26) ← ringed
27
62

12. Ring how many there are in all.
5
6
15
(51) ← ringed

Cumulative Test — Chapters 1-9, Page 3

13. Ring the number that is less.
(47) ← ringed   50

14. Ring the number that is greater.
18   (21) ← ringed

15. Write the number that comes after.
61, __62__

16. Write the number that comes between.
49, __50__, 51

17. Ring the one that is third.
first — (third cow ringed)

18. Ring the one who is second.
(second person ringed)

Cumulative Test — Chapters 1-9, Page 4

19. 4 + 5 = __9__

20. 2 + 4 = __6__

21. 9 + 3 = __12__

22. 6 − 0 = __6__

23. There are 3 red hats. There are 7 blue hats. How many hats are there in all?
__10__ hats

24. Roy has 6 books. Nancy has 5 books. How many books do they have in all?
__11__ books

25. 11 − 7 = __4__

26. There were 12 eggs. 4 were broken. How many are left?
__8__ eggs

27. There were 9 birds. Then 3 flew away. How many birds are left?
__6__ birds

28. 8¢ − 3¢ = 5¢

Cumulative Test — Chapters 1-9, Page 5

29. 6¢ + 3¢ = 9¢

30. 7¢ + 1¢ = 8¢

31. Ring the one that is longest. (longest worm ringed)

32. How long is the crayon?
__4__ long

33. Use your inch ruler. How long is the worm?
__2__ inches

34. Use your centimeter ruler. How long is the stick?
__5__ centimeters

Free Response Format • Test Answers

Cumulative Test — CHAPTERS 1–9, PAGE 6

35. Ring the one that is heavier, or .

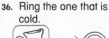

36. Ring the one that is cold.

37. Ring the one that holds about the same as a .

38. Ring the one that is about as heavy as an .

39. Ring the one that shows fair shares.

40. Ring the one that shows one third.

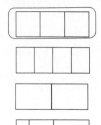

(third box ringed)

Cumulative Test — CHAPTERS 1–9, PAGE 7

41. Ring the fraction that matches this picture.

$\frac{1}{5}$ $\frac{1}{4}$

($\frac{1}{3}$) $\frac{1}{2}$

42. What fraction of the group is small? Ring the answer.

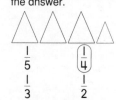

$\frac{1}{5}$ ($\frac{1}{4}$)

$\frac{1}{3}$ $\frac{1}{2}$

43. 2 ✏ for 🪙 Ring how many ✏ you can buy with 🪙🪙.

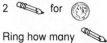

2 ✏

3 ✏

5 ✏

(6 ✏)

44. Ring the number pair. The sum is 7. The difference is 5.

2 and 7
3 and 4
5 and 2
(1 and 6)

Cumulative Test — CHAPTERS 1–9, PAGE 8

45. Ring how much the girl spent.

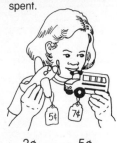

2¢   5¢

7¢   (12¢)

46. There are 3 🔔. There are 5 🗝. How many more 🗝 are there than 🔔?

__2__ more 🗝

Use the table below to answer questions 47–48.

Money to Spend

Children	Dimes	Pennies
Ali	3	2
Barb	1	5
Connie	4	0
Dave	2	3

47. Who has the most dimes? Ring the name.

Ali   Barb
(Connie)   Dave

48. Ring what you do to find how much money Ali and Barb have in all.

(Add)   Subtract

Cumulative Test — CHAPTERS 1–9, PAGE 9

Use the graph below to answer question 49.

Favorite Colors

Red — 4
Blue — 6
Yellow — 3
Green — 5

Each 👤 stands for 1 person.

49. Ring the color that is the favorite of the most people.

(Blue)   Yellow   Green   Red

Use the graph below to answer question 50.

Favorite Toys

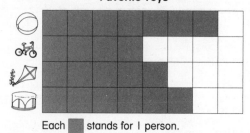

Each ▪ stands for 1 person.

50. Ring the toy that is the favorite of the most people.

(ball)   bicycle   kite   drum

Free Response Format • Test Answers

Pretest
CHAPTER 10 PAGE 1

Read each question. Find the answer.

1. Ring which happened last.

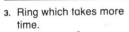

2. Ring which happened first.

3. Ring which takes more time.

4. Ring which takes less time.

Pretest
CHAPTER 10 PAGE 2

5. Ring the one that shows 2 o'clock.

6. Ring the one that shows 7:30.

7. Ring the time it is.

3:00   6:00
(9:00)   12:00

8. Ring the time it is.

(8:30)   9:30
7:30   6:30

9. Ring the time it is.

(11:00)   12:00
1:00   7:00

10. Ring the time it is.

6:00   (12:30)
1:30   not here

Pretest
CHAPTER 10 PAGE 3

Use the calendar below to answer questions 11–14.

			April			
Sunday	Monday	Tuesday	Wednesday	Thursday	Friday	Saturday
			1	2	3	4
5	6	7	8	9	10	11

11. Ring the day on which this month begins.

Friday   Monday
(Wednesday)   Thursday

12. Ring the day of the week that is the third day of this month.

Tuesday   Saturday
(Friday)   Monday

13. Ring the day that comes just after Wednesday.

Tuesday   Friday
(Thursday)   Saturday

14. Ring the day that comes just before Saturday.

(Friday)   Monday
Wednesday   Sunday

Pretest
CHAPTER 10 PAGE 4

15. Ring how much money there is.

5¢   (6¢)
30¢   60¢

16. Ring how much money there is.

3¢   5¢
(15¢) 30¢

17. Ring how much money there is.

3¢   (20¢)
25¢   30¢

18. Ring how much money there is.

25¢   27¢
35¢   (46¢)

Name _____
Pretest CHAPTER 10 PAGE 5

19. Ring how much money there is.

5¢   14¢
24¢ (29¢)

20. Ring how much money there is.

17¢   28¢
(32¢) 37¢

Use coins to answer questions 21–24.

21. Rita has 1 nickel and 2 pennies. How much money does she have?
___7___ ¢

22. Kit has 2 dimes and 2 nickels. How much money does he have?
___30___ ¢

23. Lois has 5 dimes. How much money does she have?
___50___ ¢

24. Ned has 1 quarter, 1 dime, and 3 pennies. How much money does he have?
___38___ ¢

Name _____
Pretest CHAPTER 10 PAGE 6

25. Ring the group of coins that is the same amount as .

26. Ring the group of coins that is the same amount as 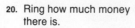.

Name _____
Pretest CHAPTER 10 PAGE 7

27. Ring the group of coins that is the same amount as .

28. Ring the group of coins that is the same amount as .

Name _____
Pretest CHAPTER 10 PAGE 8

29. The movie will be over in 1 hour. Ring the time the movie will be over.

7:00   6:00
8:30   (9:00)

30. The plane will take off in 30 minutes. Ring the time the plane will take off.

(3:00)   2:00
3:30   5:30

31. Karen is going to play for 30 minutes. Ring the time she will finish.

5:00   (4:30)
7:30   7:00

32. Art class will be over in 1 hour. Ring the time art class will be over.

(11:30) 1:00
9:30   11:00

Free Response Format • Test Answers

Posttest

CHAPTER 10 PAGE 1

Read each question. Find the answer.

1. Ring which happened last.

2. Ring which happened first.

3. Ring which takes more time.

4. Ring which takes less time.

Posttest

CHAPTER 10 PAGE 2

5. Ring the one that shows 1 o'clock.

6. Ring the one that shows 5:30.

7. Ring the time it is.

2:00   4:00
(3:00) 12:00

8. Ring the time it is.

(4:30) 3:30
8:30   6:30

9. Ring the time it is.

1:00   12:00
(11:00) 2:00

10. Ring the time it is.

5:30   (6:30)
7:30   not here

Posttest

CHAPTER 10 PAGE 3

Use the calendar below to answer questions 11–14.

May						
Sunday	Monday	Tuesday	Wednesday	Thursday	Friday	Saturday
					1	2
3	4	5	6	7	8	9

11. Ring the day on which this month begins.

Sunday   (Friday)
Saturday Monday

12. Ring the day of the week that is the third day of this month.

Tuesday   (Sunday)
Saturday Thursday

13. Ring the day that comes just after Tuesday.

Monday   Friday
(Wednesday) Saturday

14. Ring the day that comes just before Friday.

(Thursday) Saturday
Sunday   Tuesday

Posttest

CHAPTER 10 PAGE 4

15. Ring how much money there is.

(4¢)   5¢
20¢   40¢

16. Ring how much money there is.

6¢   16¢
30¢   (60¢)

17. Ring how much money there is.

5¢   21¢
(30¢) 50¢

18. Ring how much money there is.

5¢   17¢
(22¢) 35¢

Free Response Format • Test Answers

Posttest — Chapter 10, Page 5

19. Ring how much money there is.

4¢   (28¢)
31¢   40¢

20. Ring how much money there is.

6¢   26¢
(41¢)   51¢

Use coins to answer questions 21–24.

21. Rae has 2 nickels and 1 penny. How much money does she have?

___11___ ¢

22. Kim has 1 dime and 3 nickels. How much money does she have?

___25___ ¢

23. Lola has 6 dimes. How much money does she have?

___60___ ¢

24. Nick has 1 quarter, 1 dime, and 1 nickel. How much money does he have?

___40___ ¢

Posttest — Chapter 10, Page 6

25. Ring the group of coins that is the same amount as [dime].

26. Ring the group of coins that is the same amount as [quarter].

Posttest — Chapter 10, Page 7

27. Ring the group of coins that is the same amount as [3 dimes].

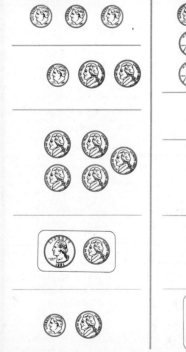

28. Ring the group of coins that is the same amount as [coins].

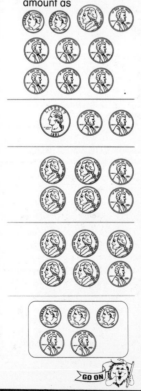

Posttest — Chapter 10, Page 8

29. The movie will be over in 1 hour. Ring the time the movie will be over.

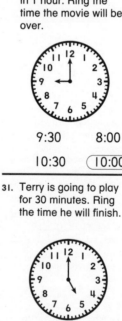

9:30   8:00
10:30   (10:00)

30. The school day will end in 30 minutes. Ring the time the school day will end.

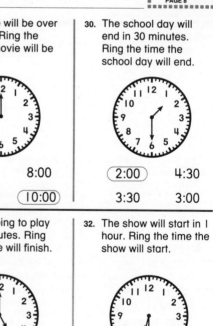

(2:00)   4:30
3:30   3:00

31. Terry is going to play for 30 minutes. Ring the time he will finish.

4:30   (5:30)
8:00   6:00

32. The show will start in 1 hour. Ring the time the show will start.

(7:30)   5:30
7:00   8:30

Free Response Format • Test Answers

Pretest — Chapter 11, Page 1

Read each question. Find the answer.

1. $3 + 8 = \underline{\ 11\ }$
2. $6 + 0 = \underline{\ 6\ }$
3. $7 - 5 = \underline{\ 2\ }$
4. $9 - 1 = \underline{\ 8\ }$

5. Ring the number of cubes in all.

   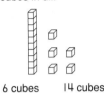

   6 cubes   14 cubes
   13 cubes   (15 cubes)

6. Ring the number of cubes in all.

   3 cubes   (30 cubes)
   20 cubes   33 cubes

7. Ring the number of cubes in all.

   10 cubes   36 cubes
   30 cubes   (37 cubes)

8. Ring the number of cubes in all.

   (46 cubes)   64 cubes
   60 cubes   66 cubes

Pretest — Chapter 11, Page 2

9. Ring the addition sentence that tells how many cubes in all.

   $4 + 1 = 5$
   $10 + 3 = 13$
   $10 + 10 = 20$
   $(13 + 10 = 23)$

10. Ring the addition sentence that tells how many cubes in all.

    $2 + 6 = 8$
    $3 + 7 = 10$
    $20 + 10 = 30$
    $(22 + 16 = 38)$

11. $\begin{array}{r} 10 \\ +60 \\ \hline 70 \end{array}$

12. $\begin{array}{r} 43 \\ +21 \\ \hline 64 \end{array}$

13. $\begin{array}{r} 32 \\ +54 \\ \hline 86 \end{array}$

14. $\begin{array}{r} 33 \\ +33 \\ \hline 66 \end{array}$

15. Chris has 10 bats. Pat has 12 bats. How many bats in all?
    $\underline{\ 22\ }$ bats

16. There are 23 girls and 24 boys playing. How many children are playing?
    $\underline{\ 47\ }$ children

Pretest — Chapter 11, Page 3

17. Tod found 35 nuts. Bob found 14 nuts. How many nuts are there?
    $\underline{\ 49\ }$ nuts

18. Sara had 20 berries. She picked 16 more. How many does she have?
    $\underline{\ 36\ }$ berries

19. Ring the subtraction sentence that tells the story.

    $(22 - 10 = 12)$
    $12 - 10 = 2$
    $12 - 2 = 10$
    $22 - 11 = 11$

20. Ring the subtraction sentence that tells the story.

    $(23 - 13 = 10)$
    $13 - 10 = 3$
    $23 - 10 = 13$
    $10 - 3 = 7$

21. $\begin{array}{r} 40 \\ -20 \\ \hline 20 \end{array}$

22. $\begin{array}{r} 68 \\ -41 \\ \hline 27 \end{array}$

Pretest — Chapter 11, Page 4

23. $\begin{array}{r} 54 \\ -20 \\ \hline 34 \end{array}$

24. $\begin{array}{r} 49 \\ -39 \\ \hline 10 \end{array}$

25. There were 19. Then 10 were sold. How many are left?
    $\underline{\ 9\ }$

26. Kate had 53. She gave away 22. How many are left?
    $\underline{\ 31\ }$

27. There were 24. Then 11 ran away. How many are left?
    $\underline{\ 13\ }$

28. There were 17. Then 12 flew away. How many are left?
    $\underline{\ 5\ }$

29. Mona had 63¢. Then she spent 21¢. Ring about how much money she has now.
    (about 40¢)   about 50¢

30. Sasha spent 48¢. Then she spent 31¢. Ring about how much she spent in all.
    about 70¢   (about 80¢)

31. Ned had 25. He gave away 15. How many does he have now?
    $\underline{\ 10\ }$

32. Mike is 4 years old. His sister is 3 years older. How old is Mike's sister?
    $\underline{\ 7\ }$ years old

Free Response Format • Test Answers

Posttest — Chapter 11, Page 1

Read each question. Find the answer.

1. $4 + 7 = \underline{11}$
2. $0 + 7 = \underline{7}$
3. $6 - 6 = \underline{0}$
4. $11 - 8 = \underline{3}$

5. Ring the number of cubes in all.

   7 cubes   15 cubes
   14 cubes   (16 cubes)

6. Ring the number of cubes in all.

   2 cubes   (20 cubes)
   10 cubes   22 cubes

7. Ring the number of cubes in all.

   (25 cubes)   52 cubes
   26 cubes   62 cubes

8. Ring the number of cubes in all.

   (38 cubes)   83 cubes
   80 cubes   88 cubes

Posttest — Chapter 11, Page 2

9. Ring the addition sentence that tells how many cubes in all.

   $1 + 5 = 6$
   $(10 + 14 = 24)$
   $9 + 13 = 22$
   $10 + 5 = 15$

10. Ring the addition sentence that tells how many cubes in all.
    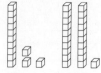
    $10 + 3 = 13$
    $(13 + 21 = 34)$
    $4 + 3 = 7$
    $13 + 20 = 33$

11. $\begin{array}{r} 70 \\ +10 \\ \hline 80 \end{array}$

12. $\begin{array}{r} 54 \\ +21 \\ \hline 75 \end{array}$

13. $\begin{array}{r} 36 \\ +12 \\ \hline 48 \end{array}$

14. $\begin{array}{r} 22 \\ +22 \\ \hline 44 \end{array}$

15. Carl has 14 tops. Pam has 10 tops. How many tops in all?
    $\underline{24}$ tops

16. There are 42 boys and 32 girls playing. How many children are playing?
    $\underline{74}$ children

Posttest — Chapter 11, Page 3

17. Toni found 40 rocks. Lee found 13 rocks. How many rocks are there?
    $\underline{53}$ rocks

18. Walt had 36 berries. He picked 23 more. How many does he have?
    $\underline{59}$ berries

19. Ring the subtraction sentence that tells the story.

    $15 - 11 = 4$
    $15 - 2 = 13$
    $26 - 10 = 16$
    $(26 - 11 = 15)$

20. Ring the subtraction sentence that tells the story.

    $13 - 10 = 3$
    $(23 - 10 = 13)$
    $23 - 1 = 22$
    $13 - 3 = 10$

21. $\begin{array}{r} 60 \\ -30 \\ \hline 30 \end{array}$

22. $\begin{array}{r} 57 \\ -31 \\ \hline 26 \end{array}$

Posttest — Chapter 11, Page 4

23. $\begin{array}{r} 45 \\ -20 \\ \hline 25 \end{array}$

24. $\begin{array}{r} 68 \\ -58 \\ \hline 10 \end{array}$

25. There were 18. Then 10 were sold. How many are left?
    $\underline{8}$

26. Tom had 45. He gave away 33. How many are left?
    $\underline{12}$

27. There were 35. Then 14 ran away. How many are left?
    $\underline{21}$

28. There were 18. Then 11 ran away. How many are left?
    $\underline{7}$

29. Erin had 52¢. Then she spent 30¢. Ring about how much money she has now.
    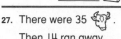
    (about 20¢)   about 30¢

30. Mark spent 52¢. Then he spent 21¢. Ring about how much he spent in all.
    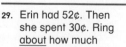
    (about 70¢)   about 80¢

31. Eric had 32. He gave away 11. How many does he have now?
    $\underline{21}$

32. Marie is 15 years old. Her brother is 3 years older. How old is Marie's brother?
    $\underline{18}$ years old

Free Response Format • Test Answers

Pretest — Chapter 12, Page 1

Read each question. Find the answer.

1. Ring the addition sentence that tells how many in all.

   (6 + 8 = 14)
   6 + 2 = 8
   2 + 4 = 6
   8 + 2 = 10

2. Ring the addition sentence that tells how many in all.

   0 + 0 = 0
   (9 + 9 = 18)
   9 + 8 = 17
   9 + 0 = 9

3. Ring the addition sentence that tells how many in all.

   1 + 7 = 8
   8 + 1 = 9
   8 + 9 = 17
   (7 + 8 = 15)

4. Ring the addition sentence that tells how many in all.

   (8 + 9 = 17)
   7 + 1 = 8
   8 + 1 = 9
   8 + 8 = 16

Pretest — Chapter 12, Page 2

5. 5 + 8 = __13__
6. 8 + 8 = __16__
7. 9 + 5 = __14__
8. 6 + 7 = __13__

9. 9
   +7

   16

10. 8 + 6 = __14__

Use counters to answer questions 11–14.

11. Mary has 9 .
    Karl has 4 .
    Write how many there are in all.

    __13__

12. There are 7 red on the rack. There are 7 blue on the rack. Write how many there are in all.

    __14__

Pretest — Chapter 12, Page 3

13. Mark has 9 🌸.
    Pam has 7 🌸.
    Write how many 🌸 there are in all.

    __16__

14. There are 7 red 🧤 on a rack. There are 6 blue 🧤 on the rack. Write how many there are in all.

    __13__

15. Ring the subtraction sentence that tells the story.

    9 − 1 = 8
    (17 − 8 = 9)
    16 − 8 = 8
    18 − 9 = 9

16. Ring the subtraction sentence that tells the story.

    7 − 1 = 6
    14 − 7 = 7
    6 − 6 = 0
    (13 − 7 = 6)

Pretest — Chapter 12, Page 4

17. Ring the subtraction sentence that tells the story.

    9 − 7 = 2
    (16 − 7 = 9)
    9 − 8 = 1
    15 − 9 = 6

18. Ring the subtraction sentence that tells the story.

    (15 − 8 = 7)
    8 − 7 = 1
    8 − 1 = 7
    14 − 8 = 6

19. 15
    − 6

     9

20. 14
    − 7

     7

21. 13 − 7 = __6__

22. 16 − 8 = __8__

Free Response Format • Test Answers

Pretest — Chapter 12, Page 5

Use counters to answer questions 23 and 24.

23. Sally baked 18 🍪.
    Her friends ate 9 🍪.
    How many are left?

    __9__ 🍪

24. There were 15 🎈.
    Then 7 🎈 popped.
    How many are left?

    __8__

25. Ring the number sentence that does not belong to the fact family for 5, 9, and 14.
    - (5 + 4 = 9)
    - 9 + 5 = 14
    - 14 − 9 = 5
    - 14 − 5 = 9

26. Ring the number sentence that does not belong to the fact family for 8, 7, and 15.
    - 15 − 7 = 8
    - 8 + 7 = 15
    - 15 − 8 = 7
    - (8 − 7 = 1)

27. Ring the number sentence that belongs to the same fact family as 13 − 8 = 5.
    - 5 + 3 = 8
    - (8 + 5 = 13)
    - 5 − 3 = 2
    - 8 − 5 = 3

28. Ring the number sentence that belongs to the same fact family as 16 − 7 = 9.
    - 7 + 2 = 9
    - 9 − 7 = 2
    - 14 − 7 = 7
    - (9 + 7 = 16)

Pretest — Chapter 12, Page 6

Use the table below to answer questions 29–30.

FAVORITE SNACKS	Boys	Girls
Popcorn	IIII	HHT III
Fruit	HHT	HHT IIII
Cheese	HHT IIII	HHT I

29. How many boys and girls like cheese best?

    __15__

30. How many boys and girls like fruit best?

    __14__

31.
    There are 9 balloons. There are 3 children. How many balloons does each child get?

    __3__ balloons

32.
    There are 16 stars. There are 2 children. How many stars does each child get?

    __8__ stars

Posttest — Chapter 12, Page 1

Read each question. Find the answer.

1. Ring the addition sentence that tells how many in all.
   (stars image)
   - (8 + 7 = 15)
   - 7 + 1 = 8
   - 8 + 2 = 10
   - 5 + 2 = 7

2. Ring the addition sentence that tells how many in all.
   (circles image)
   - 0 + 0 = 0
   - 7 + 0 = 7
   - 6 + 7 = 13
   - (7 + 7 = 14)

3. Ring the addition sentence that tells how many in all.

   - 4 + 5 = 9
   - 9 + 6 = 15
   - 8 + 5 = 13
   - (9 + 5 = 14)

4. Ring the addition sentence that tells how many in all.

   - 5 + 3 = 8
   - (5 + 8 = 13)
   - 3 + 8 = 11
   - 5 + 9 = 14

Posttest — Chapter 12, Page 2

5. 6 + 7 = __13__
6. 9 + 9 = __18__
7. 8 + 6 = __14__
8. 9 + 8 = __17__
9.
   $$\begin{array}{r} 6 \\ + 9 \\ \hline 15 \end{array}$$
10. 8 + 8 = __16__

Use counters to answer questions 11–14.

11. Mark has 7 🌼.
    Pam has 9 🌼.
    Write how many there are in all.

    __16__

12. There are 6 red 🧤 on a rack. There are 7 blue 🧤 on the rack. Write how many there are in all.

    __13__

Free Response Format • Test Answers

Posttest — Chapter 12, Page 3

13. Mary has 9 .
Karl has 5 .
Write how many there are in all.

___14___

14. There are 7 red  on the rack. There are 7 blue  on the rack. Write how many  there are in all.

___14___

15. Ring the subtraction sentence that tells the story.

8 − 8 = 0
(13 − 5 = 8)
14 − 5 = 9
8 − 5 = 3

16. Ring the subtraction sentence that tells the story.

9 − 6 = 3
(15 − 6 = 9)
10 − 4 = 6
15 − 9 = 6

Posttest — Chapter 12, Page 4

17. Ring the subtraction sentence that tells the story.

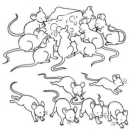

(14 − 6 = 8)
8 − 6 = 2
15 − 7 = 8
13 − 7 = 6

18. Ring the subtraction sentence that tells the story.

(14 − 5 = 9)
9 − 5 = 4
15 − 9 = 6
5 − 5 = 0

19. 16
   − 8
   ─────
      8

20. 15
   − 7
   ─────
      8

21. 17 − 8 = ___9___

22. 13 − 7 = ___6___

Posttest — Chapter 12, Page 5

Use counters to answer questions 23 and 24.

23. Mac picked 16 🍑. He gave 9 to his friend. How many 🍑 are left?

___7___

24. There were 18 🥚. Then 9 🥚 were broken. How many 🥚 are left?

___9___

25. Ring the number sentence that does <u>not</u> belong to the fact family for 4, 9, and 13.

(9 − 4 = 5)
13 − 9 = 4
4 + 9 = 13
9 + 4 = 13

26. Ring the number sentence that does <u>not</u> belong to the fact family for 7, 8, and 15.

(8 − 7 = 1)
15 − 8 = 7
7 + 8 = 15
8 + 7 = 15

27. Ring the number sentence that belongs to the same fact family as 14 − 6 = 8.

8 − 6 = 2
(14 − 8 = 6)
4 + 4 = 8
2 + 4 = 6

28. Ring the number sentence that belongs to the same fact family as 17 − 8 = 9.

9 − 1 = 8
8 + 1 = 9
(8 + 9 = 17)
18 − 9 = 9

Posttest — Chapter 12, Page 6

Use the table below to answer questions 29–30.

THINGS CHILDREN LIKE TO DO	Girls	Boys
Jumping	IIII	HHT IIII
Running	HHT	HHT IIII
Swinging	HHT III	HHT I

29. How many girls and boys like running best?

___14___

30. How many girls and boys like jumping best?

___13___

31.

There are 8 books.
There are 2 children.
How many books does each child get?

___4___ books

32.

There are 15 boats.
There are 5 children.
How many boats does each child get?

___3___ boats

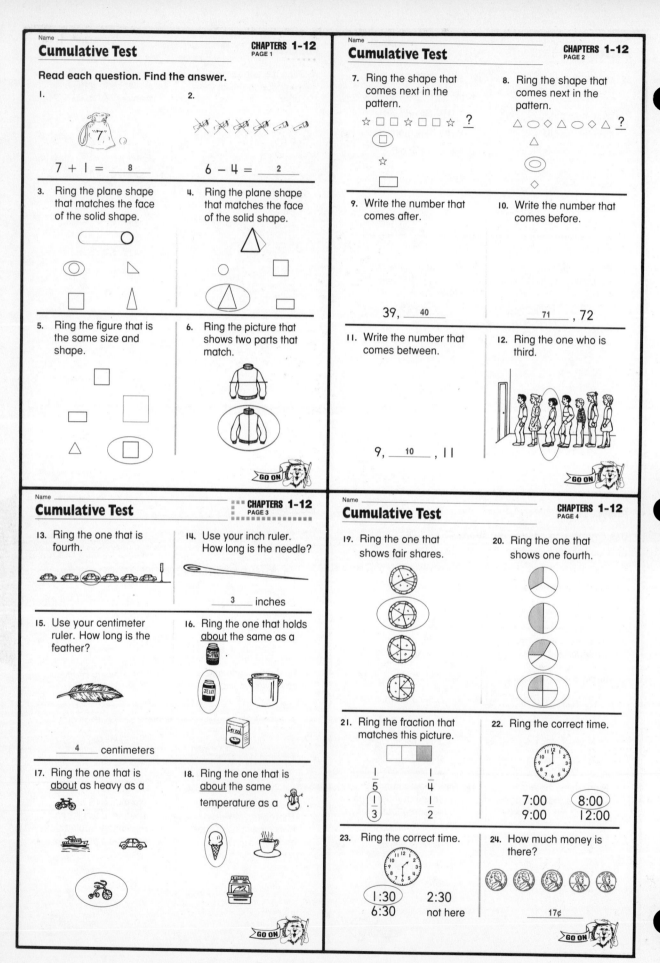

Cumulative Test — CHAPTERS 1-12, PAGE 5

25. How much money is there?

23¢

26. How much money is there?
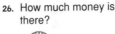
45¢

27. Ring how many cubes there are in all.

- 3 cubes
- 4 cubes
- 10 cubes
- (13 cubes)

28. Ring how many cubes there are in all.

- 4 cubes
- 6 cubes
- 24 cubes
- (42 cubes)

29. Ring the addition sentence that tells how many cubes in all.

- (31 + 24 = 55)
- 4 + 6 = 10
- 30 + 1 = 31
- 30 + 20 = 50

30.
```
  70
+ 14
-----
  84
```

Cumulative Test — CHAPTERS 1-12, PAGE 6

31.
```
  22
+ 22
-----
  44
```

32. Ella has 46 beads. Gus has 21 beads. How many beads are there in all?

____67____ beads

33. Ring the subtraction sentence that tells the story.
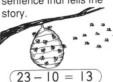
- (23 − 10 = 13)
- 13 − 10 = 3
- 13 − 3 = 10
- 23 − 0 = 23

34.
```
  73
− 21
-----
  52
```

35.
```
  54
− 34
-----
  20
```

36. There were 41 balloons. Then 30 popped. How many are left?

____11____ balloons

Cumulative Test — CHAPTERS 1-12, PAGE 7

37. 9 + 8 = ___17___

38. 7 + 7 = ___14___

39. There are 5 red balls. There are 7 blue balls. How many balls are there in all?

____12____ balls

40. 16 − 9 = ___7___

41.
14 − 6 = ___8___

42. There were 14 grapes. Then Tim ate 5. How many are left?

____9____ grapes

Cumulative Test — CHAPTERS 1-12, PAGE 8

43.

Mother will be home in 1 hour. Ring what time she will be home.
- 4:00
- 5:00
- (6:00)
- 7:00

44.
There are 20 🍪.
There are 10 children.
How many 🍪 will each child get?

____2____ 🍪

45. 3 🔔 for 🪙
Ring how many 🔔 you can buy with 🪙🪙🪙.

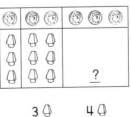

3 🔔   4 🔔
7 🔔   (9 🔔)

46. Ring the number pair. The sum is 8. The difference is 0.
- 1 and 7
- 0 and 8
- 8 and 8
- (4 and 4)

Name _____

Cumulative Test
CHAPTERS 1-12
PAGE 9

Use the graph below to answer question 47.

GLASSES OF MILK IN ONE DAY

Hal	🥛 🥛
Terri	🥛 🥛 🥛 🥛
Ira	🥛 🥛
Nan	🥛 🥛 🥛

Each 🥛 stands for 1 glass.

47. Ring how many glasses of milk Ira drank.

(2 glasses)   3 glasses   4 glasses   5 glasses

Use the graph below to answer question 48.

FAVORITE JUICES

Orange					
Grape					
Apple					
Pineapple					

Each ■ stands for 1 person.

48. Ring the juice that is the favorite of the most people.

Orange   (Grape)   Apple   Pineapple

Name _____

Cumulative Test
CHAPTERS 1-12
PAGE 10

Use the table below to answer questions 49–50.

EATING TIMES (in minutes)			
	Breakfast	Lunch	Dinner
Amy	10	15	25
Bob	10	20	30

49. Ring how long it took Bob to eat lunch.

  10 minutes
  15 minutes
  (20 minutes)
  30 minutes

50. Ring what you do to find how much longer it took Amy to eat dinner than to eat lunch.

  Add   (Subtract)

Name _____

End-of-Book Test
PAGE 1

Read each question. Find the answer.

1. 🐟 🐟 🐟 ❌   2. 💰 ● ●

   $7 - 1 = \underline{6}$ $5 + 2 = \underline{7}$

3. Ring the plane shape that matches the face of the solid shape.

   ▭ ○
   □ △

4. Ring the plane shape that matches the face of the solid shape.

   △ □
   □ ○

5. Ring the figure that is the same size and shape.

   □
   △ □
   ▭ ⬭

6. Ring the picture that shows two parts that match.

   🦋
   🦋

Name _____

End-of-Book Test
PAGE 2

7. Ring the shape that comes next in the pattern.

   ◇ ○ ◇ ○ ?
   ◈ △ ○

8. Ring the shape that comes next in the pattern.

   □ ○ ◇ □ ○ ◇ ?
   ◇ ○ ▭

9. Write the number that comes just before.

   $\underline{30}$, 31

10. Write the number that comes just after.

    68, $\underline{69}$

11. Write the number that comes between.

    44, $\underline{45}$, 46

12. Ring the one who is fifth.

13. Ring the one that is third.

14. Use your centimeter ruler. How long is the key?

    $\underline{3}$ centimeters

Free Response Format • Test Answers

End-of-Book Test — PAGE 3

15. Use your inch ruler. How long is the leaf?

____2____ inches

16. Ring the one that holds about the same as

17. Ring the one that is about as heavy as a .

18. Ring the one that is about the same temperature as .

19. Ring the one that shows fair shares.

20. Ring the one that shows one half.

End-of-Book Test — PAGE 4

21. Ring the fraction that matches this picture.

$\frac{1}{5}$ $\frac{1}{4}$
$\frac{1}{3}$ $\frac{1}{2}$

22. Ring the correct time.

3:00   (4:00)
5:00   12:00

23. Ring the correct time.

6:30   (10:30)
11:30   9:30

24. Ring how much money there is.

5¢   35¢
(40¢)   50¢

25. Ring how much money there is.

5¢   17¢
(27¢)   32¢

26. Ring how much money there is.

10¢   20¢
25¢   (35¢)

27. Ring how many cubes there are in all.

6 cubes   7 cubes
10 cubes   (16 cubes)

28. Ring how many cubes there are in all.

10 cubes   17 cubes
(37 cubes) 73 cubes

End-of-Book Test — PAGE 5

29. Ring the addition sentence that tells how many cubes there are in all.

(23 + 14 = 37)
3 + 4 = 7
5 + 5 = 10
20 + 10 = 30

30.
```
  45
+ 30
----
  75
```

31.
```
  44
+ 44
----
  88
```

32. Tina has 25 beans. Bert has 14 beans. How many beans are there in all?

____39____ beans

33. Ring the subtraction sentence that tells the story.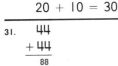

(26 − 12 = 14)
14 − 10 = 4
14 − 2 = 12
12 − 2 = 10

34.
```
  56
− 31
----
  25
```

End-of-Book Test — PAGE 6

35.
```
  63
− 23
----
  40
```

36. Fred won 24 stars. Amy won 10 stars. How many more stars did Fred win?

____14____ more stars

37. 8 + 5 = __13__

38. 9 + 0 = __9__

39. There are 4 white . There are 7 yellow . How many in all?

__11__

40. 17 − 8 = __9__

41. 12 − 5 = __7__

42. Use pennies to solve. Pam has 15 pennies. She spends 8 of them. How many pennies does Pam have left?

____7____ pennies

Free Response Format • Test Answers

End-of-Book Test

PAGE 7

43. Story time will start in 1 hour. Ring the time it will start.

1:00   2:00
(3:00) 12:00

44. There are 12 eggs. There are 3 children. How many eggs does each child get?

____4____ eggs

45. 4 🧶 for 🪙
Ring how many 🧶 you can buy with 🪙 🪙.

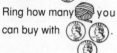

4 🧶   9 🧶
10 🧶   (12 🧶)

46. Ring the number pair. The sum is 6. The difference is 2.

4 and 6
3 and 3
2 and 2
(2 and 4)

End-of-Book Test

PAGE 8

Use the graph below to answer question 47.

PINE CONES FOUND

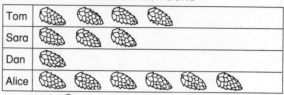

Each 🌲 stands for one pine cone.

47. Ring how many 🌲 Dan found.

(1 pine cone)   3 pine cones
 4 pine cones   6 pine cones

Use the graph below to answer question 48.

FLOWERS PICKED

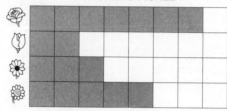

Each ▩ stands for one flower.

48. Ring how many 🌼 were picked.

2 🌼   3 🌼   (5 🌼)   7 🌼

End-of-Book Test

PAGE 9

Use the table below to answer questions 49–50.

Number of Children

	School Bus	City Bus
Boys	14	4
Girls	12	5

49. How many girls are on the school bus?

____12____ girls

50. Ring what you do to find how many children are on the city bus.

(Add)   Subtract

Free Response Format • Test Answers

III
Management Forms

Answer Sheet

This Copying Master is an individual recording sheet for up to 50 items on the standardized-format test.

Grading Made Easy

This percent converter can be used for all quizzes and tests. The percents given are based on all problems having equal value. Percents are rounded to the nearest whole percent giving the benefit of 0.5 percent.

Individual Record Form

One Copying Master for each chapter is provided. Criterion scores for each tested objective are given for the Pretest and Posttest. The student's total scores are recorded at the top of the page. The scores for each objective can also be recorded. You can use the Review Options in the Pupil's Edition, Teacher's Edition, and Workbooks that are listed on the form to assign additional review for the student unable to pass the test.

Formal Assessment Class Record Form

The scores for all tests can be recorded for your class on these record forms. The Criterion Score for each test is given.

Cumulative Record Form

Individual student progress can be recorded on the Copying Masters that are provided. Copies of each grade's tested objectives and criteria are provided. Using these forms, you can find the objectives that will be covered in the next or previous grade.

Name _____ Date _____

Test Answer Sheet

MATHEMATICS PLUS

Test Title _____

1. Ⓐ Ⓑ Ⓒ Ⓓ
2. Ⓐ Ⓑ Ⓒ Ⓓ
3. Ⓐ Ⓑ Ⓒ Ⓓ
4. Ⓐ Ⓑ Ⓒ Ⓓ
5. Ⓐ Ⓑ Ⓒ Ⓓ

6. Ⓐ Ⓑ Ⓒ Ⓓ
7. Ⓐ Ⓑ Ⓒ Ⓓ
8. Ⓐ Ⓑ Ⓒ Ⓓ
9. Ⓐ Ⓑ Ⓒ Ⓓ
10. Ⓐ Ⓑ Ⓒ Ⓓ

11. Ⓐ Ⓑ Ⓒ Ⓓ
12. Ⓐ Ⓑ Ⓒ Ⓓ
13. Ⓐ Ⓑ Ⓒ Ⓓ
14. Ⓐ Ⓑ Ⓒ Ⓓ
15. Ⓐ Ⓑ Ⓒ Ⓓ

16. Ⓐ Ⓑ Ⓒ Ⓓ
17. Ⓐ Ⓑ Ⓒ Ⓓ
18. Ⓐ Ⓑ Ⓒ Ⓓ
19. Ⓐ Ⓑ Ⓒ Ⓓ
20. Ⓐ Ⓑ Ⓒ Ⓓ

21. Ⓐ Ⓑ Ⓒ Ⓓ
22. Ⓐ Ⓑ Ⓒ Ⓓ
23. Ⓐ Ⓑ Ⓒ Ⓓ
24. Ⓐ Ⓑ Ⓒ Ⓓ
25. Ⓐ Ⓑ Ⓒ Ⓓ

26. Ⓐ Ⓑ Ⓒ Ⓓ
27. Ⓐ Ⓑ Ⓒ Ⓓ
28. Ⓐ Ⓑ Ⓒ Ⓓ
29. Ⓐ Ⓑ Ⓒ Ⓓ
30. Ⓐ Ⓑ Ⓒ Ⓓ

31. Ⓐ Ⓑ Ⓒ Ⓓ
32. Ⓐ Ⓑ Ⓒ Ⓓ
33. Ⓐ Ⓑ Ⓒ Ⓓ
34. Ⓐ Ⓑ Ⓒ Ⓓ
35. Ⓐ Ⓑ Ⓒ Ⓓ

36. Ⓐ Ⓑ Ⓒ Ⓓ
37. Ⓐ Ⓑ Ⓒ Ⓓ
38. Ⓐ Ⓑ Ⓒ Ⓓ
39. Ⓐ Ⓑ Ⓒ Ⓓ
40. Ⓐ Ⓑ Ⓒ Ⓓ

41. Ⓐ Ⓑ Ⓒ Ⓓ
42. Ⓐ Ⓑ Ⓒ Ⓓ
43. Ⓐ Ⓑ Ⓒ Ⓓ
44. Ⓐ Ⓑ Ⓒ Ⓓ
45. Ⓐ Ⓑ Ⓒ Ⓓ

46. Ⓐ Ⓑ Ⓒ Ⓓ
47. Ⓐ Ⓑ Ⓒ Ⓓ
48. Ⓐ Ⓑ Ⓒ Ⓓ
49. Ⓐ Ⓑ Ⓒ Ⓓ
50. Ⓐ Ⓑ Ⓒ Ⓓ

HBJ material copyrighted under notice appearing earlier in this work.

Grading Made Easy

MATHEMATICS PLUS

Total Number of Test Items

Number Wrong	4	5	6	7	8	9	10	11	12	13	14	15	16	17	18	19	20	21	22	23	24	25	26	27	28	29	30	32	50
1	75	80	83	86	88	89	90	91	92	92	93	93	94	94	94	95	95	95	95	96	96	96	96	96	96	97	97	97	98
2	50	60	67	71	75	78	80	82	83	85	86	87	88	88	89	89	90	90	91	91	92	92	92	93	93	93	93	94	96
3	25	40	50	57	63	67	70	73	75	77	79	80	81	82	83	84	85	86	86	87	88	88	88	89	89	90	90	91	94
4	0	20	33	43	50	56	60	64	67	69	71	73	75	76	78	79	80	81	82	83	83	84	85	85	86	86	87	88	92
5		0	17	29	38	44	50	55	58	62	64	67	69	71	72	74	75	76	77	78	79	80	81	81	82	83	83	84	90
6			0	14	25	33	40	45	50	54	57	60	63	65	67	68	70	71	73	74	75	76	77	78	79	79	80	81	88
7				0	13	22	30	36	42	46	50	53	56	59	61	63	65	67	68	70	71	72	73	74	75	76	77	78	86
8					0	11	20	27	33	38	43	47	50	53	56	58	60	62	64	65	67	68	69	70	71	72	73	75	84
9						0	10	18	25	31	36	40	44	47	50	53	55	57	59	61	63	64	65	67	68	69	70	72	82
10							0	9	17	23	29	33	38	41	44	47	50	52	55	57	58	60	62	63	64	66	67	69	80
11								0	8	15	21	27	31	35	39	42	45	48	50	52	54	56	58	59	61	62	63	66	78
12									0	8	14	20	25	29	33	37	40	43	45	48	50	52	54	56	57	59	60	63	76
13										0	7	13	19	24	28	32	35	38	41	43	46	48	50	52	54	55	57	59	74
14											0	7	13	18	22	26	30	33	36	39	42	44	46	48	50	52	53	56	72
15												0	6	12	17	21	25	29	32	35	38	40	42	44	46	48	50	53	70
16													0	6	11	16	20	24	27	30	33	36	38	41	43	45	47	50	68
17														0	6	11	15	19	23	26	29	32	35	37	39	41	43	47	66
18															0	5	10	14	18	22	25	28	31	33	36	38	40	44	64
19																0	5	10	14	17	21	24	27	30	32	34	37	41	62
20																	0	5	9	13	17	20	23	26	29	31	33	38	60
21																		0	5	9	13	16	19	22	25	28	30	34	58
22																			0	4	8	12	15	19	21	24	27	31	56
23																				0	4	8	12	15	18	21	23	28	54
24																					0	4	8	11	14	17	20	25	52
25																						0	4	7	11	14	17	22	50
26																							0	4	7	10	13	19	48
27																								0	4	7	10	16	46
28																									0	3	7	13	44
29																										0	3	9	42
30																											0	6	40
31																												3	38
32																												0	36

Number of Test Items Wrong

HBJ material copyrighted under notice appearing earlier in this work.

Individual Record Form

MATHEMATICS PLUS

Grade 1 • Chapter 1

Student Name: _____

	Date	Score
Inventory Test		
Pretest		
Posttest		

TESTED OBJECTIVES / CHAPTER TESTS / REVIEW OPTIONS

Obj. No.	Tested Objective	Test Items - Concept	Test Items - Skill	Problem Solving	Criterion Scores	Pre-test Score	Post-test Score	Lesson Page Numbers	Pupil's Edition	Teacher's Edition	Workbooks P	Workbooks R	Workbooks E
1-A	To sort by two attributes and to extend a simple pattern	1-2	3-4		3/4			3-4		MIN, p. 3	1.1	1.1	1.1
1-B	To identify groups of 0 to 10 objects		5-12		5/8			5-6	MP, p. H3	MIN, p. 5	1.2	1.2	1.2
								9-10		MIN, p. 9	1.4	1.4	1.4
								11-12		ATS, p. C1	1.5	1.5	1.5
								13-14		MIN, p. 13	1.6	1.6	1.6
								15-16		MIN, p. 15	1.7	1.7	1.7
								21-22		MIN, p. 21	1.9	1.9	1.9
								23-24		ATS, p. C2	1.10	1.10	1.10
								25-26	MP, p. H3	ATS, p. C2	1.11	1.11	1.11
1-C	To compare and order numbers through 10	13-14	15-16		3/4			7-8		ATS, p. C1	1.3	1.3	1.3
								27-28		MIN, p. 27	1.12	1.12	1.12
								29-30	MP, p. H4	MIN, p. 29	1.13	1.13	1.13
1-D	To solve problems by drawing a picture and reading a pictograph			17-20	3/4			17-18		MIN, p. 17	1.8	1.8	1.8
								31-32	MP, p. H4	MIN, p. 31	1.14	1.14	1.14

KEY: MP—More Practice   MIN—Meeting Individual Needs (First Activity)   ATS—Alternative Teaching Strategies
Workbooks: P—Practice   R—Reteaching   E—Enrichment

HBJ material copyrighted under notice appearing earlier in this work.

Individual Record Form

MATHEMATICS PLUS

Grade 1 • Chapter 2

Student Name: _____

	Date	Score
Pretest		
Posttest		

TESTED OBJECTIVES / CHAPTER TESTS

Obj. No.	Tested Objective	Test Items - Concept	Test Items - Skill	Test Items - Problem Solving	Criterion Scores	Pretest Score	Posttest Score
2-A	To identify an addition sentence represented by a model	1-6			4/6		
2-B	To add basic facts with sums to 6 in horizontal and vertical format		7-14		5/8		
2-C	To identify combinations of addends with sums to 6		15-20		4/6		
2-D	To solve problems by using a picture			21-24	3/4		

REVIEW OPTIONS

Lesson Page Numbers	Pupil's Edition	Teacher's Edition	Workbooks P	Workbooks R	Workbooks E
41-42		MIN, p. 41	2.1	2.1	2.1
43-44	MP, p. H5	MIN, p. 43	2.2	2.2	2.2
45-46		MIN, p. 45	2.3	2.3	2.3
47-48		MIN, p. 47	2.4	2.4	2.4
57-58	MP, p. H5	MIN, p. 57	2.8	2.8	2.8
53-54		ATS, p. C3	2.6	2.6	2.6
55-56	MP, p. H6	ATS, p. C4	2.7	2.7	2.7
49-50		ATS, p. C3	2.5	2.5	2.5

KEY: MP—More Practice MIN—Meeting Individual Needs (First Activity) ATS—Alternative Teaching Strategies
Workbooks: P—Practice R—Reteaching E—Enrichment

HBJ material copyrighted under notice appearing earlier in this work.

Individual Record Form

MATHEMATICS PLUS

Grade 1 • Chapter 3

Student Name: _____

	Date	Score
Pretest		
Posttest		
Cumulative Test		

TESTED OBJECTIVES

Obj. No.	Tested Objective	CHAPTER TESTS							REVIEW OPTIONS						
		Test Items			Criterion Scores	Pre-test Score	Post-test Score	Lesson Page Numbers	Pupil's Edition	Teacher's Edition	Workbooks		
		Concept	Skill	Problem Solving							P	R	E
3–A	To identify a subtraction sentence represented by a model	1–4	5–8		5/8			69–70		ATS, p. C5	3.1	3.1	3.1
								71–72		ATS, p. C5	3.2	3.2	3.2
								73–74	MP, p. H7	MIN, p. 73	3.3	3.3	3.3
3–B	To subtract basic facts to 6 in horizontal and vertical format		9–16		5/8			75–76		MIN, p. 75	3.4	3.4	3.4
								81–82		ATS, p. C6	3.6	3.6	3.6
								83–84		ATS, p. C6	3.7	3.7	3.7
								85–86	MP, p. H7	MIN, p. 85	3.8	3.8	3.8
3–C	To identify families of facts	17–18	19–20		3/4			87–88	MP, p. H8	MIN, p. 87	3.9	3.9	3.9
3–D	To solve problems by choosing the operation and by choosing an appropriate question			21–24	3/4			77–78		MIN, p. 77	3.5	3.5	3.5
								89–90	MP, p. H8	MIN, p. 89	3.10	3.10	3.10

KEY: MP—More Practice MIN—Meeting Individual Needs (First Activity) ATS—Alternative Teaching Strategies
Workbooks: P—Practice R—Reteaching E—Enrichment

HBJ material copyrighted under notice appearing earlier in this work.

Individual Record Form

MATHEMATICS PLUS

Grade 1 • Chapter 4

Student Name: _____

	Date	Score
Pretest		
Posttest		

TESTED OBJECTIVES

CHAPTER TESTS

REVIEW OPTIONS

Obj. No.	Tested Objective	Test Items — Concept	Test Items — Skill	Test Items — Problem Solving	Criterion Scores	Pre-test Score	Post-test Score	Lesson Page Numbers	Pupil's Edition	Teacher's Edition	Workbooks P	Workbooks R	Workbooks E
4-A	To add basic facts to 10	1–6	7–16		11/16			101–102		MIN, p. 101	4.1	4.1	4.1
								103–104		MIN, p. 103	4.2	4.2	4.2
								105–106		ATS, p. C7	4.3	4.3	4.3
								111–112		MIN, p. 111	4.5	4.5	4.5
								113–114		MIN, p. 113	4.6	4.6	4.6
								115–116	MP, p. H9	ATS, p. C8	4.7	4.7	4.7
4-B	To add three addends with sums to 10	17–18	19–20		3/4			117–118	MP, p. H9	ATS, p. C8	4.8	4.8	4.8
4-C	To solve problems by using pictures and identifying irrelevant information			21–24	3/4			107–108	MP, p. H10	ATS, p. C7	4.4	4.4	4.4
								121–122	MP, p. H10	MIN, p. 121	4.10	4.10	4.10

KEY: MP—More Practice MIN—Meeting Individual Needs (First Activity) ATS—Alternative Teaching Strategies
Workbooks: P—Practice R—Reteaching E—Enrichment

HBJ material copyrighted under notice appearing earlier in this work.

Individual Record Form

MATHEMATICS PLUS

Grade 1 • Chapter 5

Student Name: _____

	Date	Score
Pretest		
Posttest		

TESTED OBJECTIVES / CHAPTER TESTS

Obj. No.	Tested Objective	Test Items - Concept	Test Items - Skill	Test Items - Problem Solving	Criterion Scores	Pretest Score	Posttest Score
5–A	To subtract basic facts through 10	1–6, 15–16	7–14		11/16		
5–B	To use inverse operations and identify families of facts to recall sums and differences through 10	17–18	19–20		3/4		
5–C	To solve problems by guessing and checking and by using models			21–24	3/4		

REVIEW OPTIONS

Lesson Page Numbers	Pupil's Edition	Teacher's Edition	Workbooks P	Workbooks R	Workbooks E
131–132		MIN, p. 131	5.1	5.1	5.1
133–134		MIN, p. 133	5.2	5.2	5.2
135–138		ATS, p. C9	5.3	5.3	5.3
141–142	MP, p. H11	ATS, p. C9	5.5	5.5	5.5
143–144		ATS, p. C10	5.6	5.6	5.6
145–146	MP, p. H11	ATS, p. C10	5.7	5.7	5.7
137–138	MP, p. H12	MIN, p. 137	5.4	5.4	5.4
149–150	MP, p. H12	ATS, p. C10	5.9	5.9	5.9

KEY: MP—More Practice   MIN—Meeting Individual Needs (First Activity)   ATS—Alternative Teaching Strategies
Workbooks: P—Practice   R—Reteaching   E—Enrichment

HBJ material copyrighted under notice appearing earlier in this work.

Individual Record Form

MATHEMATICS PLUS

Grade 1 • Chapter 6

Student Name: _____

	Date	Score
Pretest		
Posttest		
Cumulative Test		

TESTED OBJECTIVES / CHAPTER TESTS / REVIEW OPTIONS

Obj. No.	Tested Objective	Test Items: Concept	Test Items: Skill	Test Items: Problem Solving	Criterion Scores	Pretest Score	Posttest Score	Lesson Page Numbers	Pupil's Edition	Teacher's Edition	Workbooks P	Workbooks R	Workbooks E
6–A	To identify plane and solid figures	1–2	3–4		3/4			159–160		ATS, p. C11	6.1	6.1	6.1
								161–162		MIN, p. 161	6.2	6.2	6.2
								163–164	MP, p. H13	ATS, p. C11	6.3	6.3	6.3
6–B	To identify whether an object is inside, outside, or on a plane figure; to identify open and closed figures	5–6	7–8		3/4			165–166		MIN, p. 165	6.4	6.4	6.4
								167–168	MP, p. H13	MIN, p. 167	6.5	6.5	6.5
6–C	To count the number of sides and corners of plane figures		9–12		3/4			169–170	MP, p. H13	ATS, p. C12	6.6	6.6	6.6
6–D	To identify congruent figures and to identify lines of symmetry	13–16			3/4			175–176		MIN, p. 175	6.8	6.8	6.8
								177–178	MP, p. H14	MIN, p. 177	6.9	6.9	6.9
6–E	To identify and extend patterns	17–20			3/4			179–180		ATS, p. C12	6.10	6.10	6.10
								181–182	MP, p. H14	MIN, p. 181	6.11	6.11	6.11
6–F	To solve problems by using graphs and by identifying a pattern			21–24	3/4			171–172		MIN, p. 171	6.7	6.7	6.7
								183–184	MP, p. H14	MIN, p. 183	6.12	6.12	6.12

KEY: MP—More Practice MIN—Meeting Individual Needs (First Activity) ATS—Alternative Teaching Strategies
Workbooks: P—Practice R—Reteaching E—Enrichment

HBJ material copyrighted under notice appearing earlier in this work.

Individual Record Form

MATHEMATICS PLUS

Grade 1 • Chapter 7

Student Name: _____

	Date	Score
Pretest		
Posttest		

TESTED OBJECTIVES / CHAPTER TESTS

Obj. No.	Tested Objective	Test Items - Concept	Test Items - Skill	Test Items - Problem Solving	Criterion Scores	Pretest Score	Posttest Score
7-A	To identify numbers to 99	1–4	5–6		4/6		
7-B	To compare and order numbers to 100	7–8	9–12		4/6		
7-C	To identify ordinal numbers first through tenth		13–16		3/4		
7-D	To skip-count by twos, fives, and tens	17–18	19–20		3/4		
7-E	To solve problems by using estimation and by using patterns			21–24	3/4		

REVIEW OPTIONS

Lesson Page Numbers	Pupil's Edition	Teacher's Edition	Workbooks P	Workbooks R	Workbooks E
195–196		MIN, p. 195	7.1	7.1	7.1
197–198		ATS, p. C13	7.2	7.2	7.2
199–200		MIN, p. 199	7.3	7.3	7.3
201–202		MIN, p. 201	7.4	7.4	7.4
203–204		MIN, p. 203	7.5	7.5	7.5
209–210	MP, p. H15		7.7	7.7	7.7
213–214		ATS, p. C13	7.9	7.9	7.9
215–216		ATS, p. C14	7.10	7.10	7.10
217–218		ATS, p. C14	7.11	7.11	7.11
219–220	MP, p. H15	MIN, p. 219	7.12	7.12	7.12
223–224	MP, p. H16	MIN, p. 223	7.14	7.14	7.14
221–222	MP, p. H16	MIN, p. 221	7.13	7.13	7.13
205–206		MIN, p. 205	7.6	7.6	7.6
225–226	MP, p. H16	MIN, p. 225	7.15	7.15	7.15

KEY: MP—More Practice   MIN—Meeting Individual Needs (First Activity)   ATS—Alternative Teaching Strategies
Workbooks: P—Practice   R—Reteaching   E—Enrichment

Individual Record Form

MATHEMATICS PLUS

Grade 1 • Chapter 8

Student Name: _____

	Date	Score
Pretest		
Posttest		

TESTED OBJECTIVES / CHAPTER TESTS / REVIEW OPTIONS

Obj. No.	Tested Objective	Test Items - Concept	Test Items - Skill	Test Items - Problem Solving	Criterion Scores	Pretest Score	Posttest Score	Lesson Page Numbers	Pupil's Edition	Teacher's Edition	Workbooks P	Workbooks R	Workbooks E
8-A	To recall basic facts with sums to 12	1–4	5–8	9–10	5/8			235–236		ATS, p. C15	8.1	8.1	8.1
								237–238		MIN, p. 237	8.2	8.2	8.2
								239–240		ATS, p. C15	8.3	8.3	8.3
								241–242		MIN, p. 241	8.4	8.4	8.4
								243–244		MIN, p. 243	8.5	8.5	8.5
								253–254	MP, p. H17	ATS, p. C16	8.9	8.9	8.9
8-B	To recall basic facts with differences from 12	11–14	15–18	19–20	7/10			235–236		ATS, p. C15	8.1	8.1	8.1
								249–250		ATS, p. C16	8.7	8.7	8.7
								251–252		ATS, p. C16	8.8	8.8	8.8
								253–254	MP, p. H17	MIN, p. 253	8.9	8.9	8.9
8-C	To add and subtract pennies; to count pennies to find the total amount and count change		21–24		3/4			257–258	MP, p. H17	MIN, p. 257	8.11	8.11	8.11
8-D	To use inverse operations and to identify families of facts	25–26	27–28		3/4			255–256	MP, p. H18	MIN, p. 255	8.10	8.10	8.10
8-E	To solve problems by using bar graphs and choosing the operation			29–32	3/4			245–246		MIN, p. 245	8.6	8.6	8.6
								261–262	MP, p. H18	MIN, p. 261	8.13	8.13	8.13

KEY: MP—More Practice MIN—Meeting Individual Needs (First Activity) ATS—Alternative Teaching Strategies
Workbooks: P—Practice R—Reteaching E—Enrichment

HBJ material copyrighted under notice appearing earlier in this work.

Individual Record Form

Grade 1 • Chapter 9

MATHEMATICS PLUS

Student Name: _____

	Date	Score
Pretest		
Posttest		
Cumulative Test		

TESTED OBJECTIVES

| Obj. No. | Tested Objective | CHAPTER TESTS ||||| Lesson Page Numbers | REVIEW OPTIONS || Workbooks |||
| | | Test Items ||| Criterion Scores | Pre-test Score | Post-test Score | | Pupil's Edition | Teacher's Edition | P | R | E |
		Concept	Skill	Problem Solving									
9-A	To estimate and measure length using nonstandard units, inches, and centimeters	1–5	6–10		7/10			271–272		MIN, p. 271	9.1	9.1	9.1
								273–274		MIN, p. 273	9.2	9.2	9.2
								275–276		ATS, p. C17	9.3	9.3	9.3
								277–278		MIN, p. 277	9.4	9.4	9.4
								279–280	MP, p. H19	MIN, p. 279	9.5	9.5	9.5
9-B	To estimate weight, capacity, and temperature	11–14	15–18		5/8			281–282	MP, p. H19	MIN, p. 281	9.6	9.6	9.6
								283–284		ATS, p. C17	9.7	9.7	9.7
								285–286		MIN, p. 285	9.8	9.8	9.8
9-C	To identify equal parts, halves, thirds, and fourths of a region or group of objects	19–23	24–28		7/10			291–292		MIN, p. 291	9.10	9.10	9.10
								293–294		ATS, p. C18	9.11	9.11	9.11
								295–296		MIN, p. 295	9.12	9.12	9.12
								297–298		ATS, p. C18	9.13	9.13	9.13
								299–300	MP, p. H20	MIN, p. 299	9.14	9.14	9.14
9-D	To solve problems by using tables			29–32	3/4			287–288	MP, p. H20	MIN, p. 287	9.9	9.9	9.9

KEY: MP—More Practice MIN—Meeting Individual Needs (First Activity) ATS—Alternative Teaching Strategies
Workbooks: P—Practice R—Reteaching E—Enrichment

Individual Record Form

MATHEMATICS PLUS

Grade 1 • Chapter 10

Student Name: _____

	Date	Score
Pretest		
Posttest		

TESTED OBJECTIVES

Obj. No.	Tested Objective	CHAPTER TESTS					REVIEW OPTIONS			Workbooks		
		Test Items			Criterion Scores	Pre-test Score	Post-test Score	Lesson Page Numbers	Pupil's Edition	Teacher's Edition	P	R	E
		Concept	Skill	Problem Solving								
10-A	To sequence events; to estimate which event takes more or less time	1–2	3–4		3/4			313–314		ATS, p. C19	10.1	10.1	10.1
								315–316	MP, p. H21	MIN, p. 315	10.2	10.2	10.2
10-B	To tell time to the hour and half hour	5–6	7–10		5/8			317–318		MIN, p. 317	10.3	10.3	10.3
								319–320		MIN, p. 319	10.4	10.4	10.4
								321–322		ATS, p. C19	10.5	10.5	10.5
								323–324	MP, p. H21	MIN, p. 323	10.6	10.6	10.6
10-C	To read a calendar		11–14		3/4			325–326		ATS, p. C20	10.7	10.7	10.7
10-D	To find the value of a group of coins: pennies, nickels, dimes and quarters	15–16	17–20	21–24	7/10			331–332		ATS, p. C20	10.9	10.9	10.9
								333–334		MIN, p. 333	10.10	10.10	10.10
								335–336		MIN, p. 335	10.11	10.11	10.11
								337–338		MIN, p. 337	10.12	10.12	10.12
								339–340	MP, p. H22	MIN, p. 339	10.13	10.13	10.13
10-E	To determine equivalent groups of coins	25–26	27–28		3/4			343–344		MIN, p. 343	10.15	10.15	10.15
								341–342		MIN, p. 341	10.14	10.14	10.14
10-F	To solve problems by using a picture and by using a model			29–32	3/4			327–328	MP, p. H22	MIN, p. 327	10.8	10.8	10.8
								343–344		MIN, p. 343	10.15	10.15	10.15

KEY: MP—More Practice MIN—Meeting Individual Needs (First Activity) ATS—Alternative Teaching Strategies
Workbooks: P—Practice R—Reteaching E—Enrichment

HBJ material copyrighted under notice appearing earlier in this work.

Individual Record Form

MATHEMATICS PLUS

Grade 1 • Chapter 11

Student Name: _____

	Date	Score
Pretest		
Posttest		

TESTED OBJECTIVES / CHAPTER TESTS / REVIEW OPTIONS

Obj. No.	Tested Objective	Test Items – Concept	Test Items – Skill	Test Items – Problem Solving	Criterion Scores	Pre-test Score	Post-test Score	Lesson Page Numbers	Pupil's Edition	Teacher's Edition	Workbooks P	Workbooks R	Workbooks E
11-A	To add and subtract basic facts with sums to 12		1–4		3/4			353–354	MP, p. H23	ATS, p. C21	11.1	11.1	11.1
11-B	To identify a two-digit number represented by groups of tens and ones	5–8			3/4			355–356	MP, p. H23	ATS, p. C21	11.2	11.2	11.2
11-C	To add two-digit numbers without regrouping	9–10	11–14	15–18	7/10			357–358		MIN, p. 357	11.3	11.3	11.3
								359–360		ATS, p. C22	11.4	11.4	11.4
								361–362	MP, p. H24	MIN, p. 361	11.5	11.5	11.5
								371–376		MIN, p. 371	11.9	11.9	11.9
11-D	To subtract two-digit numbers without regrouping	19–20	21–24	25–28	7/10			365–366		MIN, p. 365	11.6	11.6	11.6
								367–368		ATS, p. C22	11.7	11.7	11.7
								369–370	MP, p. H24	MIN, p. 369	11.8	11.8	11.8
								371–376		MIN, p. 371	11.9	11.9	11.9
11-E	To solve problems by using estimation and by choosing a sensible answer			29–32	3/4			373–374		MIN, p. 373	11.10	11.10	11.10
								375–376	MP, p. H24	MIN, p. 375	11.11	11.11	11.11

KEY: MP—More Practice MIN—Meeting Individual Needs (First Activity) ATS—Alternative Teaching Strategies
Workbooks: P—Practice R—Reteaching E—Enrichment

HBJ material copyrighted under notice appearing earlier in this work.

Individual Record Form

MATHEMATICS PLUS

Grade 1 • Chapter 12

Student Name: _____

	Date	Score
Pretest		
Posttest		
Cumulative Test		
End-of-Book Test		

TESTED OBJECTIVES

Obj. No.	Tested Objective	Chapter Tests — Test Items			Criterion Scores	Pre-test Score	Post-test Score
		Concept	Skill	Problem Solving			
12–A	To add basic facts with sums to 18	1–4	5–10	11–14	9/14		
12–B	To subtract basic facts with sums to 18	15–18	19–22	23–24	7/10		
12–C	To identify fact families with sums to 18	25–26	27–28		3/4		
12–D	To solve problems by using a model			29–32	3/4		

REVIEW OPTIONS

Lesson Page Numbers	Pupil's Edition	Teacher's Edition	Workbooks		
			P	R	E
385–386		MIN, p. 385	12.1	12.1	12.1
387–388		MIN, p. 387	12.2	12.2	12.2
389–390		ATS, p. C23	12.3	12.3	12.3
391–392		ATS, p. C23	12.4	12.4	12.4
393–394		MIN, p. 393	12.5	12.5	12.5
395–396		MIN, p. 395	12.6	12.6	12.6
401–402	MP, p. H25	MIN, p. 401	12.8	12.8	12.8
403–404	MP, p. H25	MIN, p. 403	12.9	12.9	12.9
407–408		MIN, p. 407	12.11	12.11	12.11
411–412		MIN, p. 411	12.13	12.13	12.13
385–386		MIN, p. 385	12.1	12.1	12.1
387–388		MIN, p. 387	12.2	12.2	12.2
401–402		ATS, p. C24	12.8	12.8	12.8
403–404		MIN, p. 403	12.9	12.9	12.9
405–406	MP, p. H26	MIN, p. 405	12.10	12.10	12.10
397–398		MIN, p. 397	12.7	12.7	12.7
409–410	MP, p. H26	MIN, p. 409	12.12	12.12	12.12

KEY: MP—More Practice   MIN—Meeting Individual Needs (First Activity)   ATS—Alternative Teaching Strategies
Workbooks: P—Practice   R—Reteaching   E—Enrichment

HBJ material copyrighted under notice appearing earlier in this work.

Formal Assessment

MATHEMATICS PLUS
Grade 1

Class Record Form

School / Teacher / NAMES Date	Inventory	CHAPTER 1		CHAPTER 2		CHAPTER 3		Cumulative	CHAPTER 4		CHAPTER 5		CHAPTER 6		Cumulative	CHAPTER 7	
		Pretest	Posttest	Pretest	Posttest	Pretest	Posttest		Pretest	Posttest	Pretest	Posttest	Pretest	Posttest		Pretest	Posttest
Criterion Score	35/50	15/20	15/20	18/24	18/24	18/24	18/24	35/50	18/24	18/24	18/24	18/24	18/24	18/24	35/50	18/24	18/24

continued

Formal Assessment

MATHEMATICS PLUS
Grade 1

Class Record Form (continued)

School / Teacher / Criterion Score / NAMES	CHAPTER 8 Pretest	CHAPTER 8 Posttest	CHAPTER 9 Pretest	CHAPTER 9 Posttest	Cumulative	CHAPTER 10 Pretest	CHAPTER 10 Posttest	CHAPTER 11 Pretest	CHAPTER 11 Posttest	CHAPTER 12 Pretest	CHAPTER 12 Posttest	Cumulative	End-of-Book			
Criterion Score	24/32	24/32	24/32	24/32	35/50	24/32	24/32	24/32	24/32	24/32	24/32	35/50	35/50			
Date																

HBJ material copyrighted under notice appearing earlier in this work.

Cumulative Record Form

MATHEMATICS PLUS
Grade 1

Student _____ Teacher _____

INVENTORY TEST
Criteria 35/50   Test Score _____

CHAPTER 1

Test Scores:	Criteria 15/20	Pretest	Posttest	Needs More Work	Accomplished
☐☐ 1–A To sort by two attributes and to extend a simple pattern					
☐☐ 1–B To identify groups of 0 to 10 objects					
☐☐ 1–C To compare and order numbers through 10					
☐☐ 1–D To solve problems by drawing a picture and reading a pictograph					

CHAPTER 2

Test Scores:	Criteria 18/24	Pretest	Posttest	Needs More Work	Accomplished
☐☐ 2–A To identify an addition sentence represented by a model					
☐☐ 2–B To add basic facts with sums to 6 in horizontal and vertical format					
☐☐ 2–C To identify combinations of addends with sums to 6					
☐☐ 2–D To solve problems by using a picture and acting out problems					

CHAPTER 3

Test Scores:	Criteria 18/24	Pretest	Posttest	Needs More Work	Accomplished
☐☐ 3–A To identify a subtraction sentence represented by a model					
☐☐ 3–B To subtract basic facts to 6 in horizontal and vertical format					
☐☐ 3–C To identify families of facts					
☐☐ 3–D To solve problems by choosing the operation or choosing an appropriate question					

CUMULATIVE TEST CHAPTERS 1–3
Criteria 35/50   Test Score _____

CHAPTER 4

Test Scores:	Criteria 18/24	Pretest	Posttest	Needs More Work	Accomplished
☐☐ 4–A To add basic facts to 10					
☐☐ 4–B To add three addends with sums to 10					
☐☐ 4–C To solve problems by using pictures and identifying irrelevant information					

CHAPTER 5

Test Scores:	Criteria 18/24	Pretest	Posttest	Needs More Work	Accomplished
☐☐ 5–A To subtract basic facts through 10					
☐☐ 5–B To use inverse operations and identify families of facts to recall sums and differences through 10					
☐☐ 5–C To solve problems by guessing and checking and using models					

continued

Cumulative Record Form

MATHEMATICS PLUS
Grade 1

Student _____   Teacher _____

CHAPTER 6

Test Scores:	Criteria 18/24	Pretest	Posttest	Needs More Work	Accomplished
☐☐ 6–A	To identify plane and solid figures				
☐☐ 6–B	To identify whether an object is inside, outside, or on a plane figure; to identify open and closed figures				
☐☐ 6–C	To count the number of sides and corners of plane figures				
☐☐ 6–D	To identify congruent figures and to identify lines of symmetry				
☐☐ 6–E	To identify and extend patterns				
☐☐ 6–F	To solve problems by using graphs and identifying a pattern				

CUMULATIVE TEST CHAPTERS 1–6
Criteria ___35/50___   Test Score _____

CHAPTER 7

Test Scores:	Criteria 18/24	Pretest	Posttest	Needs More Work	Accomplished
☐☐ 7–A	To identify numbers to 99				
☐☐ 7–B	To compare and order numbers to 100				
☐☐ 7–C	To identify ordinal numbers first through tenth				
☐☐ 7–D	To skip count by twos, fives, and tens				
☐☐ 7–E	To solve problems by using estimation and by using patterns				

CHAPTER 8

Test Scores:	Criteria 24/32	Pretest	Posttest	Needs More Work	Accomplished
☐☐ 8–A	To recall basic facts with sums to 12				
☐☐ 8–B	To recall basic facts with differences from 12				
☐☐ 8–C	To add and subtract pennies; to count pennies to find the total amount and count change				
☐☐ 8–D	To use inverse operations and to identify families of facts				
☐☐ 8–E	To solve problems by using bar graphs and choosing the operation				

CHAPTER 9

Test Scores:	Criteria 24/32	Pretest	Posttest	Needs More Work	Accomplished
☐☐ 9–A	To estimate and measure length using nonstandard units, inches, and centimeters				
☐☐ 9–B	To estimate weight, capacity, and temperature				
☐☐ 9–C	To identify equal parts, halves, thirds, and fourths of a region or group of objects				
☐☐ 9–D	To solve problems by using tables and visualizing results				

CUMULATIVE TEST CHAPTERS 1–9
Criteria ___35/50___   Test Score _____

continued

Cumulative Record Form

MATHEMATICS PLUS
Grade 1

Student _____   Teacher _____

CHAPTER 10

Test Scores:	Criteria Pretest Posttest 24/32 ___ ___	Needs More Work	Accomplished
☐ ☐ 10–A	To sequence events; to estimate which event takes more or less time		
☐ ☐ 10–B	To tell time to the hour and half hour		
☐ ☐ 10–C	To read a calendar		
☐ ☐ 10–D	To find the value of a group of coins: pennies, nickels, dimes, and quarters		
☐ ☐ 10–E	To determine equivalent groups of coins		
☐ ☐ 10–F	To solve problems by using a picture and by using a model		

CHAPTER 11

Test Scores:	Criteria Pretest Posttest 24/32 ___ ___	Needs More Work	Accomplished
☐ ☐ 11–A	To add and subtract basic facts with sums to 12		
☐ ☐ 11–B	To identify a two-digit number represented by groups of tens and ones		
☐ ☐ 11–C	To add two-digit numbers without regrouping		
☐ ☐ 11–D	To subtract two-digit numbers without regrouping		
☐ ☐ 11–E	To solve problems by using estimation and by choosing a sensible answer		

CHAPTER 12

Test Scores:	Criteria Pretest Posttest 24/32 ___ ___	Needs More Work	Accomplished
☐ ☐ 12–A	To add basic facts with sums to 18		
☐ ☐ 12–B	To subtract basic facts with sums to 18		
☐ ☐ 12–C	To identify fact families with sums to 18		
☐ ☐ 12–D	To solve problems by using a model and by choosing the strategy		

CUMULATIVE TEST CHAPTERS 1–12
Criteria 35/50   Test Score _____

END-OF-BOOK TEST
Criteria 35/50   Test Score _____

Cumulative Record Form

MATHEMATICS PLUS
Grade 2

Student _____   Teacher _____

INVENTORY TEST
Criteria 35/50   Test Score _____

CHAPTER 1

Test Scores:	Criteria Pretest Posttest 18/24 _____ _____	Needs More Work	Accomplished
☐☐ 1–A	To add basic facts with sums to 10		
☐☐ 1–B	To subtract basic facts with minuends to 10		
☐☐ 1–C	To use inverse operations and to identify families of facts to 10		
☐☐ 1–D	To solve problems by using a model or writing a number sentence		

CHAPTER 2

Test Scores:	Criteria Pretest Posttest 18/24 _____ _____	Needs More Work	Accomplished
☐☐ 2–A	To add basic facts with sums to 18		
☐☐ 2–B	To subtract basic facts with minuends to 18		
☐☐ 2–C	To use inverse operations and to identify families of facts to 18		
☐☐ 2–D	To solve problems by choosing the operation and identifying irrelevant information		

CHAPTER 3

Test Scores:	Criteria Pretest Posttest 24/32 _____ _____	Needs More Work	Accomplished
☐☐ 3–A	To identify groups of ten; to identify tens and ones to 99		
☐☐ 3–B	To compare and order numbers to 100		
☐☐ 3–C	To identify ordinal numbers first through twentieth		
☐☐ 3–D	To count by twos, threes, fives, and tens		
☐☐ 3–E	To identify odd and even numbers		
☐☐ 3–F	To solve problems by finding a pattern and using a table		

CUMULATIVE TEST CHAPTERS 1–3
Criteria 35/50   Test Score _____

CHAPTER 4

Test Scores:	Criteria Pretest Posttest 18/24 _____ _____	Needs More Work	Accomplished
☐☐ 4–A	To count by tens, fives, and ones		
☐☐ 4–B	To identify amounts of money made up of pennies, nickels, dimes, quarters, and half-dollars		
☐☐ 4–C	To identify the same amount of money using different combinations of coins; to determine the change left when purchasing an item		
☐☐ 4–D	To solve problems by using a model and making a decision		

continued

Cumulative Record Form

MATHEMATICS PLUS
Grade 2

Student _____   Teacher _____

CHAPTER 5

Test Scores:	Criteria Pretest Posttest 18/24 ____ ____	Needs More Work	Accomplished
☐ ☐ 5-A	To tell time to the hour, half-hour, and minute		
☐ ☐ 5-B	To estimate the amount of time needed to complete an activity		
☐ ☐ 5-C	To read a calendar		
☐ ☐ 5-D	To solve problems by using data and by using a picture		

CHAPTER 6

Test Scores:	Criteria Pretest Posttest 18/24 ____ ____	Needs More Work	Accomplished
☐ ☐ 6-A	To add two-digit numbers with and without regrouping		
☐ ☐ 6-B	To estimate sums to 99		
☐ ☐ 6-C	To add money amounts to 99¢		
☐ ☐ 6-D	To solve problems by using data and to identify whether a problem has enough information to solve		

CUMULATIVE TEST CHAPTERS 1–6

Criteria 35/50   Test Score _____

CHAPTER 7

Test Scores:	Criteria Pretest Posttest 18/24 ____ ____	Needs More Work	Accomplished
☐ ☐ 7-A	To subtract two-digit numbers with and without regrouping		
☐ ☐ 7-B	To use addition to check subtraction		
☐ ☐ 7-C	To add and subtract money amounts to 99¢; to add and subtract two-digit numbers		
☐ ☐ 7-D	To solve problems by making and using graphs, by choosing the operation, and by using data		

CHAPTER 8

Test Scores:	Criteria Pretest Posttest 24/32 ____ ____	Needs More Work	Accomplished
☐ ☐ 8-A	To identify plane and solid figures		
☐ ☐ 8-B	To identify and extend patterns		
☐ ☐ 8-C	To identify congruent and symmetric figures		
☐ ☐ 8-D	To identify equal parts; to identify fractions that represent part of a whole and part of a group; to compare fractions		
☐ ☐ 8-E	To solve problems by using a table and by using a picture		

CHAPTER 9

Test Scores:	Criteria Pretest Posttest 24/32 ____ ____	Needs More Work	Accomplished
☐ ☐ 9-A	To estimate and measure length to the nearest inch, foot, and centimeter; to measure perimeter in centimeters		
☐ ☐ 9-B	To measure weight with nonstandard units, pounds, and kilograms		
☐ ☐ 9-C	To measure capacity with quarts, pints, and cups; to estimate and measure the volume of a cube		
☐ ☐ 9-D	To use a thermometer to measure temperature		
☐ ☐ 9-E	To solve problems by choosing a reasonable answer and by guessing and checking the area of a rectangle		

CUMULATIVE TEST CHAPTERS 1–9

Criteria 35/50   Test Score _____

continued

Cumulative Record Form

MATHEMATICS PLUS
Grade 2

Student _____   Teacher _____

CHAPTER 10

Test Scores:	Criteria Pretest Posttest 24/32 ____ ____	Needs More Work	Accomplished
☐☐ 10–A	To identify numbers to 1,000; to identify place value		
☐☐ 10–B	To compare and order numbers to 1,000		
☐☐ 10–C	To count by ones, fives, tens, and hundreds		
☐☐ 10–D	To identify combinations of coins that equal $1.00; to identify money amounts expressed as dollars and cents		
☐☐ 10–E	To solve problems by using a bar graph and by using a table to choose an appropriate question		

CHAPTER 11

Test Scores:	Criteria Pretest Posttest 18/24 ____ ____	Needs More Work	Accomplished
☐☐ 11–A	To add three-digit numbers with and without regrouping		
☐☐ 11–B	To subtract three-digit numbers with and without regrouping		
☐☐ 11–C	To estimate sums and differences by rounding		
☐☐ 11–D	To solve problems by using a graph and by identifying reasonable results		

CHAPTER 12

Test Scores:	Criteria Pretest Posttest 24/32 ____ ____	Needs More Work	Accomplished
☐☐ 12–A	To find products using objects, pictures, and repeated addition		
☐☐ 12–B	To multiply basic facts with factors of 0–5		
☐☐ 12–C	To find quotients using objects, pictures, and repeated subtraction		
☐☐ 12–D	To solve problems by using a graph and by using a drawing		

CUMULATIVE TEST CHAPTERS 1–12

Criteria 35/50 Test Score _____

END-OF-BOOK TEST

Criteria 35/50 Test Score _____

Cumulative Record Form

MATHEMATICS PLUS
Grade 3

Student _____ Teacher _____

INVENTORY TEST
Criteria 35/50   Test Score _____

CHAPTER 1

Test Scores:	Criteria 18/24	Pretest ___	Posttest ___	Needs More Work	Accomplished
☐☐ 1-A	To add and subtract basic facts with sums to 18 and differences from 18				
☐☐ 1-B	To use inverse operations to identify fact families and to find missing addends				
☐☐ 1-C	To choose the correct operation to solve the problem				
☐☐ 1-D	To solve a problem by acting it out or guessing and checking				

CHAPTER 2

Test Scores:	Criteria 24/32	Pretest ___	Posttest ___	Needs More Work	Accomplished
☐☐ 2-A	To identify odd and even numbers				
☐☐ 2-B	To read and identify numbers through hundred thousands; to identify place value				
☐☐ 2-C	To compare and order numbers through hundreds				
☐☐ 2-D	To round numbers to the nearest ten or nearest hundred				
☐☐ 2-E	To identify ordinal numbers first through fiftieth				
☐☐ 2-F	To solve problems by using a table				

CHAPTER 3

Test Scores:	Criteria 18/24	Pretest ___	Posttest ___	Needs More Work	Accomplished
☐☐ 3-A	To estimate sums and differences by rounding				
☐☐ 3-B	To add and subtract one- and two-digit numbers with and without regrouping				
☐☐ 3-C	To find the sum of more than two addends				
☐☐ 3-D	To solve problems with too much or too little information and to solve problems by organizing data				

CUMULATIVE TEST CHAPTERS 1-3
Criteria 35/50   Test Score _____

CHAPTER 4

Test Scores:	Criteria 18/24	Pretest ___	Posttest ___	Needs More Work	Accomplished
☐☐ 4-A	To estimate sums and differences by rounding				
☐☐ 4-B	To add three- and four-digit numbers and money amounts with regrouping				
☐☐ 4-C	To subtract three- and four-digit numbers and money amounts with regrouping				
☐☐ 4-D	To solve problems by choosing the operation and by choosing a number sentence				

continued

Cumulative Record Form

MATHEMATICS PLUS
Grade 3

Student _____ Teacher _____

CHAPTER 5

Test Scores:	Criteria Pretest Posttest 24/32 ____ ____	Needs More Work	Accom- plished
☐☐ 5-A	To use a calendar		
☐☐ 5-B	To tell time to the hour, half hour, quarter hour, and minute		
☐☐ 5-C	To use a schedule		
☐☐ 5-D	To count money amounts to $9.99; to compare amounts of money; to solve problems using money		
☐☐ 5-E	To solve problems by working backward and by acting them out		

CHAPTER 6

Test Scores:	Criteria Pretest Posttest 15/20 ____ ____	Needs More Work	Accom- plished
☐☐ 6-A	To relate multiplication to skip counting, repeated addition, and a number line		
☐☐ 6-B	To multiply basic facts when one of the factors is 0, 1, 2, 3, 4, or 5		
☐☐ 6-C	To solve problems by finding a pattern or by using a model		

CUMULATIVE TEST CHAPTERS 1-6
Criteria ___35/50___ Test Score _____

CHAPTER 7

Test Scores:	Criteria Pretest Posttest 18/24 ____ ____	Needs More Work	Accom- plished
☐☐ 7-A	To multiply basic facts when one of the factors is 6, 7, 8, or 9		
☐☐ 7-B	To multiply 3 factors		
☐☐ 7-C	To solve problems with too much or too little informa- tion and to solve problems by making choices		

CHAPTER 8

Test Scores:	Criteria Pretest Posttest 18/24 ____ ____	Needs More Work	Accom- plished
☐☐ 8-A	To demonstrate an understanding of division using models and repeated subtraction		
☐☐ 8-B	To use inverse operations to identify fact families and to find factors		
☐☐ 8-C	To divide basic facts with 0, 1, 2, 3, 4, and 5 as factors		
☐☐ 8-D	To solve multistep problems and to solve problems by writing a number sentence		

CHAPTER 9

Test Scores:	Criteria Pretest Posttest 18/24 ____ ____	Needs More Work	Accom- plished
☐☐ 9-A	To divide basic facts		
☐☐ 9-B	To divide a two-digit number by a one-digit number to get a quotient with a remainder		
☐☐ 9-C	To solve problems by choosing the operation and by choosing the method of computation		

CUMULATIVE TEST CHAPTERS 1-9
Criteria ___35/50___ Test Score _____

continued

8 • Chapters 5-9 Cumulative Record Form Grade 3

Cumulative Record Form

MATHEMATICS PLUS
Grade 3

Student _____   Teacher _____

CHAPTER 10

Test Scores:	Criteria 24/32	Pretest	Posttest	Needs More Work	Accomplished
☐☐ 10–A	To identify solid and plane figures and compare their attributes				
☐☐ 10–B	To identify lines, line segments, angles, and right angles				
☐☐ 10–C	To identify congruent figures and symmetrical figures				
☐☐ 10–D	To find the perimeter, area, and volume of figures using nonstandard units				
☐☐ 10–E	To use ordered pairs to locate points on a grid				
☐☐ 10–F	To solve problems by finding a pattern and by using a pictograph				

CHAPTER 11

Test Scores:	Criteria 24/32	Pretest	Posttest	Needs More Work	Accomplished
☐☐ 11–A	To estimate and measure length using customary and metric units				
☐☐ 11–B	To estimate and measure capacity using customary and metric units				
☐☐ 11–C	To estimate and measure weight using customary and metric units				
☐☐ 11–D	To estimate and measure temperature using customary and metric units				
☐☐ 11–E	To solve problems by using a picture and by using a bar graph				

CHAPTER 12

Test Scores:	Criteria 24/32	Pretest	Posttest	Needs More Work	Accomplished
☐☐ 12–A	To identify and find part of a whole and part of a group				
☐☐ 12–B	To identify equivalent fractions				
☐☐ 12–C	To compare fractions				
☐☐ 12–D	To identify mixed numbers				
☐☐ 12–E	To read and write decimals to hundredths				
☐☐ 12–F	To add and subtract decimals				
☐☐ 12–G	To solve multistep problems; to solve problems by using a picture				

CHAPTER 13

Test Scores:	Criteria 24/32	Pretest	Posttest	Needs More Work	Accomplished
☐☐ 13–A	To multiply and divide by tens and hundreds				
☐☐ 13–B	To multiply two- and three-digit numbers				
☐☐ 13–C	To estimate products				
☐☐ 13–D	To divide with one- and two-digit numbers to find quotients with and without remainders				
☐☐ 13–E	To solve problems by using estimation and by choosing a strategy				

CUMULATIVE TEST CHAPTERS 1–13
Criteria 35/50   Test Score _____

END-OF-BOOK TEST
Criteria 35/50   Test Score _____

Cumulative Record Form   Grade 3

Cumulative Record Form

MATHEMATICS PLUS
Grade 4

Student _____ Teacher _____

INVENTORY TEST
Criteria 35/50   Test Score _____

CHAPTER 1

Test Scores:	Criteria Pretest Posttest 18/24 ___ ___	Needs More Work	Accomplished
☐ ☐ 1–A	To read and identify numbers in expanded and standard form through hundred millions; to identify place value		
☐ ☐ 1–B	To compare and order numbers through millions		
☐ ☐ 1–C	To estimate numbers by rounding to the nearest 10, 100, or 1,000		
☐ ☐ 1–D	To identify ordinal numbers through hundredth		
☐ ☐ 1–E	To solve problems by using a table and finding a pattern		

CHAPTER 2

Test Scores:	Criteria Pretest Posttest 18/24 ___ ___	Needs More Work	Accomplished
☐ ☐ 2–A	To add or subtract basic facts		
☐ ☐ 2–B	To estimate sums or differences for reasonableness, using front-end digits and rounding		
☐ ☐ 2–C	To add or subtract whole numbers and money amounts with and without regrouping		
☐ ☐ 2–D	To solve problems by using a picture or using a table to analyze data		

CHAPTER 3

Test Scores:	Criteria Pretest Posttest 24/32 ___ ___	Needs More Work	Accomplished
☐ ☐ 3–A	To multiply basic facts		
☐ ☐ 3–B	To divide basic facts		
☐ ☐ 3–C	To use inverse operations and to identify fact families		
☐ ☐ 3–D	To solve problems by writing a number sentence and identifying needed or extraneous information		

CUMULATIVE TEST CHAPTERS 1–3
Criteria 35/50   Test Score _____

CHAPTER 4

Test Scores:	Criteria Pretest Posttest 24/32 ___ ___	Needs More Work	Accomplished
☐ ☐ 4–A	To tell time		
☐ ☐ 4–B	To compute elapsed time using an analog clock, a digital clock, and a calendar		
☐ ☐ 4–C	To use and interpret tally tables, frequency tables, pictographs, bar graphs, and line graphs		
☐ ☐ 4–D	To use ordered pairs to find points on a coordinate grid		
☐ ☐ 4–E	To solve problems by using a table or schedule and analyzing data to make decisions		

continued

10 • Chapters 1–4   Cumulative Record Form Grade 4

Cumulative Record Form

MATHEMATICS PLUS
Grade 4

Student _____   Teacher _____

CHAPTER 5

Test Scores:	Criteria 18/24	Pretest ___	Posttest ___	Needs More Work	Accomplished
☐☐ 5-A	To multiply by multiples of 10, 100, and 1,000				
☐☐ 5-B	To estimate products				
☐☐ 5-C	To multiply two-, three-, and four-digit numbers by one-digit numbers				
☐☐ 5-D	To solve problems by working backward and choosing a method of computation				

CHAPTER 6

Test Scores:	Criteria 24/32	Pretest ___	Posttest ___	Needs More Work	Accomplished
☐☐ 6-A	To use mental math to multiply by multiples of 10				
☐☐ 6-B	To estimate products				
☐☐ 6-C	To multiply numbers through thousands, or money amounts by a two-digit number				
☐☐ 6-D	To solve problems by using a graph and by guessing and checking				

CUMULATIVE TEST CHAPTERS 1-6
Criteria 35/50   Test Score _____

CHAPTER 7

Test Scores:	Criteria 18/24	Pretest ___	Posttest ___	Needs More Work	Accomplished
☐☐ 7-A	To choose and use appropriate metric units of length, capacity, and mass				
☐☐ 7-B	To choose and use appropriate customary units of length, capacity, and weight				
☐☐ 7-C	To find the perimeter				
☐☐ 7-D	To convert from smaller to larger customary units and vice versa				
☐☐ 7-E	To solve multistep problems and to solve problems by using a picture				

CHAPTER 8

Test Scores:	Criteria 24/32	Pretest ___	Posttest ___	Needs More Work	Accomplished
☐☐ 8-A	To estimate quotients with one-digit divisors				
☐☐ 8-B	To identify multiplication and division as inverse operations				
☐☐ 8-C	To divide one-, two-, and three-digit numbers and money amounts by a one-digit number				
☐☐ 8-D	To find the median, range, and average of a set of data				
☐☐ 8-E	To choose appropriate strategies to solve problems and to choose the method of computation				

CHAPTER 9

Test Scores:	Criteria 24/32	Pretest ___	Posttest ___	Needs More Work	Accomplished
☐☐ 9-A	To identify and distinguish between plane and solid figures and their properties				
☐☐ 9-B	To measure area in square units and multiply to find area; to measure volume in cubic units and multiply to find volume				
☐☐ 9-C	To identify lines, line segments, rays, perpendicular and parallel lines, angles, circles, and parts of a circle				
☐☐ 9-D	To identify congruent and similar figures; to identify lines of symmetry; to identify a slide, flip, and turn of a figure				
☐☐ 9-E	To solve multistep problems and to solve problems by using a model				

CUMULATIVE TEST CHAPTERS 1-9
Criteria 35/50   Test Score _____

continued

Cumulative Record Form Grade 4

Cumulative Record Form

MATHEMATICS PLUS
Grade 4

Student _____ Teacher _____

CHAPTER 10

Test Scores:	Criteria Pretest Posttest 24/32 ___ ___	Needs More Work	Accomplished
☐☐ 10–A	To identify a fractional part of a group; to identify the fractional part of a number		
☐☐ 10–B	To identify equivalent fractions; to simplify fractions		
☐☐ 10–C	To compare fractions with like and unlike denominators		
☐☐ 10–D	To identify fractions as mixed numbers		
☐☐ 10–E	To solve problems by acting them out and by choosing appropriate strategies		

CHAPTER 11

Test Scores:	Criteria Pretest Posttest 24/32 ___ ___	Needs More Work	Accomplished
☐☐ 11–A	To estimate the fractional part of a number and sums and differences of fractions		
☐☐ 11–B	To add and subtract fractions with like and unlike denominators; to add and subtract mixed numbers with like denominators		
☐☐ 11–C	To identify the length of an object to a fractional part of an inch		
☐☐ 11–D	To identify possible combinations and arrangements using tree diagrams; to find the probability of an event		
☐☐ 11–E	To solve problems by using a model and by making an organized list		

CHAPTER 12

Test Scores:	Criteria Pretest Posttest 24/32 ___ ___	Needs More Work	Accomplished
☐☐ 12–A	To read and identify decimals with tenths and hundredths; to relate fractions and decimals		
☐☐ 12–B	To identify equivalent decimals; to compare and order decimals		
☐☐ 12–C	To estimate decimals by rounding to the nearest whole number; to estimate decimal sums and differences		
☐☐ 12–D	To add and subtract decimals		
☐☐ 12–E	To solve problems by working backward and by using estimation		

CHAPTER 13

Test Scores:	Criteria Pretest Posttest 18/24 ___ ___	Needs More Work	Accomplished
☐☐ 13–A	To estimate the quotient of a number when dividing by a two-digit divisor		
☐☐ 13–B	To divide by multiples of 10 with and without remainders		
☐☐ 13–C	To divide with two-digit divisors with and without remainders		
☐☐ 13–D	To solve problems by finding the hidden question and by interpreting the remainder		

CUMULATIVE TEST CHAPTERS 1–13
Criteria 35/50   Test Score _____

END-OF-BOOK TEST
Criteria 35/50   Test Score _____

Cumulative Record Form

MATHEMATICS PLUS
Grade 5

Student _____ Teacher _____

INVENTORY TEST
Criteria 35/50   Test Score _____

CHAPTER 1

Test Scores:	Criteria Pretest Posttest 18/24 _____ _____	Needs More Work	Accomplished
☐☐ 1–A	To read and identify numbers in expanded and standard form from hundred millions to thousandths; to identify place value		
☐☐ 1–B	To compare and order whole numbers and decimals		
☐☐ 1–C	To estimate whole numbers and decimals by rounding		
☐☐ 1–D	To solve problems by using a table or a drawing		

CHAPTER 2

Test Scores:	Criteria Pretest Posttest 24/32 _____ _____	Needs More Work	Accomplished
☐☐ 2–A	To use addition and subtraction as inverse operations		
☐☐ 2–B	To estimate whole-number and decimal sums and differences		
☐☐ 2–C	To add and subtract whole numbers		
☐☐ 2–D	To add and subtract decimals to thousandths		
☐☐ 2–E	To solve problems by identifying and solving hidden questions and choosing a number sentence		

CHAPTER 3

Test Scores:	Criteria Pretest Posttest 18/24 _____ _____	Needs More Work	Accomplished
☐☐ 3–A	To estimate products		
☐☐ 3–B	To multiply with multiples of 10		
☐☐ 3–C	To multiply whole numbers		
☐☐ 3–D	To solve multistep problems and to solve problems by finding a pattern		

CUMULATIVE TEST CHAPTERS 1–3
Criteria 35/50   Test Score _____

CHAPTER 4

Test Scores:	Criteria Pretest Posttest 24/32 _____ _____	Needs More Work	Accomplished
☐☐ 4–A	To identify and classify lines, rays, angles, polygons, triangles, quadrilaterals, and circles		
☐☐ 4–B	To identify congruent, similar, and symmetrical figures		
☐☐ 4–C	To identify slides, flips, and turns		
☐☐ 4–D	To identify solid figures and their attributes		
☐☐ 4–E	To solve problems by guessing and checking and by solving a simpler problem		

continued

Cumulative Record Form Grade 5

Cumulative Record Form

MATHEMATICS PLUS
Grade 5

Student _____   Teacher _____

CHAPTER 5

Test Scores:	Criteria 24/32	Pretest	Posttest	Needs More Work	Accomplished
☐ ☐ 5–A	To determine divisibility by 2, 3, 5, and 10				
☐ ☐ 5–B	To estimate quotients with one-digit divisors				
☐ ☐ 5–C	To divide whole numbers by a one-digit divisor with and without a remainder				
☐ ☐ 5–D	To find the average of a set of data				
☐ ☐ 5–E	To solve problems by choosing the operation and working backward				

CHAPTER 6

Test Scores:	Criteria 18/24	Pretest	Posttest	Needs More Work	Accomplished
☐ ☐ 6–A	To estimate quotients with a whole-number divisor				
☐ ☐ 6–B	To divide whole numbers by two-digit divisors				
☐ ☐ 6–C	To solve problems by using estimation and by choosing a strategy to solve a problem				

CUMULATIVE TEST CHAPTERS 1–6
Criteria 35/50   Test Score _____

CHAPTER 7

Test Scores:	Criteria 18/24	Pretest	Posttest	Needs More Work	Accomplished
☐ ☐ 7–A	To find the mean, median, mode, and range of a set of data				
☐ ☐ 7–B	To read and interpret bar graphs, pictographs, line graphs, and circle graphs				
☐ ☐ 7–C	To graph ordered pairs				
☐ ☐ 7–D	To solve problems by using a table and using logical reasoning				

CHAPTER 8

Test Scores:	Criteria 24/32	Pretest	Posttest	Needs More Work	Accomplished
☐ ☐ 8–A	To estimate decimal products				
☐ ☐ 8–B	To multiply decimals by whole numbers and decimals by decimals				
☐ ☐ 8–C	To divide decimals by whole numbers				
☐ ☐ 8–D	To solve problems by choosing the method of computation and identifying relevant and irrelevant information				

CHAPTER 9

Test Scores:	Criteria 24/32	Pretest	Posttest	Needs More Work	Accomplished
☐ ☐ 9–A	To identify fractions; to identify equivalent fractions; to express fractions in simplest form				
☐ ☐ 9–B	To identify numbers as prime or composite; to find common factors and multiples of whole numbers				
☐ ☐ 9–C	To rename mixed numbers and improper fractions as mixed numbers or whole numbers				
☐ ☐ 9–D	To compare and order fractions				
☐ ☐ 9–E	To solve problems by using a picture and by acting it out				

CUMULATIVE TEST CHAPTERS 1–9
Criteria 35/50   Test Score _____

continued

HBJ material copyrighted under notice appearing earlier in this work.

Cumulative Record Form

MATHEMATICS PLUS
Grade 5

Student _____ Teacher _____

CHAPTER 10

Test Scores:	Criteria 24/32	Pretest	Posttest	Needs More Work	Accomplished
☐ ☐ 10–A	To estimate the sums and differences of fractions				
☐ ☐ 10–B	To add and subtract like and unlike fractions and mixed numbers				
☐ ☐ 10–C	To multiply fractions and whole numbers; fractions and fractions; fractions and mixed numbers				
☐ ☐ 10–D	To divide fractions				
☐ ☐ 10–E	To solve problems by using tables and diagrams				

CHAPTER 11

Test Scores:	Criteria 24/32	Pretest	Posttest	Needs More Work	Accomplished
☐ ☐ 11–A	To choose appropriate metric and customary units of length, capacity, and mass				
☐ ☐ 11–B	To compute using metric and customary measures and intervals of time				
☐ ☐ 11–C	To identify the length of an object to the nearest $\frac{1}{16}$ of an inch and to the nearest 1 millimeter				
☐ ☐ 11–D	To solve problems by evaluating answers for reasonableness and by using a schedule				

CHAPTER 12

Test Scores:	Criteria 24/32	Pretest	Posttest	Needs More Work	Accomplished
☐ ☐ 12–A	To find the perimeter of a polygon and the circumference of a circle				
☐ ☐ 12–B	To find the area of parallelograms, triangles, and complex and curved figures				
☐ ☐ 12–C	To estimate and find the volume of a rectangular prism				
☐ ☐ 12–D	To solve problems by using a model and using a formula				

CHAPTER 13

Test Scores:	Criteria 18/24	Pretest	Posttest	Needs More Work	Accomplished
☐ ☐ 13–A	To identify ratios and equivalent ratios; to use scale drawings to find actual distances				
☐ ☐ 13–B	To express a ratio as a percent, and a percent as a fraction and as a decimal				
☐ ☐ 13–C	To determine possible outcomes and to find the probability of an event				
☐ ☐ 13–D	To solve problems by making an organized list and by conducting a simulation				

CUMULATIVE TEST CHAPTERS 1–13
Criteria 35/50   Test Score _____

END-OF-BOOK TEST
Criteria 35/50   Test Score _____

Cumulative Record Form

MATHEMATICS PLUS
Grade 6

Student _____   Teacher _____

INVENTORY TEST
Criteria 35/50   Test Score _____

CHAPTER 1

Test Scores:	Criteria Pretest Posttest 24/32 _____ _____	Needs More Work	Accomplished
☐ ☐ 1–A	To read and identify numbers in standard and expanded form from hundred billions to ten-thousandths; to identify place value		
☐ ☐ 1–B	To compare and order whole numbers and decimals		
☐ ☐ 1–C	To round whole numbers through billions; to round money amounts		
☐ ☐ 1–D	To solve problems using the order of operations		
☐ ☐ 1–E	To identify powers, exponents, squares, and square roots		
☐ ☐ 1–F	To solve problems by finding a pattern or by using a table		

CHAPTER 2

Test Scores:	Criteria Pretest Posttest 15/20 _____ _____	Needs More Work	Accomplished
☐ ☐ 2–A	To estimate sums and differences of whole numbers and decimals		
☐ ☐ 2–B	To find sums and differences of whole numbers and decimals		
☐ ☐ 2–C	To identify and evaluate algebraic expressions		
☐ ☐ 2–D	To solve problems by guessing and checking and by choosing the method of computation		

CHAPTER 3

Test Scores:	Criteria Pretest Posttest 15/20 _____ _____	Needs More Work	Accomplished
☐ ☐ 3–A	To estimate whole-number and/or decimal products		
☐ ☐ 3–B	To multiply a whole number by a one-, two-, or three-digit number		
☐ ☐ 3–C	To multiply a decimal by a decimal or by a whole number		
☐ ☐ 3–D	To solve multistep problems and to solve problems by making decisions		

CHAPTER 4

Test Scores:	Criteria Pretest Posttest 24/32 _____ _____	Needs More Work	Accomplished
☐ ☐ 4–A	To determine divisibility by 2, 3, 5, 9, or 10		
☐ ☐ 4–B	To estimate whole-number quotients; to round decimal quotients		
☐ ☐ 4–C	To divide a whole number by a whole number		
☐ ☐ 4–D	To divide a decimal by a whole number and a decimal by a decimal		
☐ ☐ 4–E	To identify and evaluate algebraic expressions		
☐ ☐ 4–F	To solve problems by choosing an equation and by working backward		

CUMULATIVE TEST CHAPTERS 1–4
Criteria 35/50   Test Score _____

continued

Cumulative Record Form

MATHEMATICS PLUS
Grade 6

Student _____   Teacher _____

CHAPTER 5

Test Scores:	Criteria 24/32	Pretest ___	Posttest ___	Needs More Work	Accomplished
☐☐ 5-A	To collect and organize data				
☐☐ 5-B	To analyze data using bar graphs, histograms, line graphs, circle graphs, and stem-and-leaf plots				
☐☐ 5-C	To find the range, mean, median, and mode for a collection of data				
☐☐ 5-D	To use the counting principle and tree diagrams to identify the number of possible outcomes				
☐☐ 5-E	To determine the probability of events				
☐☐ 5-F	To solve problems by finding a pattern and choosing an appropriate graph				

CHAPTER 6

Test Scores:	Criteria 24/32	Pretest ___	Posttest ___	Needs More Work	Accomplished
☐☐ 6-A	To identify factors, prime and composite numbers, prime factors, greatest common factor, and least common multiple				
☐☐ 6-B	To find equivalent fractions; to identify least common denominator; to identify a fraction in simplest form				
☐☐ 6-C	To compare and order fractions				
☐☐ 6-D	To identify a mixed number as a fraction and vice versa				
☐☐ 6-E	To solve problems by acting them out and by choosing an appropriate strategy				

CHAPTER 7

Test Scores:	Criteria 18/24	Pretest ___	Posttest ___	Needs More Work	Accomplished
☐☐ 7-A	To estimate sums and differences of fractions and mixed numbers				
☐☐ 7-B	To add and subtract like and unlike fractions				
☐☐ 7-C	To add and subtract mixed numbers with like and unlike denominators, with and without renaming				
☐☐ 7-D	To solve problems by using a diagram and by choosing an appropriate strategy				

CHAPTER 8

Test Scores:	Criteria 24/32	Pretest ___	Posttest ___	Needs More Work	Accomplished
☐☐ 8-A	To multiply fractions, mixed numbers, and whole numbers; to simplify fractions before multiplying				
☐☐ 8-B	To estimate the product of fractions by fractions and fractions by whole numbers				
☐☐ 8-C	To divide fractions, whole numbers, and mixed numbers by fractions or mixed numbers				
☐☐ 8-D	To change a fraction to a decimal and vice versa				
☐☐ 8-E	To solve problems by writing an equation and solving a simpler problem				

CUMULATIVE TEST CHAPTERS 1-8
Criteria   35/50   Test Score _____

continued

Cumulative Record Form

MATHEMATICS PLUS
Grade 6

Student _____   Teacher _____

CHAPTER 9

Test Scores:	Criteria Pretest Posttest 24/32 ___ ___		Needs More Work	Accomplished
☐☐	9–A	To choose and use appropriate metric units of length, capacity, and mass		
☐☐	9–B	To determine the more precise measure		
☐☐	9–C	To compute using customary and metric measure and intervals of time		
☐☐	9–D	To choose and use appropriate customary units of length, capacity, and mass		
☐☐	9–E	To solve problems by using estimation and using a schedule		

CHAPTER 10

Test Scores:	Criteria Pretest Posttest 24/32 ___ ___		Needs More Work	Accomplished
☐☐	10–A	To identify a ratio comparing two numbers; to identify equivalent ratios and unit rates		
☐☐	10–B	To write and solve proportions		
☐☐	10–C	To find the decimal or fraction equal to a percent or vice versa		
☐☐	10–D	To estimate or find the percent of a number		
☐☐	10–E	To solve problems by using a scale drawing		

CHAPTER 11

Test Scores:	Criteria Pretest Posttest 24/32 ___ ___		Needs More Work	Accomplished
☐☐	11–A	To identify and classify lines, angles, polygons, and solid figures		
☐☐	11–B	To identify constructions of congruent line segments, line bisectors, and congruent angles		
☐☐	11–C	To identify similar, congruent, and symmetrical figures		
☐☐	11–D	To identify translations of figures		
☐☐	11–E	To solve problems by using a circle graph and by finding a pattern		

CUMULATIVE TEST CHAPTERS 1–11
Criteria   35/50   Test Score _____

continued

Cumulative Record Form

MATHEMATICS PLUS
Grade 6

Student _____   Teacher _____

CHAPTER 12

Test Scores:	Criteria Pretest Posttest 24/32	Needs More Work	Accomplished
☐☐ 12–A	To find perimeter of regular and irregular polygons		
☐☐ 12–B	To find the area of rectangles, parallelograms, triangles, and irregular figures; to relate perimeter and area		
☐☐ 12–C	To identify parts of a circle; to find circumference and area of a circle		
☐☐ 12–D	To find the surface area of a rectangular prism; to find the volume of a rectangular prism		
☐☐ 12–E	To solve problems by using a formula and using a model		

CHAPTER 13

Test Scores:	Criteria Pretest Posttest 24/32	Needs More Work	Accomplished
☐☐ 13–A	To identify integers; to compare and order integers		
☐☐ 13–B	To add and subtract integers		
☐☐ 13–C	To identify and graph ordered pairs on a coordinate plane		
☐☐ 13–D	To choose strategies to solve problems		

CHAPTER 14

Test Scores:	Criteria Pretest Posttest 18/24	Needs More Work	Accomplished
☐☐ 14–A	To identify and evaluate algebraic expressions		
☐☐ 14–B	To identify rational numbers; to solve equations involving rational numbers		
☐☐ 14–C	To solve inequalities		
☐☐ 14–D	To use relations to extend patterns; to graph relations		
☐☐ 14–E	To solve problems by writing an equation and by using logical reasoning		

CUMULATIVE TEST CHAPTERS 1–14
Criteria 35/50   Test Score _____

END-OF-BOOK TEST
Criteria 35/50   Test Score _____

Cumulative Record Form

MATHEMATICS PLUS
Grade 7

Student _____ Teacher _____

INVENTORY TEST
Criteria 35/50   Test Score _____

CHAPTER 1

Test Scores:	Criteria Pretest Posttest 24/32	Needs More Work	Accomplished
☐☐ 1-A	To estimate sums, differences, products, and quotients of whole numbers and decimals		
☐☐ 1-B	To use addition, subtraction, multiplication, and division to solve whole-number and decimal problems		
☐☐ 1-C	To identify and evaluate numbers in standard form and exponent form		
☐☐ 1-D	To express and evaluate numbers in scientific notation		
☐☐ 1-E	To solve problems by using a picture and by guessing and checking		

CHAPTER 2

Test Scores:	Criteria Pretest Posttest 18/24	Needs More Work	Accomplished
☐☐ 2-A	To read and interpret bar graphs, line graphs, circle graphs, histograms, and stem-and-leaf plots		
☐☐ 2-B	To use frequency distribution tables; to use unbiased surveys		
☐☐ 2-C	To analyze sets of data using range, quartiles, mean, median, and mode; to read and interpret box-and-whisker graphs		
☐☐ 2-D	To solve problems by using tables and graphs and by selecting different ways to graph a set of data		

CHAPTER 3

Test Scores:	Criteria Pretest Posttest 24/32	Needs More Work	Accomplished
☐☐ 3-A	To find the prime factorization of a composite number		
☐☐ 3-B	To find the greatest common factor and the least common multiple of two or more numbers		
☐☐ 3-C	To identify equivalent fractions; fractions in simplest form; to compare and order two or more fractions and mixed numbers; to identify one or two fractions between any two given fractions		
☐☐ 3-D	To identify one or two fractions between any two given fractions		
☐☐ 3-E	To solve problems by finding a pattern and by using a picture		

CHAPTER 4

Test Scores:	Criteria Pretest Posttest 24/32	Needs More Work	Accomplished
☐☐ 4-A	To estimate sums, differences, products, and quotients of fractions		
☐☐ 4-B	To add and subtract fractions and mixed numbers with and without regrouping		
☐☐ 4-C	To multiply and divide fractions and mixed numbers		
☐☐ 4-D	To solve multistep problems and to solve problems by solving a simpler problem		

CUMULATIVE TEST CHAPTERS 1-4
Criteria 35/50   Test Score _____

continued

Cumulative Record Form

MATHEMATICS PLUS
Grade 7

Student _____   Teacher _____

CHAPTER 5

Test Scores:	Criteria Pretest Posttest	Needs More Work	Accom-plished
	24/32 ___ ___		
☐☐ 5–A	To identify and evaluate expressions		
☐☐ 5–B	To solve problems using the order of operations		
☐☐ 5–C	To solve a one-step inequality involving addition		
☐☐ 5–D	To solve one-step equations		
☐☐ 5–E	To solve problems by using a formula and by working backward		

CHAPTER 6

Test Scores:	Criteria Pretest Posttest	Needs More Work	Accom-plished
	18/24 ___ ___		
☐☐ 6–A	To identify, measure, and classify lines, angles, triangles, polygons, and quadrilaterals; to identify congruent figures and geometric relationships		
☐☐ 6–B	To identify constructions of congruent angles and triangles, and line bisectors		
☐☐ 6–C	To solve problems using logical reasoning and Venn diagrams		
☐☐ 6–D	To solve problems by finding a pattern and by using a formula		

CHAPTER 7

Test Scores:	Criteria Pretest Posttest	Needs More Work	Accom-plished
	24/32 ___ ___		
☐☐ 7–A	To identify ratios and rates; to complete equal ratios and solve proportions		
☐☐ 7–B	To use ratios and proportions to compute unit prices and to solve problems		
☐☐ 7–C	To use scale drawings to solve problems		
☐☐ 7–D	To identify corresponding parts of similar figures; to find the ratio of corresponding sides; to use proportions to find the missing measures of similar figures		
☐☐ 7–E	To solve problems by using a map and by using a formula		

CHAPTER 8

Test Scores:	Criteria Pretest Posttest	Needs More Work	Accom-plished
	24/32 ___ ___		
☐☐ 8–A	To identify fraction, decimal, ratio, and percent equivalencies		
☐☐ 8–B	To estimate and find the percent of a number, the percent one number is of another; to find a number when a percent of it is known by using estimation, proportions, and equations		
☐☐ 8–C	To analyze a circle graph		
☐☐ 8–D	To solve sales tax, discount, and simple-interest problems		
☐☐ 8–E	To solve problems by using a graph and by guessing and checking		

CUMULATIVE TEST CHAPTERS 1–8

Criteria 35/50   Test Score _____

continued

Cumulative Record Form Grade 7   Chapters 5–8 • 21

Cumulative Record Form

MATHEMATICS PLUS
Grade 7

Student _____   Teacher _____

CHAPTER 9

Test Scores:	Criteria Pretest Posttest 24/32 ___ ___ ___		Needs More Work	Accomplished
☐ ☐	9–A	To define or identify integers; to determine the absolute value of an integer		
☐ ☐	9–B	To add, subtract, multiply, and divide integers		
☐ ☐	9–C	To write and solve algebraic expressions and equations using integers		
☐ ☐	9–D	To graph ordered pairs on a coordinate plane		
☐ ☐	9–E	To solve problems by working backward and writing an equation		

CHAPTER 10

Test Scores:	Criteria Pretest Posttest 24/32 ___ ___ ___		Needs More Work	Accomplished
☐ ☐	10–A	To identify rational, irrational, and real numbers; to identify terminating and repeating decimals; to change decimals to percents, and fractions to decimals		
☐ ☐	10–B	To compare and order rational numbers; to apply the Density Property		
☐ ☐	10–C	To write numbers using integers as exponents and using scientific notation		
☐ ☐	10–D	To identify perfect squares and to find the square roots of positive rational numbers		
☐ ☐	10–E	To solve problems by acting them out and by using a table		

CHAPTER 11

Test Scores:	Criteria Pretest Posttest 24/32 ___ ___ ___		Needs More Work	Accomplished
☐ ☐	11–A	To identify sample spaces and use tree diagrams to find outcomes		
☐ ☐	11–B	To identify the number of combinations and permutations		
☐ ☐	11–C	To identify the mathematical probability that an event will occur; to use mathematical probability to predict expected outcomes		
☐ ☐	11–D	To find the probability of independent and dependent events		
☐ ☐	11–E	To solve problems by making an organized list		

CUMULATIVE TEST CHAPTERS 1–11

Criteria 35/50   Test Score _____

CHAPTER 12

Test Scores:	Criteria Pretest Posttest 24/32 ___ ___ ___		Needs More Work	Accomplished
☐ ☐	12–A	To choose an appropriate metric or customary unit of measure; to choose between an estimate or an exact measurement for a given situation		
☐ ☐	12–B	To determine the more precise unit of measure		
☐ ☐	12–C	To find the perimeter of a polygon; to find the circumference of a circle		
☐ ☐	12–D	To find the area of parallelograms, triangles, circles, and trapezoids		
☐ ☐	12–E	To identify figures with line symmetry and turn symmetry; to identify translations, reflections, and rotation images		
☐ ☐	12–F	To solve problems by using a formula and using a model		

continued

Cumulative Record Form

MATHEMATICS PL...
Grade 7

Student _____ Teacher _____

CHAPTER 13

Test Scores:	Criteria 18/24	Pretest	Posttest	Needs More Work	Accomplished
☐☐ 13–A	To identify solid figures				
☐☐ 13–B	To find the surface area of a solid figure				
☐☐ 13–C	To find the volume, capacity, and mass of a solid figure				
☐☐ 13–D	To solve problems by using estimation and using a model				

CHAPTER 14

Test Scores:	Criteria 24/32	Pretest	Posttest	Needs More Work	Accomplished
☐☐ 14–A	To identify patterns to solve problems				
☐☐ 14–B	To identify relations and use relations to solve problems				
☐☐ 14–C	To identify and use the Pythagorean Property to solve problems				
☐☐ 14–D	To identify whether or not a relation is a function				
☐☐ 14–E	To solve problems by solving a simpler problem and to choose a strategy to solve a problem				

CUMULATIVE TEST CHAPTERS 1–14
Criteria   35/50   Test Score _____

END-OF-BOOK TEST
Criteria   35/50   Test Score _____

...tive Record Form

MATHEMATICS PLUS
Grade 8

Teacher _____

...NTORY TEST
...iteria 35/50   Test Score _____

CHAPTER 1

Test Scores:	Criteria 18/24	Pretest	Posttest			Needs More Work	Accomplished
☐ ☐	1–A	To interpret rounded numbers and to identify overestimates and underestimates					
☐ ☐	1–B	To use addition, subtraction, multiplication, and division with whole numbers and decimals to solve problems					
☐ ☐	1–C	To express and evaluate numbers in exponent form					
☐ ☐	1–D	To evaluate numerical expressions using the rules for the order of operations					
☐ ☐	1–E	To solve multistep problems and to solve problems by finding a pattern					

CHAPTER 2

Test Scores:	Criteria 18/24	Pretest	Posttest			Needs More Work	Accomplished
☐ ☐	2–A	To identify algebraic expressions for word expressions; to evaluate algebraic expressions					
☐ ☐	2–B	To solve one- and two-step equations involving whole numbers and decimals; to identify an equation used to solve a problem					
☐ ☐	2–C	To solve one- and two-step inequalities using whole number replacements; to identify an inequality used to solve a problem					
☐ ☐	2–D	To solve problems by guessing and checking and by choosing an equation					

CHAPTER 3

Test Scores:	Criteria 24/32	Pretest	Posttest			Needs More Work	Accomplished
☐ ☐	3–A	To identify prime and composite numbers and to write the prime factorization of a number					
☐ ☐	3–B	To use the GCF and the LCM to solve problems					
☐ ☐	3–C	To write equivalent fractions; to express fractions in simplest form, fractions as whole numbers or mixed numbers and vice versa; to compare and order fractions and mixed numbers					
☐ ☐	3–D	To estimate sums and differences of fractions and mixed numbers					
☐ ☐	3–E	To add and subtract fractions and mixed numbers					
☐ ☐	3–F	To solve problems by making choices and by using pictures					

CHAPTER 4

Test Scores:	Criteria 18/24	Pretest	Posttest			Needs More Work	Accomplished
☐ ☐	4–A	To estimate products and quotients of fractions and mixed numbers					
☐ ☐	4–B	To multiply and divide fractions or mixed numbers					
☐ ☐	4–C	To solve one- and two-step equations using fractions					
☐ ☐	4–D	To identify a decimal equivalent to a fraction and vice versa					
☐ ☐	4–E	To solve problems by solving a simpler problem and by choosing a strategy to solve a problem					

CUMULATIVE TEST CHAPTERS 1–4
Criteria 35/50   Test Score _____

continued

Cumulative Record Form   Grade 8

Cumulative Record Form

MATHEMATICS PL[US]
Grade 8

Student _____   Teacher _____

CHAPTER 5

Test Scores:	Criteria 24/32	Pretest ___	Posttest ___	Needs More Work	Accomplished
☐ ☐ 5–A	To identify points, lines, line segments, rays, angles, intersecting lines, and angle relationships between types of angle pairs				
☐ ☐ 5–B	To identify angles and angle relationships between types of angle pairs				
☐ ☐ 5–C	To identify constructions of congruent segments and angles, parallel and perpendicular lines, and bisections of segments and angles				
☐ ☐ 5–D	To identify the relationships of angles formed by parallel lines and transversals				
☐ ☐ 5–E	To identify properties of polygons: quadrilaterals and triangles				
☐ ☐ 5–F	To identify congruent figures and their corresponding parts				
☐ ☐ 5–G	To solve problems by finding a pattern and by making choices				

CHAPTER 6

Test Scores:	Criteria 24/32	Pretest ___	Posttest ___	Needs More Work	Accomplished
☐ ☐ 6–A	To read, write, and simplify ratios and rates				
☐ ☐ 6–B	To identify and write proportions to solve problems; to use a scale drawing to find actual measurement or scale measurement				
☐ ☐ 6–C	To identify an equivalent ratio, decimal, or percent when one form is given				
☐ ☐ 6–D	To use estimation in solving percent problems; to solve percent problems				
☐ ☐ 6–E	To solve simple-interest problems				
☐ ☐ 6–F	To solve problems by using a map and by using estimation				

CHAPTER 7

Test Scores:	Criteria 24/32	Pretest ___	Posttest ___	Needs More Work	Accomplished
☐ ☐ 7–A	To compare and order integers				
☐ ☐ 7–B	To add, subtract, multiply, and divide integers				
☐ ☐ 7–C	To write numbers using negative exponents, and to evaluate powers with negative exponents				
☐ ☐ 7–D	To multiply and divide powers				
☐ ☐ 7–E	To use scientific notation to name a number and vice versa				
☐ ☐ 7–F	To solve problems by using a table and by writing an equation				

continued

Cumulative Record Form

MATHEMATICS PLUS
Grade 8

Teacher _____

CHAPTER 8

Test Scores:	Criteria 24/32	Pretest	Posttest		Needs More Work	Accomplished
☐ ☐	8-A	To identify rational and irrational numbers; to compare and order rational numbers				
☐ ☐	8-B	To find the square and square root of a number				
☐ ☐	8-C	To add, subtract, multiply, and divide rational numbers				
☐ ☐	8-D	To solve one- and two-step equations involving rational numbers				
☐ ☐	8-E	To solve and graph inequalities using real numbers				
☐ ☐	8-F	To solve problems by using a table and using a formula				

CUMULATIVE TEST CHAPTERS 1–8

Criteria ___35/50___   Test Score _____

CHAPTER 9

Test Scores:	Criteria 24/32	Pretest	Posttest		Needs More Work	Accomplished
☐ ☐	9-A	To locate points on a coordinate plane; to graph points, linear equations, and inequalities on a coordinate plane				
☐ ☐	9-B	To identify and graph relations and functions				
☐ ☐	9-C	To find the slope of a line				
☐ ☐	9-D	To solve a system of equations by graphing				
☐ ☐	9-E	To identify transformations on a coordinate plane				
☐ ☐	9-F	To solve problems by guessing and checking and by using a graph to estimate				

CHAPTER 10

Test Scores:	Criteria 24/32	Pretest	Posttest		Needs More Work	Accomplished
☐ ☐	10-A	To analyze sets of data using range, quartiles, extremes, mean, median, and mode; to read and interpret box-and-whisker graphs				
☐ ☐	10-B	To read and interpret frequency tables, histograms, line graphs, bar graphs, circle graphs, and stem-and-leaf plots				
☐ ☐	10-C	To use extrapolation and interpolation to make estimates				
☐ ☐	10-D	To identify the type of correlation between two variables using a scattergram; to read and interpret scattergrams				
☐ ☐	10-E	To distinguish between sufficient and insufficient data to solve problems and to choose an appropriate graph				

CHAPTER 11

Test Scores:	Criteria 24/32	Pretest	Posttest		Needs More Work	Accomplished
☐ ☐	11-A	To use factorials to find the number of choices, permutations, and combinations				
☐ ☐	11-B	To find simple probabilities; to find probabilities of independent and dependent events				
☐ ☐	11-C	To use random numbers to simulate probability experiments and to make predictions using experimental probability				
☐ ☐	11-D	To use Venn diagrams to show relationships among groups				
☐ ☐	11-E	To solve problems by using a diagram and to solve problems by acting them out				

continued

Cumulative Record Form

MATHEMATICS PLU[S]
Grade 8

Student _____   Teacher _____

CUMULATIVE TEST CHAPTERS 1–11
Criteria 35/50   Test Score _____

CHAPTER 12

Test Scores:	Criteria Pretest Posttest 24/32 ___ ___	Needs More Work	Accomplished
☐☐ 12–A	To choose an appropriate metric or customary unit of measure; to estimate measures using non-standard and standard units of measure		
☐☐ 12–B	To identify the precision, the greatest possible error, and the significant digits of a measurement		
☐☐ 12–C	To find the perimeter of polygons; to find the circumference of a circle		
☐☐ 12–D	To find the area of rectangles, parallelograms, triangles, trapezoids, and circles		
☐☐ 12–E	To solve problems by making a model and by using a formula		

CHAPTER 13

Test Scores:	Criteria Pretest Posttest 24/32 ___ ___	Needs More Work	Accomplished
☐☐ 13–A	To identify and describe solid figures		
☐☐ 13–B	To find the surface area of prisms, pyramids, cylinders, and cones		
☐☐ 13–C	To find the volume of prisms, pyramids, cylinders, and cones		
☐☐ 13–D	To identify relationships between metric units of volume, capacity, and mass		
☐☐ 13–E	To solve problems by using a table and by using a formula		

CHAPTER 14

Test Scores:	Criteria Pretest Posttest 24/32 ___ ___	Needs More Work	Accomplished
☐☐ 14–A	To identify right triangles and to use properties of 45–45 right triangles and 30–60 right triangles to find unknown lengths		
☐☐ 14–B	To use proportions to find unknown lengths in similar figures		
☐☐ 14–C	To identify and use tangent, sine, and cosine ratios to find unknown lengths in right triangles		
☐☐ 14–D	To solve problems by making choices and by using a picture		

CUMULATIVE TEST CHAPTERS 1–14
Criteria 35/50   Test Score _____

END-OF-BOOK TEST
Criteria 35/50   Test Score _____

Cumulative Record Form   Grade 8